Multiscale Theory of Composites and Random Media

Multiscale Theory of Composites and Random Media

Xi Frank Xu

CRC Press
Taylor & Francis Group
Boca Raton London New York

CRC Press is an imprint of the
Taylor & Francis Group, an **informa** business

CRC Press
Taylor & Francis Group
6000 Broken Sound Parkway NW, Suite 300
Boca Raton, FL 33487-2742

First issued in paperback 2020

ISBN-13: 978-1-4822-5624-6 (hbk)
ISBN-13: 978-0-367-65702-4 (pbk)

Library of Congress Cataloging-in-Publication Data

Names: Xu, Xi Frank, author.
Title: Multiscale theory of composites and random media / Xi Frank Xu.
Description: Boca Raton : Taylor & Francis, a CRC title, part of the Taylor & Francis imprint, a member of the Taylor & Francis Group, the academic division of T&F Informa, plc, [2019] | Includes bibliographical references and index. |
Identifiers: LCCN 2018015084 (print) | LCCN 2018030719 (ebook) | ISBN 9780429894381 (Adobe PDF) | ISBN 9780429894374 (ePub) | ISBN 9780429894367 (Mobipocket) | ISBN 9781482256246 (hardback : acid-free paper) | ISBN 9780429470653 (ebook)
Subjects: LCSH: Composite materials--Mathematical models. | Multiscale modeling.
Classification: LCC TA418.9.C6 (ebook) | LCC TA418.9.C6 X824 2019 (print) |
DDC 620.1/18015118--dc23
LC record available at https://lccn.loc.gov/2018015084

Visit the Taylor & Francis Web site at
http://www.taylorandfrancis.com

and the CRC Press Web site at
http://www.crcpress.com

To Xin and Hannah

Contents

List of symbols and acronyms

Brief definitions of frequently used symbols are listed below, with meanings of acronyms provided at the end. Note that a few symbols have different definitions in different sections. In such cases, the definitions are given locally and used consistently within sections.

SYMBOLS

$\boldsymbol{A}_n, \tilde{\boldsymbol{A}}_n$: concentration tensor and normalized concentration tensor of Phase-n or Component-n

a: semi-axis of a spheroid, or radius of a sphere

a_1, a_2, a_3: semi-axis of an ellipsoid

[B]: strain displacement matrix

C, [C]: covariance

$\hat{C}$, $\hat{C}_f$, $\hat{C}_Z$: power spectral density

C_V: coefficient of variation

c^*: percolation threshold

c_{ij}, c_{ijk}: auto- and triple-correlations among phases i, j, k

c_0, c_i, $\tilde{c}_n^{(i)}$, $\tilde{c}_n$: volumes of the matrix, Phase-i, the ith inclusion in Component-n, and Component-n

$\tilde{c}_{nm}$, $\tilde{c}_{h,mn}$: intercomponent and interpoint auto-correlations

D, D_n, D_c: domains of a BVP, Phase-n, and a crack

$\tilde{D}_n$: domain of an ellipsoid in Component-n

$\mathcal{D}$: set of grey levels

∂D_u, ∂D_t: parts of boundary imposed with displacement and traction conditions, respectively

$D_n^{(i)}$: domain of the ith inclusion of Component-n

d: dimensionality

d_q: short-range-correlation energy function

E: strain function space

E: Young's modulus

{E_0}: strain vector in finite elements

e_0: hydrostatic strain
{F}, {F_0}, {f}: force vector in finite elements
$\boldsymbol{f}$, f, f_n: body force or excitation force
f_k: fraction of Type-k fillers
f_X, F_X: probability density function and cumulative distribution of the random variable X
$f_H(\eta)$: probability density function of the aspect ratio
$f_{HN}(\eta, n)$: bivariate probability density function of the aspect ratio and the contrast ratio
$\boldsymbol{G},\hat{\boldsymbol{G}}$: Green functions in real and Fourier space
$G,\hat{G}$: transfer functions in real and Fourier space
$\boldsymbol{G}^{\infty},\hat{\boldsymbol{G}}^{\infty}$: free space Green functions in real and Fourier space
[G]: Green function matrix in finite elements
g: mapping between two coefficients of correlation
g_K: function of random variables
H_n: nth Hermite polynomial
$\mathcal{H},\mathcal{H}^+,\mathcal{H}^-$: Hashin-Shtrikman functional, and its upper and lower bounds
h_n: random point field for the centers of all the inclusions in Component-n
$\boldsymbol{I}$: Identity tensor
I: intensity of a white noise
$\boldsymbol{K}$: nonlocal kernel
K: conductivity or permeability
[K], [K_0], [K_m], [K_{mn}]: stiffness matrix in finite elements
K^{EB}: ellipsoidal bound of the effective conductivity
$K^e,K_+^{(n)},K_-^{(n)}$: exact, nth order upper bound, nth-order lower bound of the effective conductivity
$K_-^{(3)*}$: best possible lower bound of the effective conductivity
K_r: reference permeability
$\boldsymbol{L}$, $\boldsymbol{L}_0$, $\boldsymbol{L}_n$: elastic moduli tensors of a composite, a reference medium, and Phase-n
L, L_1: RVE or window size
$\mathcal{L}$: BVP domain size
$[L],[\bar{L}],[L_0],[L_\sigma]$: matrix of elastic moduli in finite elements
$\boldsymbol{L}^e,\boldsymbol{L}_-^{(n)},\boldsymbol{L}_+^{(n)}$: exact, nth-order lower bound, nth-order upper bound of the effective elastic moduli
$\boldsymbol{L}'$, $\boldsymbol{L}''$: higher order constitutive tensor
$\boldsymbol{L}^{EB}$: ellipsoidal bound of the effective elastic moduli
$\boldsymbol{L}^A,\boldsymbol{L}_m^A$: polarization moduli
$\boldsymbol{L}_\rho^{(2)}$: size-dependent HS bound
[L_C]: lower triangular matrix of Cholesky decomposition about [C]
l_1, l_2: lattice coordinate
ℓ_c: correlation length
ℓ_{i,k_i}, $\mathcal{L}_k$: Lagrange interpolation polynomial and basis function
$\boldsymbol{M},\boldsymbol{M}_0,\boldsymbol{M}_r$: compliance tensors of a composite, a reference medium, and Phase-r

M_K: number of terms in a polynomial chaos expansion
$\mathcal{N}$: neighborhood set
N, N_c, N_e, N_f, N_K: numbers of phases, cracks, finite elements, fillers, and random variables
$\tilde{N}$, $\tilde{N}_n$: numbers of components, and inclusions in Component-n
N_i, {N}, [N]: shape function, the vector, and the matrix
n, n_k: contrast ratios of the inclusion phase and Phase-k
n_{pc}: order of a polynomial chaos expansion
$n_p^{(e)}$: number of elemental nodes or shape functions
P: probability
P_n: nth Legendre polynomial
$\boldsymbol{p}$, $\boldsymbol{p}_r$: stress polarization functions of a composite and Phase-r
$\{\mathrm{P_r}\}$: vector of the stress polarization of Phase-r in finite elements
$\boldsymbol{P}_{ri}^{(e)}$, $\{\mathrm{P_r}\}^{(e)}$: polarization value of Phase-r on the nodes of an element
$\boldsymbol{p}_r^{(e)}$, $\{\mathrm{p_r}\}^{(e)}$: polarization value of Phase-r in an element
$\boldsymbol{Q}$: transformation matrix
Q_m: set of orthogonal polynomials
$\boldsymbol{q}$: heat flux
q: exponent
R_2, R_3, R_n: two-, three-, and n-point correlations in random morphology
R^d: d-dimensional Euclidean space
$\Re$: set of real numbers
$\boldsymbol{S}$, $\boldsymbol{S}_0$, $\boldsymbol{S}_c$: Eshelby tensors of an ellipsoid, a circles/sphere, and a crack
$\boldsymbol{S}_n$, $\bar{\boldsymbol{S}}_n$: Eshelby tensor and average Eshelby tensor of Component- or Phase-n
$\boldsymbol{S}_{nm}$: generalized Eshelby tensor of Components n and m
$\boldsymbol{S}_0^\rho$: size-dependent spherical Eshelby tensor
dS: surface differential
$\mathcal{S}$: set of indices
^{T}: superscript as the transpose of a matrix or a vector
T: temperature in random morphology
$\mathcal{T}$: operator of translation
$\tilde{\boldsymbol{t}}, \tilde{\boldsymbol{t}}_0, \tilde{\boldsymbol{t}}'$: total, slow-scale, and fast-scale stress tractions
$\tilde{t}$: heat flux boundary condition
$\{\tilde{\mathrm{t}}\}$: stress traction or boundary heat flux in finite elements
U: energy in random morphology
$\mathcal{U}$, $\mathcal{U}^+$, $\mathcal{U}^-$: exact, upper bound, and lower bound of potential energy
$\mathcal{U}^c$, $\mathcal{U}^{c+}$, $\mathcal{U}^{c-}$: exact, upper bound, and lower bound of complementary energy
U_n: uniformly distributed random variable in [0,1]
$\{\mathrm{U}\}$, $\{\mathrm{U_0}\}$: vector of displacement or temperature in finite elements
$U_{0ik}^{(e)}$, $\{\mathrm{U_0}\}^{(e)}$: nodal displacement in an element
$\boldsymbol{u}$, $\boldsymbol{u}_0$, $\boldsymbol{u}'$: total, slow-scale, and fast-scale displacements

$\tilde{\boldsymbol{u}}, \tilde{\boldsymbol{u}}_0, \tilde{\boldsymbol{u}}'$: total, slow-scale, and fast-scale displacement boundary conditions
u: temperature, or 1D displacement in dynamics
$\tilde{u}$: temperature boundary condition
$\dot{u}, \dot{u}_0, \dot{u}'$: total, slow-scale, and fast-scale velocities
$u_{0k}^{(e)}, \{u_0\}^{(e)}$: displacement in an element
dV: volume differential
V_n: nth-order energy in random morphology
V_c: volume of a crack
dW: Gaussian white noise
$\{\mathbf{w}\}$: sample vector of a Gaussian white noise
X, x: morphological set and a sample
$\boldsymbol{x}_n^{(i)}$: center of the ith inclusion in Component-n
Y: RVE domain
Y, $\tilde{\mathrm{Y}}$: non-Gaussian random variable or field, and the normalized one
Y_σ: yield criterion
Z, $\tilde{Z}$: Gaussian random variable or field, and the normalized one
$\{\mathbf{z}\}$: sample vector of a Gaussian field
α: density of inclusions or cracks
a^*: critical density
β: nonlinear parameter
$\boldsymbol{\Gamma}, \hat{\boldsymbol{\Gamma}}$: modified Green functions in real and Fourier space
$\boldsymbol{\Gamma}^\infty, \hat{\boldsymbol{\Gamma}}^\infty$: free space modified Green functions in real and Fourier space
$\tilde{\boldsymbol{\Gamma}}^\infty$: free space modified Green function characterized with a conductivity of unity
$[\Gamma]$: matrix of the modified Green function in finite elements
$[\Gamma_{rs}]$: matrix for the integral of the modified Green function and a correlation function c_{rs}
γ: angle between O′O″ and O′O‴ in the tetrahedron of Figure 4.5
γ_0, γ_{012}: engineering shear strain
γ_s: crack surface energy density
$\gamma_{max}, \gamma_{min}$: contrast ratio between the elastic moduli of the inclusions and the matrix
δ: variation
δ_{ij}: Kronecker-delta function
$\boldsymbol{\varepsilon}, \boldsymbol{\varepsilon}_0, \boldsymbol{\varepsilon}'$: total, slow-, and fast-scale strains
$\boldsymbol{\varepsilon}_r$: strain in Phase-$r$
ε: nonlinearity parameter of a Duffing oscillator
$\varepsilon_{0kj}^{(e)}, \{\varepsilon_0\}^{(e)}$: elemental strain
ζ: damping ratio
ζ_1, ζ_2: geometric parameter of the bulk modulus
η, η_k: aspect ratios of the inclusion phase and Phase-k

$\boldsymbol{\eta}$, η_i: vector of random variables, and the i-th component
η_1, η_2: geometric parameter of the shear modulus
$\tilde{\eta}$: clustering parameter
θ, θ', θ_1, θ_2: spherical coordinate
$\boldsymbol{\theta}$: vector of Euler angles
$\vartheta \in \Theta$: sample in random space
κ_0, κ_n: bulk moduli of a reference medium and Phase-n
κ^{EB}: ellipsoidal bound of the effective bulk modulus
κ_A, κ_n^A: bulk moduli of the concentration tensor and the polarization moduli tensor
κ_S,κ_S^ρ: bulk moduli of the Eshelby tensor and the size-dependent Eshelby tensor
$\kappa_+^{(n)}$, $\kappa_-^{(n)}$: nth-order upper and lower bounds of the effective bulk modulus
$\kappa_\rho^{(2)}$: size-dependent HS bound of the effective bulk modulus
κ_0^ε, κ_0^σ: reference bulk moduli in the plane strain and plain stress cases
$\boldsymbol{\Lambda}$: first derivative of the Green function
λ: Lame's constant
λ_n: eigenvalue
μ_0, μ_n: shear moduli of a reference medium and Phase-n
μ_A, μ_n^A: shear moduli of the concentration tensor and the polarization moduli tensor
μ_S,μ_S^ρ: "shear moduli" of the Eshelby tensor and the size-dependent Eshelby tensor
μ^{EB}: ellipsoidal bound of the effective shear modulus
$\mu_+^{(n)}$, $\mu_-^{(n)}$: nth-order upper and lower bounds of the effective shear modulus
$\mu_\rho^{(2)}$: size-dependent HS bound of the effective shear modulus
v_0, v_n: Poisson's ratios of a reference medium and Phase-n
ξ_n: nth-wave vector in Fourier space
ρ, $\hat{\rho}$: correlation coefficients in real and Fourier space
ρ_0,$\hat{\rho}_0$: isotropic correlation coefficients in real and Fourier space
ρ_{nm}, $\tilde{\rho}_{nm}$: correlation coefficients between Phases m and n, and between Components m and n
ρ_n: correlation coefficient of the nth orthonormal polynomial Q_n
ρ_Y, ρ_Z: correlation coefficients of a non-Gaussian field Y and a Gaussian field Z
Σ: stress function space
σ, σ_z, σ_0, σ_λ, σ_μ: standard deviation
$\boldsymbol{\sigma}$, $\boldsymbol{\sigma}_0$, $\boldsymbol{\sigma}'$: total, slow-scale, and fast-scale stresses
$\boldsymbol{\sigma}_r$: stress in Phase-r
$\boldsymbol{\sigma}^e$: von Mises effective stress
σ_{11}^c,σ_{12}^c: critical tensile and shear stresses of a cracked body
ς: Gaussian random variable

ϕ, Φ, ϕ_k, Φ_k: Gaussian PDF and CDF and the k-variate pair
$\tilde{\phi}$: PDF of a standard Gaussian variable
$\varphi, \varphi', \varphi_1, \tilde{\varphi}$: spherical coordinate
φ_n: eigenfunction
χ_c, χ_n: indicator functions of a crack, and Phase- or Component-n
$\chi_n^{(i)}$: indicator function of the ith inclusion of Component-n
$\tilde{\chi}_n$: indicator function of an ellipsoid of Component-n centered at the origin
Ψ: fourth-order geometric parameter
ω, ω_n: frequency and natural frequency

ACRONYMS

BVP: boundary value problem
CDF: cumulative distribution function
CNT: carbon nanotube
FEM: finite element method
FFT: fast Fourier transform
HS: Hashin-Shtrikman
K-L: Karhunen-Loeve
MSFEM: multiscale stochastic finite element method
PCE: polynomial chaos expansion
PDF: probability density function
PSD: power spectral density
RFB: random field based
RVE: representative volume element
SFEM: stochastic finite element method
SRC: short-range-correlation

Preface

This book stems from a graduate course on composites first taught at Stevens Institute of Technology in 2007 and later at Beijing Jiaotong University. A strong impulse is to emphasize criticality of emerging scale-coupling mechanics by presenting relevant progress made over the past decade on mechanics of random composites and percolation threshold.

With regard to fundamental theory of micromechanics, there have been two major stages of development. The first foundation stage was laid in the late 1950s and early 1960s by two important representative works, Eshelby's solution (1957) and Hashin-Shtrikman principle (1962). The second stage was completed around the turn of the 21st century, signified by several monographs published at the time (Nemat-Nasser & Hori, 1999; Milton, 2002; Torquato, 2002). In this second stage, mechanics of random composites were formulated by Willis (1981, 2002) in a rigorous probabilistic framework. Several essential variational bounds were derived by Beran (1965), Milton (2002), and others. A variety of analytical approximations and failure criteria were also heuristically developed especially to deal with complexity of failure problems. With the two stages accomplished, micro–macro theory of micromechanics is established based on a fundamental assumption of scale separation. There are, however, three critical topics remaining uncompleted: percolation, nonlinear homogenization (or micro–macro bridging of plasticity/softening/fracture), and dynamical homogenization. All the three topics are closely related to scale-coupling phenomena in that local and global behaviors strongly interact with each other.

This book makes attempts to tackle scale-coupling mechanics starting from fundamental variational principles and to address the uncompleted critical topics starting from percolation. Chapter 1 describes the background and basic idea of scale-coupling mechanics. Chapter 2 provides some preliminaries on random field theory and correlation functions. The remaining chapters are divided into two distinctive parts: Part I, covering Chapters 3 through 6, on analytical homogenization of scale-separation problems, and Part II, from Chapters 7 through 9, on computational analysis of scale-coupling problems. In Part I, variational bounds are derived in detail, from which a peculiar phenomenon of fluid–antifluid annihilation is predicted. To further take the inclusion shape into account, ellipsoidal bound is formulated leading to new and much improved formulas of percolation threshold. Various schemes of micromechanics, such as self-consistent, Mori-Tanaka, and effective medium, are commented on in passing as special cases in the adopted variational framework. In Part II, Green-function-based variational principles are formulated on a finite body composite, and numerically implemented with multiscale stochastic finite element method (MSFEM). The unconventional scale-coupling effect is first quantified as size effect of representative volume element (RVE) in Chapter 7, then further demonstrated in stress analysis of Chapter 8 and soil settlement of Chapter 9. To circumvent the curse-of-dimensionality issue, a new random field based orthogonal expansion is introduced in Chapter 9 with the applications discussed in both random media and stochastic dynamics.

Part I can serve as a text for a one-semester course of micromechanics and Part II for another on stochastic finite elements. In the text there are a number of exercises discussed that can be used as homework or projects. Nonlinearity and dynamics of composites are two big topics deliberately missing from this book, each of which probably needs a whole book to cover when the time comes. Micromechanics and composites have an immense existing literature, and the citations on these topics made in this book are far from being complete. The reader is recommended to further explore reference lists of those articles and books cited by this book.

I first wish to thank Lori Graham-Brady and Pizhong Qiao, who advised and guided me into the field of composites when I was a graduate student. I am grateful to Irene Beyerlein, Lihua Shen, George Stefanou, Zhenjun Yang, Xi Chen, and Keqiang Hu as wonderful research collaborators. I am also thankful to all the students in the United States and China taking the courses that this book stems from, especially Nelson Pinilla and Kunpeng Wang, who provided helpful feedback. My final thanks goes to Tony Moore, senior editor of Taylor & Francis, without whose support this would not have been possible.

Xi Frank Xu

Author

Xi Frank Xu, Ph.D. was a professor in civil engineering at Beijing Jiaotong University where he left in 2017, and was formerly an assistant professor at Stevens Institute of Technology, USA. He received the 2010 K.J. Bathe Award for the Best Paper by a Young Researcher in the field of computational engineering.

Chapter 1

Introduction

Emerging scale-coupling mechanics

Micromechanics and multiscale modeling began to emerge in science and engineering in the 1950s and 1990s, not by coincidence, around the times when microfibers and nanotubes were fabricated leading to the births of composites technology and nanotechnology, respectively. Micromechanics is focused on understanding of the micro–macro relationship between macroscopic behavior and microscopic information of a composite. By extending the micro–macro bridging concept of micromechanics into a new territory occupied by multiple scales, multiscale modeling of materials has introduced many new tools, and certainly many challenges as well, to deal with phenomena newly observed or not well understood. A most critical challenge confronting multiscale modeling is to fit so-called scale-coupling phenomena into the current theoretical framework of mechanics, or alternatively, as exhorted by this book, to develop new theory of scale-coupling mechanics.

In this chapter, a conceptual distinction is first made between two fundamental modeling strategies, scale separation vs. scale coupling. Next, the idea of scale-coupling mechanics is introduced through a simplistic formulation of the nonlocal theory and the strain gradient theory, which is first noted in Xu, Dui, and Ren (2014). The arrangement of the following chapters of this book is explained and commented on throughout this chapter.

1.1 PHENOMENOLOGICAL METHODOLOGY VS. MICRO-MACRO METHODOLOGY

In the field of mechanics of materials, there are two classical and fundamental modeling methodologies, phenomenological and micro–macro, as shown in Figure 1.1. A *phenomenological model* is defined as a mechanics model that describes the empirical relationships of macroscopic phenomena to each other, in a way that directly applies or is at least consistent with fundamental laws of mechanics, but has no use of microscopic information including physical mechanisms and statistics. Note that, according to this definition, a model traditionally falling into the category of so-called mechanistic models would still be a phenomenological one when no microscopic

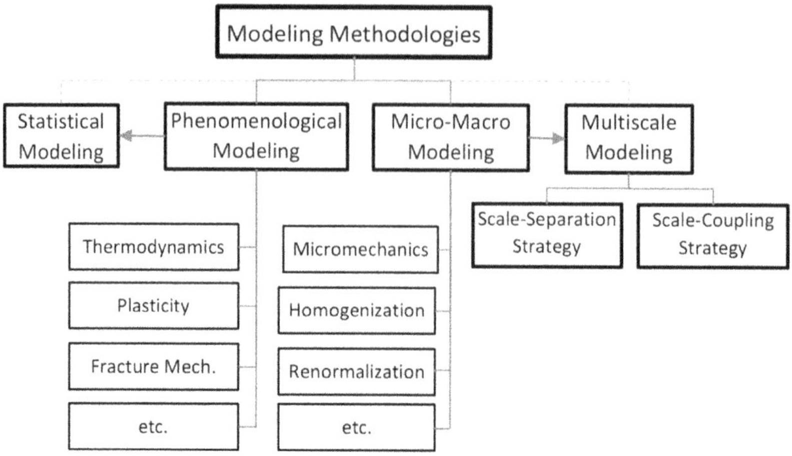

Figure 1.1 Diagram of modeling methodologies in mechanics of materials. (With kind permission from Springer Science+Business Media: *Multiscale modeling and uncertainty quantification of materials and structures*, A note on scale-coupling mechanics, 2014, 159–169, Xu X.F., G. Dui, and Q. Ren. Permission conveyed through Copyright Clearance Center, Inc.)

information is explicitly taken into account. A phenomenological model in mechanics of materials typically consists of the following three basic steps:

i. Define a number of state variables and parameters (e.g. constitutive parameters),
ii. Formulate a mathematical model using the defined variables and parameters to describe an observed phenomenon, and
iii. Measure the parameters and validate the model experimentally.

An important assumption is implicitly made that the macroscopic relationship explained by a phenomenological model can be used beyond the range of those values of parameters that have been measured or validated.

Most models developed in mechanics of materials are phenomenological, e.g. thermodynamics, plasticity, fracture mechanics, etc. A major drawback of the methodology lies in its inherent lack of incorporation of microscopic mechanisms and statistics, and thereby in its lack of in-depth physical insights in why and how variables and parameters interact with each other in such a way that yields a particular macroscopic behavior. When dealing with a complex phenomenon, a typical pitfall in a phenomenological model is that the number of variables and parameters can rise to a level that is hardly amenable. In an extreme case, by purely using curve-fitting parameters, a phenomenological model degenerates to a statistical model with no input or understanding of underlying physical mechanisms (see Figure 1.1). In damage and fracture mechanics, an overly complicated model of fatigue life prediction, or in soil mechanics, a constitutive model of an unsaturated soil, can become such a victim when too many variables or parameters with unclear physical meanings are employed. Such a model is practically close to a statistical one, with the applicability and generality severely restricted. It should be noted that, on the other hand, statistical modeling remains the most essential approach to tackle highly complex phenomena of which only primitive understanding exists (e.g. life science, social behavior, etc.).

Micro–macro methodology pertains to an analytical, numerical, or combined derivation of macroscopic behavior based on validated microscopic information, such as constitutive laws, physical properties, statistics, and data. When a micro–macro model needs fundamental microscopic laws and properties only with no other special assumptions or parameters, it is also called a *first principle* or *ab initio model.* For example, given Newton's laws of motion and an interatomic potential, macroscopic thermodynamics of an ideal gas can be numerically simulated by using molecular dynamics.Compared with a phenomenological model, a micro–macro model has two major advantages:

- An explicit micro–macro relation becomes known, thereby allowing optimization of microscopic information in geometry, physics, or biochemistry to achieve certain desired macroscopic behavior or properties; and
- An analytical, numerical, or combined approach can be frequently implemented as virtual testing to replace much more expensive physical testing, and to resolve many problems otherwise experimentally challenging or even prohibitive (see Figure 1.2).

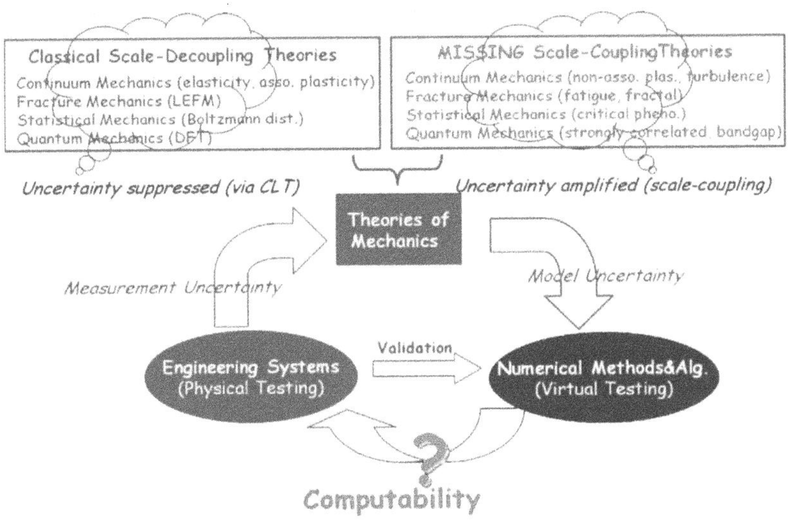

Figure 1.2 Missing scale-coupling theories across scales. (With kind permission from Springer Science+Business Media: *Multiscale modeling and uncertainty quantification of materials and structures*, A note on scale-coupling mechanics, 2014, 159–169, Xu X.F., G. Dui, and Q. Ren. Permission conveyed through Copyright Clearance Center, Inc.)

At this point, it should be remarked that *micro* and *macro* are used as two relative terms when there is a wide-scale separation between two scales in space or time. For instance, the grain scale is a microscale with respect to a specimen of steel bar, but a macroscale with respect to the iron atoms. Two famous micro–macro models in the early stage of modern science are statistical mechanics and Brownian motion. Due to technological limitation at that time in microscopic measurement, early micro–macro models usually adopt a top-down strategy (i.e. derivation of microscopic information from macroscopic measurement). A simple example is that we can estimate one of the most essential microscopic statistics, the variance of molecular velocity in the air, by measuring a mercury thermometer (i.e. macroscopic thermal expansion behavior of mercury). Since the 1950s, a bottom-up strategy has emerged, and there are three representative micro–macro theories developed in mechanics, mathematics, and physics, respectively, described as follows (refer to Figure 1.1):

i. Micromechanics theory has been formulated so far based on a fundamental concept of representative volume element (RVE) (e.g. Nemat-Nasser & Hori, 1999). In a rigorous variational framework, Part I of this book covers micromechanics theory in the regime of scale separation, and in Part II, the theory is extended to the unconventional scale-coupling regime;
ii. Homogenization theory has been mathematically developed since the 1970s based on a periodic unit cell concept (e.g. Bensoussan, Lions, & Papanicolaou, 1978; Sanchez-Palencia, 1980), which was further extended to stochastic homogenization and stochastic boundary value problems (BVPs) by assuming scale separation (Papanicolaou & Varadhan, 1981). Stochastic homogenization in the context of random composites is equivalently the RVE-based micromechanics theory as mentioned in i). In Part II of this book, stochastic homogenization and stochastic BVPs are addressed beyond the assumption of scale separation; and
iii. Renormalization group theory has been developed in physics based on a scale-invariance or self-similarity concept (e.g. Sinai, 1982). To describe a macrosystem independent of types of microinteractions, in statistical mechanics the Boltzmann distribution is obtained as a fixed point of infinite transformations by assuming a short range correlation of microscopic interactions (i.e. separation of scales). When dealing with a critical phenomenon with a long range correlation of microinteractions, i.e. scale-coupling, renormalization group theory can be similarly viewed as identification of a fixed point of transformations by using a concept of self-similar random fields (Jona-Lasinio, 2001). It is worth noting that in chaos theory and extreme value theory, there are similar ideas about a fixed point of probability distributions under infinite transformations. In most engineering practices,

the scale separation assumption is invalid and the three asymptotic distributions of extreme value theory are subjected to modifications. In mechanics, one such modified strength theory has been proposed to reinterpret the Weibull distribution and rebuild the weakest link model (Xu & Beyerlein, 2016). Nonlinearity and dynamics, the two most crucial parts of scale-coupling mechanics, unfortunately are not ready to be included in this book.

1.2 MULTISCALE METHODOLOGY

With the rapid growth of computing technology and the arising of nanotechnology since the 1990s, multiscale modeling has emerged as a new methodology. The term *multiscale modeling* has been widely used with quite different meanings by different scientific and engineering communities. While a rigorous definition of *multiscale modeling* has not been given with universal consensus, a distinction from classical micro–macro modeling should be clearly made. As shown in Figure 1.1, multiscale modeling can be alternatively treated as an extension of classical micro–macro modeling, but it does and should contain one or more of the following important new features:

i. A conventional micro–macro model deals with a problem of scale separation or scale invariance, with some essential asymptotic results obtained analytically. In a scale-coupling problem where the scales are not well separated, in general it is difficult to find an analytical solution, and the problem has to be resolved numerically by using a certain multiscale computational model such as those described in Part II of this book. It is therefore not surprising that most of the unresolved, complicated problems in mechanics are scale-coupling ones, from classical challenges, such as turbulence, boundary layer issues, and fatigue cracks to novel applications like microelectromechanical systems, metamaterials, etc.;
ii. A conventional micro–macro model deals with two scales only with no consideration of intermediate scales, except for renormalization theory treating an infinite number of scales self-similarly. A typical multiscale model enables modeling of microscopic effects propagating across multiple discrete scales explicitly. For example, a microcrack initiated from atomistic defects gradually extends into a macrocrack with its length comparable to the width dimension of a beam specimen. During the crack-propagating process initialized from point defects, there are three scales involved, namely, an atomistic scale, a crack length scale, and a specimen size scale. A multiscale model based on a hybrid of molecular dynamics and finite element method can be used to simulate such a crack propagation;

iii. A multiscale model cutting across diverse scales is also a multiphysics model as different physical laws are used at different scales, from quantum mechanics and molecular dynamics all the way up to subregimes of continuum mechanics with different levels of resolution, such as granular matter mechanics, geomechanics, and geophysics; and
iv. A conventional micro–macro model is normally developed in an individual scientific or engineering discipline. Many such models in different scientific and engineering fields are actually quite similar, overlapping, or complementary to each other. The prevalently observed isolation among these models is simply due to disciplinary barriers. New multiscale models have been developed with synergistic efforts crossing disciplinary boundaries.

According to the condition of scale separation or scale coupling, there are two distinctive multiscale strategies, i.e. scale separation and scale coupling. Classical theories of mechanics are formulated based on scale separation, e.g. elasticity, thermodynamics, and density functional theory, in which the effect of uncertainty is minimized through energy summation according to the central limit theorem. There are however many unresolved challenging problems, e.g. turbulence, fractal fracture, fatigue failure, critical phenomena, etc., which are typically characterized with complexity of amplified uncertainty amid scale-coupling interactions, even subjected to questioning of computability. A major hindrance to resolving these challenging problems, in the author's opinion, is continuous use of classical scale-separation theories in conflict with scale-coupling phenomena that dominantly characterize these problems. A most impressive result of such a cart-before-the-horse issue is demonstrated as negative mass and negative elastic moduli derived in elastodynamics and metamaterials. A lack of rigorous scale-coupling mechanics theory is considered to be the most critical issue in multiscale modeling. In Figure 1.2, this critical issue is illustrated along with uncertainty propagating through a loop consisting of theory, physical testing, and virtue testing of engineering systems.

1.3 SCALE-COUPLING MECHANICS

1.3.1 Existing models on scale-coupling effects

Little theoretical work has been devoted to addressing scale-coupling issues. A major relevant theoretical work is so-called generalized continua models originated from the idea of Cosserat continua and developed by Eringen (1999), Toupin (1962), Mindlin (1964), Germain (1973) and others from a variety of perspectives. Generalized continua models attempt to explicitly account for scale-coupling effects of microstructure by introducing into a continuum framework additional degrees of freedom or internal variables (e.g. a spin field in micropolar continua, a microdisplacement field in micromorphic continua, etc.).

While generalized continua models are mathematically sound, their engineering applications remain quite limited and the theory itself has been subjected to considerable debates. The main reasons for this are attributed to a lack of clarity and experimental methods about unconventional higher order variables; issues related to boundary conditions for higher order tensors, such as couple forces and couple stresses; and difficulties of resolving boundary layer effects (see e.g. Kunin, 1983). Many studies on generalized continua degenerate into purely mathematical endeavors, except for the topic of the strain gradient plasticity. The strain gradient plasticity theory, with its fundamental root in generalized continua, has been developed with a close connection to the phenomenological size effect of plasticity. Following the first proposal of the theory (Aifantis, 1984), further development has been made with phenomenological modeling (Fleck & Hutchinson, 1997; Nix & Gao, 1998). As microstructure information has not yet been directly taken into account, major issues of the theory center on the interpretation of an internal length scale (Evans & Hutchinson, 2009).

Relevant to the previously discussed strain gradient theory, nonlocal constitutive laws have been proposed in elasticity, damage, and plasticity (Rogula, 1982; Eringen, 1983; Bazant & Pijaudier-Cabot, 1988). As shown next, the nonlocal formulation and the strain gradient formulation are convertible to each other. A major issue of the nonlocal theory is a lack of physical ground in determination of a nonlocal kernel, while a choice of the latter affects modeling results significantly.

1.3.2 Nonlocal formulation of scale-coupling mechanics

In Xu (2009), variational principles are formulated for a scale-coupling composite body in elasticity, based on which a Green-function-based multiscale method is developed to numerically solve a scale-coupling boundary value problem (Xu, Chen, & Shen, 2009). In this subsection, scale-coupling mechanics is simplistically formulated in the following example in a nonlocal format.

A boundary value problem of elasticity characterized with spatially heterogeneous moduli $\boldsymbol{L}(\boldsymbol{x})$ in domain D reads

$$\nabla \cdot \boldsymbol{\sigma} + \boldsymbol{f} = 0$$

$$\boldsymbol{\sigma} = \boldsymbol{L}\boldsymbol{\varepsilon} \qquad (1.1a, b, c, d)$$

$$\text{BC}\ \boldsymbol{u} = \tilde{\boldsymbol{u}} \quad \partial D_u$$

$$\text{BC}\ \boldsymbol{\sigma} \cdot \boldsymbol{n} = \tilde{\boldsymbol{t}} \quad \partial D_t$$

which can be decomposed into a slow-scale BVP with spatially homogeneous moduli $\boldsymbol{L}_0$

$$\nabla \cdot \boldsymbol{\sigma}_0 + \boldsymbol{f} = 0$$

$$\boldsymbol{\sigma}_0 = \boldsymbol{L}_0 \boldsymbol{\varepsilon}_0 \qquad (1.2\text{a, b, c, d})$$

$$\text{BC}\ \boldsymbol{u}_0 = \tilde{\boldsymbol{u}} \quad \partial D_u$$

$$\text{BC}\ \boldsymbol{\sigma}_0 \cdot \boldsymbol{n} = \tilde{\boldsymbol{t}} \quad \partial D_t$$

and a fast-scale BVP

$$\nabla \cdot \boldsymbol{\sigma}' + \nabla \cdot \boldsymbol{p} = 0$$

$$\boldsymbol{\sigma}' = \boldsymbol{L}_0 \boldsymbol{\varepsilon}' \qquad (1.3\text{a, b, c, d})$$

$$\text{BC}\ \boldsymbol{u}' = 0 \quad \partial D_u$$

$$\text{BC}\ \boldsymbol{\sigma}' \cdot \boldsymbol{n} = 0 \quad \partial D_t$$

with the stress polarization

$$\boldsymbol{p} = (\boldsymbol{L} - \boldsymbol{L}_0)\boldsymbol{\varepsilon} \qquad (1.4)$$

The solution of the fast-scale BVP is given in terms of the Green function G as

$$\boldsymbol{u}'(\boldsymbol{x}) = \int_D \boldsymbol{G}(\boldsymbol{x}, \boldsymbol{x}') \nabla \cdot \boldsymbol{p}(\boldsymbol{x}') \, d\boldsymbol{x}' \qquad (1.5)$$

which, by using Gauss' divergence theorem, yields

$$\boldsymbol{\varepsilon}'(\boldsymbol{x}) = -\int_D \boldsymbol{\Gamma}(\boldsymbol{x}, \boldsymbol{x}') \boldsymbol{p}(\boldsymbol{x}') \, d\boldsymbol{x}' \qquad (1.6)$$

with the modified Green function

$$\Gamma_{ijkl}(\boldsymbol{x},\boldsymbol{x}') = \frac{1}{2}\left(\frac{\partial^2 G_{ik}(\boldsymbol{x},\boldsymbol{x}')}{\partial x_l \partial x_j'} + \frac{\partial^2 G_{jk}(\boldsymbol{x},\boldsymbol{x}')}{\partial x_l \partial x_i'}\right) \tag{1.7}$$

By substituting (1.6) into (1.4), the stress polarization is thus expanded as an infinite series

$$\boldsymbol{p} = \left(\boldsymbol{L} - \boldsymbol{L}_0 - (\boldsymbol{L} - \boldsymbol{L}_0)\boldsymbol{\Gamma}(\boldsymbol{L} - \boldsymbol{L}_0) + (\boldsymbol{L} - \boldsymbol{L}_0)\boldsymbol{\Gamma}(\boldsymbol{L} - \boldsymbol{L}_0)\boldsymbol{\Gamma}(\boldsymbol{L} - \boldsymbol{L}_0) - \cdots\right)\boldsymbol{\varepsilon}_0 \tag{1.8}$$

The stress therefore can be expressed in terms of the slow-scale strain $\boldsymbol{\varepsilon}_0$ as

$$\begin{aligned}
\boldsymbol{\sigma} &= \boldsymbol{L}_0\boldsymbol{\varepsilon}_0 + \boldsymbol{L}_0\boldsymbol{\varepsilon}' + \boldsymbol{p} \\
&= \boldsymbol{L}\boldsymbol{\varepsilon}_0 - \boldsymbol{L}_0\boldsymbol{\Gamma}\boldsymbol{p} + \boldsymbol{p} \\
&= \big(\boldsymbol{L} - \boldsymbol{L}\boldsymbol{\Gamma}(\boldsymbol{L} - \boldsymbol{L}_0) + \boldsymbol{L}\boldsymbol{\Gamma}(\boldsymbol{L} - \boldsymbol{L}_0)\boldsymbol{\Gamma}(\boldsymbol{L} - \boldsymbol{L}_0) - \boldsymbol{L}\boldsymbol{\Gamma}(\boldsymbol{L} - \boldsymbol{L}_0)\boldsymbol{\Gamma}(\boldsymbol{L} - \boldsymbol{L}_0) \\
&\quad \boldsymbol{\Gamma}(\boldsymbol{L} - \boldsymbol{L}_0) - \cdots\big)\boldsymbol{\varepsilon}_0
\end{aligned} \tag{1.9}$$

which indicates that the stress is nonlocal with respect to the slow-scale strain $\boldsymbol{\varepsilon}_0$. In the classical nonlocal formulation the stress is given in terms of a nonlocal kernel K, as

$$\boldsymbol{\sigma}(\boldsymbol{x}) = \int_D \boldsymbol{K}(\boldsymbol{x},\boldsymbol{x}')\boldsymbol{\varepsilon}_0(\boldsymbol{x}')d\boldsymbol{x}' \tag{1.10}$$

where the nonlocal kernel is usually assumed empirically (e.g. an exponential function). This kernel issue is considered to be a major drawback of the nonlocal formulation. By comparing (1.9) with (1.10), a rigorous mathematical expression for the nonlocal kernel is actually obtained as

$$\begin{aligned}
\boldsymbol{K} &= \boldsymbol{L} - \boldsymbol{L}\boldsymbol{\Gamma}(\boldsymbol{L} - \boldsymbol{L}_0) + \boldsymbol{L}\boldsymbol{\Gamma}(\boldsymbol{L} - \boldsymbol{L}_0)\boldsymbol{\Gamma}(\boldsymbol{L} - \boldsymbol{L}_0) - \boldsymbol{L}\boldsymbol{\Gamma}(\boldsymbol{L} - \boldsymbol{L}_0)\boldsymbol{\Gamma}(\boldsymbol{L} - \boldsymbol{L}_0) \\
&\quad \boldsymbol{\Gamma}(\boldsymbol{L} - \boldsymbol{L}_0) - \cdots
\end{aligned} \tag{1.11}$$

In engineering practices, microstructure information in a BVP domain is always incomplete, subject to randomness and uncertainty of heterogeneities. Deformation strain of a BVP is conventionally analyzed by using certain "smeared" homogeneous elastic moduli, corresponding to the slow-scale strain $\boldsymbol{\varepsilon}_0$ of the problem (1.2). With such an "apparent" strain to represent the true strain at a particular location, as Eq. (1.9) shows, the constitutive law at any individual location becomes nonlocal.

The result in (1.11) explains partly why certain regularization, such as the nonlocal or the strain gradient formulation, should be necessarily taken to prevent spurious mesh dependence of a material softening problem. In a BVP even initially characterized with approximately homogeneous elastic moduli, mechanical and environmental loading upon the body eventually yields microscopic cracks, flaws, cavities, etc., resulting in a heterogeneous field of elastic moduli $\boldsymbol{L}(\boldsymbol{x})$ and the nonlocal kernel (1.11).

1.3.3 Strain gradient formulation of scale-coupling mechanics

By employing a Taylor expansion of the slow-scale strain as

$$\boldsymbol{\varepsilon}_0(\boldsymbol{x}') = \boldsymbol{\varepsilon}_0(\boldsymbol{x}) + (\boldsymbol{x}' - \boldsymbol{x}) \cdot \nabla \boldsymbol{\varepsilon}_0(\boldsymbol{x}) + \frac{1}{2}(\boldsymbol{x}' - \boldsymbol{x})(\boldsymbol{x}' - \boldsymbol{x})\nabla^2 \boldsymbol{\varepsilon}_0(\boldsymbol{x}) + \cdots \tag{1.12}$$

and substituting the expansion into the nonlocal expression (1.10), it yields a strain gradient form

$$\boldsymbol{\sigma}(\boldsymbol{x}) = \boldsymbol{L}(\boldsymbol{x})\boldsymbol{\varepsilon}_0(\boldsymbol{x}) + \boldsymbol{L}'(\boldsymbol{x})\nabla \boldsymbol{\varepsilon}_0(\boldsymbol{x}) + \boldsymbol{L}''(\boldsymbol{x})\nabla^2 \boldsymbol{\varepsilon}_0(\boldsymbol{x}) + \cdots \tag{1.13}$$

with the first two higher order constitutive tensors given as

$$\boldsymbol{L}'(\boldsymbol{x}) = \int_D \boldsymbol{K}(\boldsymbol{x}, \boldsymbol{x}')(\boldsymbol{x}' - \boldsymbol{x})\, d\boldsymbol{x}' \tag{1.14}$$

$$\boldsymbol{L}''(\boldsymbol{x}) = \frac{1}{2}\int_D \boldsymbol{K}(\boldsymbol{x}, \boldsymbol{x}')(\boldsymbol{x}' - \boldsymbol{x})(\boldsymbol{x}' - \boldsymbol{x}) d\boldsymbol{x}' \tag{1.15}$$

As shown in Eq. (1.13), the main idea of generalized continua is recovered from the strain gradient formulation. In fact, the fast-scale and slow-scale variables employed in subsection 1.3.2 correspond to a microvolume and a macrovolume, respectively, defined in Mindlin's work (1964), while the decomposition of a BVP and the use of a Green function here are considered to be theoretically simple and straightforward.

1.3.4 Representative volume element

Denote $\mathcal{L}$, L, ℓ_c characteristic lengths of a BVP domain size, an intermediate window size, and constituent heterogeneities, respectively. The BVP size $\mathcal{L}$ is characterized by the most critical loading dimension of a BVP, e.g. the thickness of a beam. The window size L is proportional to the wavelength of a slow-scale strain $\boldsymbol{\varepsilon}_0$. The window size L is chosen such that within the window domain Y the slow-scale strain is approximately constant (e.g. the domain of a linear finite element). The characteristic length of heterogeneities, ℓ_c, is proportional to the average size of the inclusions in a particulate composite, which is more rigorously defined as a correlation length in Chapter 2. Among these three length scales, there are two types of interscale separation or coupling to be clarified:

i. *Type-I scale separation* refers to the case in which a BVP size is significantly larger than the window size, i.e. $L << \mathcal{L}$. When boundary effects of a BVP are negligible, the modified Green function $\boldsymbol{\Gamma}$ in the nonlocal formula (1.9) can be approximated as $\boldsymbol{\Gamma}^{\infty}$, the one in infinite space. By applying volume averaging over the window domain on the both sides of (1.9), it yields

$$\langle \boldsymbol{\sigma} \rangle = \left(\langle \boldsymbol{L} \rangle - \left\langle \boldsymbol{L}\boldsymbol{\Gamma}^{\infty}(\boldsymbol{L}-\boldsymbol{L}_0) \right\rangle + \left\langle \boldsymbol{L}\boldsymbol{\Gamma}^{\infty}(\boldsymbol{L}-\boldsymbol{L}_0)\boldsymbol{\Gamma}^{\infty}(\boldsymbol{L}-\boldsymbol{L}_0) \right\rangle - \cdots \right) \boldsymbol{\varepsilon}_0 \tag{1.16}$$

where the bracket denotes the volume averaging over Y. Note that the $(n+1)$th term on the right-hand side of Eq. (1.16) involves n-fold convolution of $\boldsymbol{\Gamma}^{\infty}$ operating on the (n+1)-point correlation function;

ii. *Type-II scale separation* refers to the scale separation between a window size L and a characteristic length ℓ_c of microstructure, i.e. $\ell_c << L$. In this case, each individual term on the right-hand side of Eq. (1.16) converges to a result independent of the ratio between ℓ_c and L. For example, the third term $\left\langle \boldsymbol{L}\boldsymbol{\Gamma}^{\infty}(\boldsymbol{L}-\boldsymbol{L}_0)\boldsymbol{\Gamma}^{\infty}(\boldsymbol{L}-\boldsymbol{L}_0) \right\rangle$ involving the triple correlation can be analytically simplified into an explicit result in terms of a pair of geometric parameters of the bulk or shear modulus between 0 and 1, as detailed in Chapter 4.

When the assumption of Type-II scale separation is satisfied, a formal name for window Y is well known as representative volume element (RVE). It should be emphasized that only when both Type-I and Type-II scale separation are established can such an RVE concept become applicable to a mechanics analysis. Since all materials are composed of microscopic heterogeneities, it is not an exaggeration to state that all conventional mechanics analyses based on "smeared" material constants follow such an RVE approach, most typically in a finite element analysis. Part I of this book covering Chapters 3 to 6 is devoted to homogenization of such scale separation problems.

When Type-I scale separation is not established (i.e. an RVE size L and a BVP size $\mathcal{L}$ are not well separated), such a BVP is called *scale-coupling*, onto which an RVE-based approach becomes inapplicable. Examples of scale-coupling BVPs are a concrete beam or a thin film with its thickness comparable to the RVE size, and a boundary layer existing beneath the surface of almost any material with the depth comparable to the characteristic size of microstructure. Part II of this book covering Chapters 7 to 9 introduces some recently developed models to tackle such scale-coupling problems.

It is especially remarked that, when Type-II scale separation is not satisfied, i.e. an RVE size or a strain wavelength L is not sufficiently larger than the microstructure size ℓ_c of a medium—the results for all the terms on the right-hand side of (1.16) become size-dependent, or mesh-dependent in the finite element case. A typical scale-coupling example of Type-II is that in elastodynamics there are a number of strain wavelengths coupling to the characteristic length of heterogeneities in a medium. Such a scale-coupling or size effect of RVE is specifically investigated and quantified in Chapter 7.

The previously discussed scale-coupling mechanics has been described in spatial lengths, while the concept of which is equally applicable to a dynamics problem in the temporal domain. For instance, in stochastic dynamics a white noise excitation is an essential assumption of scale separation made relative to the relaxation time or natural period of an oscillator, which leads to some most important analytical results in dynamics. When the characteristic time scale of a random excitation approaches the natural period of an oscillator, a phenomenon of so-called "probabilistic resonance" occurs (Xu, 2014). The scale-coupling effect in a one-dimensional time domain is observed to be much more strongly amplified than in a three-dimensional spatial domain, especially with regard to propagation of uncertainty, e.g. in chaotic systems.

As a concluding remark of this chapter, the scale-coupling issue is emphasized as the most critical topic of multiscale research. In a scale-coupling problem, when a conventional theory of scale separation is obstinately applied to modeling of a relevant scale-coupling phenomenon, a price often paid is that people have to end up with a physically weird model with some physically inconsistent variables or parameters (e.g. nonlocal theory in damage mechanics, negative mass, or negative moduli in elastodynamics). Therefore, in this author's viewpoint, some of the greatest forthcoming breakthroughs in mechanics are to be made on theory and models of scale-coupling mechanics in the years to come.

Chapter 2

Random morphology and correlation functions

Physical behavior of natural and synthetic heterogeneous media is a fundamental research topic in a variety of scientific and engineering disciplines, such as geophysics, material science, chemical physics, biomedical engineering, and civil and mechanical engineering (Nemat-Nessar & Hori, 1999; Milton, 2002; Torquato, 2002; Sahimi, 2003; Ostoja-Starzewski, 2008; Auriault, Boutin, & Geindreau, 2009). Examples of application include transport, electromagnetic, and mechanical properties of cellular solids, colloids, tissue, bone, rocks, soils, and manufactured composites, such as fiber reinforced plastics, foamed solids, and polymer blends, especially in the rapidly growing area of nanotechnology. To formulate a mechanics theory of random composites, it is necessary to first have some basic understanding of random field theory and statistical descriptors of morphology.

In modeling of random morphology of heterogeneous media there are two fundamental tasks (i.e. characterization and simulation):

i. Characterization: appropriate translation of morphological information that is generally perceivable by human vision into a mathematically tractable model; and
ii. Simulation: applicability of a morphological model to digital sampling, i.e., whether a model enables convenient computer generation of Monte Carlo samples with desired morphological configurations.

A straightforward mathematical tool to tackle morphological problems is the theory of stochastic processes, i.e, to view each morphological pattern as a stochastic process that is completely defined by a multivariate distribution, or approximately characterized with partial statistical information such as a hierarchical order of statistical moments or correlation functions. Preliminaries of random field theory are introduced in Section 2.1 with modeling of Gaussian random fields. There are mainly two models to simulate non-Gaussian random fields and morphologies, i.e. translation model and correlation model, which are presented in Sections 2.2 and 2.3, respectively.

2.1 GAUSSIAN RANDOM FIELDS

2.1.1 Preliminaries

A Gaussian random variable Z is characterized with the following *probability density function* (PDF)

$$\phi(z) = \frac{1}{\sigma\sqrt{2\pi}} e^{-\frac{(z-\bar{Z})^2}{2\sigma^2}} \tag{2.1a}$$

where $\bar{Z}$ and σ denote the mean value and the standard deviation, respectively (The overbar symbol in this book is exclusively used to denote statistical expectation or ensemble averaging). The corresponding *cumulative distribution function* (CDF) of (2.1a) is given as

$$\begin{aligned}\Phi(z) &= \int_{-\infty}^{z} \phi(x)\,dx \\ &= \int_{-\infty}^{z} \frac{1}{\sigma\sqrt{2\pi}} e^{-\frac{(x-\bar{Z})^2}{2\sigma^2}} dx\end{aligned} \tag{2.1b}$$

When $\bar{Z} = 0$ and $\sigma = 1$, the probability distribution is called the *standard Gaussian distribution.*

Definition

A random field $Z(\mathbf{r},\vartheta)$, $\vartheta \in \Theta$ random space, is Gaussian if and only if for every finite set of spatial points $\mathbf{r}_1,\mathbf{r}_2,\cdots,\mathbf{r}_k$ in a spatial domain D, the joint distribution of random variables $Z_1 = Z(\mathbf{r}_1)$, $Z_2 = Z(\mathbf{r}_2)$,..., $Z_k = Z(\mathbf{r}_k)$ follows a multivariate Gaussian probability density function

$$\phi_k(z) = \frac{1}{\sqrt{(2\pi)^k |C|}} e^{-\frac{1}{2}\{z-\bar{Z}\}^T [C]^{-1} \{z-\bar{Z}\}} \tag{2.2}$$

where the mean value vector $\{\bar{Z}\} = \{\bar{Z}_1,\bar{Z}_2,\cdots,\bar{Z}_k\}^T$, [C] is a symmetric positive semidefinite covariance matrix with its component $[C]_{ij} = \overline{(Z_i-\bar{Z}_i)\left(Z_j^* - \bar{Z}_j^*\right)}$ representing the covariance between random variables Z_i and Z_j, and the asterisk denotes complex conjugate.

Example

When the covariance matrix [C] in Eq. (2.2) is diagonal with all the diagonal entries equal to σ^2 and $\{\bar{Z}\} = 0$, the Gaussian random variables Z_1, Z_2, ..., Z_k are independent and identically distributed. Such a random process is called a *Gaussian white noise process.*

Example

The time integral of a Gaussian white noise process $dW(t,\vartheta)$

$$W(t,\vartheta) = \int_0^t dW(s,\vartheta) \tag{2.3}$$

results in a so-called Wiener process, or Brownian motion, which is the most widely studied in stochastic process theory. Its extension to two-dimension is called a *Gaussian free field.*

Example

When the covariance matrix [C] is singular, the multivariate PDF (2.2) does not exist, although such a Gaussian field still exists, e.g., a two-dimensional random field $Z(x_1,x_2,\vartheta) = \varsigma(\vartheta)\sin(x_1)\cos(x_2)$ with ς a Gaussian random variable. These types of Gaussian fields are called *degenerate Gaussian fields.*

Definition

A random field is strongly homogeneous or homogeneous in the strict sense when its probability distribution does not change with regard to a spatial shift. If only the mean and covariance are spatial invariant, it is called weakly homogeneous *or* homogeneous in the wide sense.

Since a Gaussian process or field is completely defined by its mean and covariance, the wide-sense stationary or homogeneity in the Gaussian case becomes equivalent to the strict-sense stationary or homogeneity. The correlation coefficient of a random field Z is defined as

$$\rho(\boldsymbol{r},\boldsymbol{s}) = \frac{\overline{\left(Z(\boldsymbol{r},\vartheta) - \bar{Z}(\boldsymbol{r})\right)\left(Z^*(\boldsymbol{s},\vartheta) - \bar{Z}^*(\boldsymbol{s})\right)}}{\sigma(\boldsymbol{r})\sigma^*(\boldsymbol{s})} \tag{2.4}$$

where $\sigma(\boldsymbol{r})$denotes the standard deviation of the random field at point $\boldsymbol{r}$. In a homogeneous random field, the correlation coefficient $\rho(\boldsymbol{r},\boldsymbol{s})$ reduces to $\rho(\boldsymbol{r}-\boldsymbol{s})$. The correlation coefficient $\rho(\boldsymbol{r})$ is linked to its *power spectral density* (PSD) $\hat{\rho}(\boldsymbol{\xi})$ via the following Wiener-Khinchin theorem

$$\hat{\rho}(\boldsymbol{\xi}) = \int \rho(\boldsymbol{r})e^{-i\boldsymbol{\xi}\cdot\boldsymbol{r}}\, d\boldsymbol{r} \tag{2.5a}$$

$$\rho(\boldsymbol{r}) = \frac{1}{(2\pi)^d}\int \hat{\rho}(\boldsymbol{\xi})e^{i\boldsymbol{\xi}\cdot\boldsymbol{r}}\, d\boldsymbol{\xi} \tag{2.5b}$$

where the hat indicates the Fourier image, $d = 2$ or 3 indicates dimensionality, and the Fourier pair $\rho(\boldsymbol{r})$ and $\hat{\rho}(\boldsymbol{\xi})$ satisfy the condition of integrability.

Example

A well-known example of nonstationary Gaussian processes is the Wiener process $W(t,\vartheta)$ (2.3) with the zero mean and the variance equal to t.

Example

An example of stationary Gaussian processes is the Ornstein-Uhlenbeck process, which is expressed in terms of the Wiener process (2.3) as

$$Z(t,\vartheta)=e^{-t}W(e^{2t},\vartheta)$$

Definition

A random field is strongly ergodic, or, simply said, ergodic, if its probability distribution can be deduced from a single and sufficiently large sample. If only the mean and covariance functions such measured converge to the true ones, the random field is called weakly ergodic *or* ergodic in the wide sense.

Example

An example of ergodic random fields is a Gaussian free field. In fact, a homogeneous Gaussian random field characterized with a continuous PSD is always ergodic.

Since an ergodic random field has its ensemble averaging identical to its spatial averaging, with the latter being spatial invariant, it implies that statistical homogeneity is a necessary but not a sufficient condition for ergodicity. In other words, an ergodic field is always homogenous, while a homogeneous field is not necessarily ergodic. For instance, a Gaussian free field plus a Gaussian random variable is homogeneous but nonergodic.

Ergodicity is an important and frequently used assumption in many theories and models of physical sciences and engineering. In some scientific and engineering fields, many researchers are even unware of the assumption usually implicitly made, not mentioning checking its validity for a particular application (e.g., molecular dynamics). In random field modeling and simulation, the ergodicity assumption is usually made. However, in a concrete application, the assumption should be carefully checked (e.g., morphology of percolation discussed at the end of this chapter).

Definition

In an isotropic random field that satisfies $\rho(\boldsymbol{r})=\rho(|\boldsymbol{r}|)$, the length scale of random fluctuation is characterized with the correlation length.

By equating to the integral $\int_{R^d} \rho(\boldsymbol{r})d\boldsymbol{r}$ *the volume of the sphere or the area of the circle with* ℓ_c *being the radius, the correlation length is defined as*

$$\ell_c = \sqrt[3]{3\int_0^{\infty} \rho(r) r^2 \, dr} \tag{2.6a}$$

in 3D, and

$$\ell_c = \sqrt{2\int_0^{\infty} \rho(r) r \, dr} \tag{2.6b}$$

in 2D.

2.1.2 Random variable representation of Gaussian random fields

There are three main random variable based expansions to represent a Gaussian random field, described as follows:

Karhunen-Loève (K-L) Expansion

A Gaussian random field can be biorthogonally decomposed into a series of orthonormal bases $\varphi_n(\boldsymbol{r})$ and a series of independent standard Gaussian variables Z_n, as follows:

$$Z(\boldsymbol{r},\vartheta) = \sum_{n=1}^{\infty} \sqrt{\lambda_n} Z_n(\vartheta)\varphi_n(\boldsymbol{r}) + \bar{Z}(\boldsymbol{r}) \tag{2.7}$$

where

$$\int_D \varphi_m(\boldsymbol{r})\varphi_n^*(\boldsymbol{r})d\boldsymbol{r} = \delta_{mn} \tag{2.8}$$

with D denoting the spatial domain, and δ_{mn} the Kronecker-delta function. By substituting the expansion (2.7) into the following definition of the covariance function

$$C(\boldsymbol{r},\boldsymbol{s}) = \overline{\left(Z(\boldsymbol{r},\vartheta) - \bar{Z}(\boldsymbol{r})\right)\left(Z^*(\boldsymbol{s},\vartheta) - \bar{Z}^*(\boldsymbol{s})\right)} \tag{2.9}$$

it yields

$$C(\boldsymbol{r},\boldsymbol{s}) = \sum_{n=1}^{\infty} \lambda_n \varphi_n(\boldsymbol{r})\varphi_n^*(\boldsymbol{s}) \tag{2.10}$$

Given that the kernel $C(\boldsymbol{r},\boldsymbol{s})$ is continuous and positive semidefinite, the validity of the K-L expansion (2.7) with a sequence of positive eigenvalues λ_n is established by Mercer's theorem, which is equivalent to spectral theorem by applying an inner product with $\varphi_n(\boldsymbol{s})$ onto both sides of (2.10) i.e.

$$\int_D C(\boldsymbol{r},\boldsymbol{s})\varphi_n(\boldsymbol{s})d\boldsymbol{s} = \lambda_n\varphi_n(\boldsymbol{r}) \tag{2.11}$$

In a random field, the multidimensional Fredholm integral equation (2.11) is usually solved numerically, except for few cases analytical solvable. An example of such a covariance function is given as (e.g. Ghanem & Spanos, 1991 and references therein)

$$C(\boldsymbol{r},\boldsymbol{s}) = \exp\left[-\left(\frac{|r_1 - s_1|}{l_1} + \frac{|r_2 - s_2|}{l_2}\right)\right] \tag{2.12}$$

It should be noted that the K-L expansion is not limited to Gaussian random fields. However, when the random variables in expansion (2.7) are not Gaussian, a resulting random field becomes statistically complicated and normally unsuitable for modeling purpose.

Fourier Series Expansion

In a homogeneous Gaussian random field the kernel in (2.11) reduces to $C(\boldsymbol{r} - \boldsymbol{s})$, and the left hand side of (2.11) becomes a convolution. When the size of a domain D is sufficiently larger than the correlation length ℓ_c, by applying convolution theorem, it yields that the power spectral density $\hat{C}$ has its value at the nth wave vector ξ_n identical to the nth eigenvalue, i.e.,

$$\hat{C}(\xi_n) = \lambda_n \tag{2.13}$$

Define the domain D as a rectangular parallelepiped $\left(-\frac{L_1}{2},\frac{L_1}{2}\right)\times\left(-\frac{L_2}{2},\frac{L_2}{2}\right)\times\left(-\frac{L_3}{2},\frac{L_3}{2}\right)$ in the Cartesian coordinate system, with the nth wave vector written as $\xi_n = \left\{\frac{2n\pi}{L_1},\frac{2n\pi}{L_2},\frac{2n\pi}{L_3}\right\}^T$.

Corresponding to the previous use of the convolution theorem, the eigenfunctions satisfying (2.8) are actually a Fourier series, i.e.:

$$\varphi_n(\boldsymbol{r}) = \sqrt{\frac{\Delta\xi_1\Delta\xi_2\Delta\xi_3}{(2\pi)^3}}e^{i\xi_n\cdot\boldsymbol{r}} \quad \text{or} \quad \varphi_n(\boldsymbol{r}) = \sqrt{\frac{1}{L_1L_2L_3}}e^{i\xi_n\cdot\boldsymbol{r}} \tag{2.14}$$

where the spacing vector of wave numbers $\Delta\boldsymbol{\xi} = \left\{ \frac{2\pi}{L_1}, \frac{2\pi}{L_2}, \frac{2\pi}{L_3} \right\}^T$. By substituting (2.13) and (2.14) into (2.7), it results in a Fourier series representation of a 3D homogeneous Gaussian random field Z, as

$$Z(\boldsymbol{r},\vartheta) = \sum_{n=1}^{\infty} \sqrt{\frac{\hat{C}(\boldsymbol{\xi}_n)}{L_1 L_2 L_3}} Z_n(\vartheta) e^{i\xi_n \cdot \boldsymbol{r}} + \bar{Z} \tag{2.15}$$

By substituting (2.15) into (2.9), the discrete version of the Wiener-Khinchin relation is recovered as follows:

$$C(\boldsymbol{r}-\boldsymbol{s}) = \frac{1}{L_1 L_2 L_3} \sum_{n=1}^{\infty} \hat{C}(\boldsymbol{\xi}_n) e^{i\xi_n \cdot (\boldsymbol{r}-\boldsymbol{s})} \tag{2.16}$$

To take advantage of fast Fourier transform (FFT) in digital generation of Gaussian random field samples, the Gaussian random variables in (2.14) are normally incorporated into the complex exponentials as

$$Z(\boldsymbol{r},\vartheta) = \sum_{n=1}^{\infty} \sqrt{\frac{\hat{C}(\boldsymbol{\xi}_n)}{L_1 L_2 L_3}} e^{i(\xi_n \cdot \boldsymbol{r} + 2\pi U_n(\vartheta))} + \bar{Z} \tag{2.17}$$

where U_n are independent standard uniformly distributed random variables in [0,1]. Equivalence between (2.17) and (2.15) is verified by substituting (2.17) into (2.9), which, by further invoking the central limit theorem, yields (2.15).

This derivation indicates that the Fourier series expansion can be treated as a special case of the K-L expansion when a domain size becomes sufficiently larger than the correlation length (i.e. Type-II scale separation defined in Subsection 1.3.4 is satisfied).

Cholesky Decomposition

Given the covariance matrix [C] of a Gaussian random field, a random sample can be generated from a Gaussian free field by using Cholesky decomposition

$$[\mathrm{C}] = [\mathrm{L_C}][\mathrm{L_C}]^T \tag{2.18}$$

where $[\mathrm{L_C}]$ is a lower triangular matrix with positive diagonal entries. Rearrange a standard Gaussian free field sample into a vector {w}.

A sample vector {z}, corresponding to a correlated zero-mean Gaussian random field, is generated as follows:

$$\{z\} = [L_C]\{w\} \tag{2.19}$$

which is simply verified by noting the identity between $[C] = \overline{\{Z\}(\vartheta)\{Z\}^T(\vartheta)}$ and (2.18).

Summary

The K-L expansion performs best in a small spatial domain with the domain size not significantly larger than the correlation length, while a random field can be either homogeneous or nonhomogeneous. Its unique advantage is that by compressing the random field information into a small number of random variables, or a so-called low dimensional setting, the K-L expansion enables resolving of such a random field problem semi-analytically, which is much more efficient than traditional Monte Carlo method.

As a special case of the K-L expansion applied to a homogenous random field with a large domain, the Fourier series expansion has its numerical advantage in digital generation of random samples by employing FFT. However, a high dimension setting of the expansion makes it unsuitable to a random-variable-based stochastic method.

While Cholesky decomposition is applicable to both homogenous and nonhomogeneous random fields, its intermediate position between the other two expansions usually makes it the least advantageous. With regard to a random field in a large domain, Cholesky decomposition cannot compete with FFT in numerical efficiency, while in the small-domain regime, the advantage of the low-dimensional K-L expansion becomes overwhelming.

2.2 NON-GAUSSIAN RANDOM FIELDS: THE TRANSLATION MODEL

2.2.1 Definition of the translation model

There is no such a universal method available to simulate a stochastic process or a random field with any given multivariate distribution. Most of existing simulation methods are based on an underlying Gaussian process or field, and the rest either present formidable computational challenges or are too restrictive to be of general interest (Johnson, 1987). Research on sample generation of non-Gaussian random processes started in the 1950s. Due to statistical complexity of non-Gaussian random processes, a non-Gaussian sample is typically generated based on the first two order statistics, the marginal PDF and the autocorrelation function. The translation model, point-wise translating a Gaussian process into a non-Gaussian one, was first studied in nonlinear systems and signal processing, and was mathematically developed in the bivariate translation system (Mardia, 1970). The translation model was later applied to multivariate systems (e.g. Grigoriu, 1984), porous media (Quiblier, 1984; Adler, 1992; Giona & Adrover, 1996), random heterogeneous materials (Cahn, 1965; Berk, 1991; Roberts & Teubner, 1995; Roberts & Knackstedt, 1996), and electrical engineering systems (Li & Hammond, 1975; Cambanis & Masry, 1978; Liu & Munson, 1982). The translation model has served as a major approach of non-Gaussian simulation. The range of its application however is limited by its inherent restrictions (i.e. a multivariate distribution with an underlying Gaussian structure, and the required positive semidefiniteness of an underlying covariance function). The translation model and its limitations are described in the context of random fields:

Definition

Given $Z(\mathbf{r},\vartheta)$ a homogeneous Gaussian random field as an input, in the translation model a non-Gaussian random field $Y(\mathbf{r},\vartheta)$ characterized with a nondegenerate marginal cumulative distribution F_Y, is obtained through a point-wise CDF-based monotonic translation

$$Y(\mathbf{r},\vartheta) = F_Y^{-1}\left(\Phi\left(Z(\mathbf{r},\vartheta)\right)\right) \tag{2.20}$$

Denote ρ_Z and ρ_Y the correlation coefficients of $Z(\mathbf{r},\vartheta)$ and $Y(\mathbf{r},\vartheta)$, respectively, and define the point-wise mapping of them as a function g, i.e. $\rho_Y(\mathbf{r}) = g\left(\rho_Z(\mathbf{r})\right)$. The following properties of the translation model are shown to be true:

Property

a. $Y(\mathbf{r},\vartheta)$ is strictly homogenous;

b. $\left|\rho_Y(\mathbf{r})\right| \le \left|\rho_Z(\mathbf{r})\right| \le 1$;

c. $\rho_Z(\mathbf{r}) = 0 \Leftrightarrow \rho_Y(\mathbf{r}) = 0$; $\rho_Z(\mathbf{r}) = 1 \Leftrightarrow \rho_Y(\mathbf{r}) = 1$;

d. $g'(\rho_Z) = \dfrac{dg(\rho_Z)}{d\rho_Z} > 0$; $\rho_Z(\mathbf{r}) > (<) 0 \Leftrightarrow \rho_Y(\mathbf{r}) > (<) 0$; Inf $\rho_Y = g(-1)$.

An N-point random field $Y(\mathbf{r},\vartheta)$ is completely defined by its multivariate distribution as follows:

$$F_Y(y_1,\cdots,y_N) = P\left(Y(\mathbf{r}_1) \le y_1,\cdots,Y(\mathbf{r}_N) \le y_N\right) \tag{2.21}$$

which, through the translation (2.20), is equivalent to

$$P\left(Z(\mathbf{r}_1) \le \Phi^{-1}\left(F_Y(y_1)\right),\cdots,Z(\mathbf{r}_N) \le \Phi^{-1}\left(F_Y(y_n)\right)\right) = \Phi_N\left(\Phi^{-1}\left(F_Y(y_1)\right),\cdots\Phi^{-1}\left(F_Y(y_N)\right)\right)$$

Since the given Gaussian multivariate distribution is strictly homogeneous, Property *a* is obtained.

We henceforth handle the translated random field $Y(\boldsymbol{r},\vartheta)$ in the normalized version

$$\tilde{Y}(\boldsymbol{r},\vartheta)=\frac{Y(\boldsymbol{r},\vartheta)-\bar{Y}}{\sqrt{\overline{Y^2}(\boldsymbol{r})-\bar{Y}^2}} \tag{2.22}$$

with the tilde indicating normalization. The translation (2.20) is equivalent to

$$Y(\boldsymbol{r},\vartheta)=F_Y^{-1}\left(\Phi\left(\tilde{Z}(\boldsymbol{r},\vartheta)\right)\right) \tag{2.23}$$

where the normalized Gaussian field $\tilde{Z}(\boldsymbol{r},\vartheta)=\frac{Z(\boldsymbol{r},\vartheta)-\bar{Z}}{\sigma_Z}$. Denote $\mathcal{T}$ the combined operation of (2.22) and (2.23), i.e. $\tilde{Y}(\boldsymbol{r},\vartheta)=\mathcal{T}\left(\tilde{Z}(\boldsymbol{r},\vartheta)\right)$, and expand the operator in the Fourier Hermite series as follows:

$$\mathcal{T}(\tilde{Z})=\sum_{n=0}^{\infty}a_n\frac{H_n(\tilde{Z})}{\sqrt{n!}} \tag{2.24}$$

where the coefficients

$$a_n=\int_{-\infty}^{+\infty}\mathcal{T}(z)\frac{H_n(z)}{\sqrt{n!}}\tilde{\phi}(z)dz \tag{2.25}$$

and the Hermite polynomials

$$H_n(z)=(-1)^n\frac{\left(\sqrt{2\pi}\right)^{n-1}}{\tilde{\phi}(z)}\frac{d^n\left(\tilde{\phi}(z)\right)}{dz^n} \tag{2.26}$$

with $\tilde{\phi}(z)$ the standard Gaussian PDF. By using the definition

$$\rho_Y(\boldsymbol{r}-\mathbf{s})=\overline{\mathcal{T}\left(\tilde{Z}(\boldsymbol{r},\vartheta)\right)\mathcal{T}\left(\tilde{Z}(\mathbf{s},\vartheta)\right)} \tag{2.27}$$

and pair orthogonality of the Hermite polynomials

$$\overline{H_m\left(\tilde{Z}(\boldsymbol{r},\vartheta)\right)H_n\left(\tilde{Z}(\mathbf{s},\vartheta)\right)}=\rho_Z^n(\boldsymbol{r}-\mathbf{s})n!\delta_{mn} \tag{2.28}$$

the correlation coefficient of $Y(\boldsymbol{r},\vartheta)$ is obtained as

$$\rho_Y(\boldsymbol{r}) = \sum_{n=1}^{\infty} a_n^2 \rho_Z^n(\boldsymbol{r}) \tag{2.29}$$

When $\boldsymbol{r} = 0$, it follows that $\rho_Z(\boldsymbol{r}) = 1 \Leftrightarrow \rho_Y(\boldsymbol{r}) = 1$ and

$$\sum_{n=1}^{\infty} a_n^2 = 1 \tag{2.30}$$

The result (2.29) directly yields Properties *b* and *c*. To obtain *d*, we need to prove the following statement:

Given a point-wise nonlinearity $\mathcal{T}$ and a Gaussian input $\tilde{Z}$, if the operator $\mathcal{T}$ is monotonic and nonconstant almost everywhere, then the correlation coefficient ρ_Z of the Gaussian image is uniquely determined by the correlation coefficient ρ_Y of the non-Gaussian translation.

Proof: The requirement of one-to-one mapping g is equivalent to having $g'(\rho_Z) > 0$ in the domain $[-1,1]$, which holds if we can show

$$g'(\rho_Z) = \int_{-\infty}^{\infty}\int_{-\infty}^{\infty} \frac{d\mathcal{T}(z_1)}{dz_1}\frac{d\mathcal{T}(z_2)}{dz_2}\tilde{\phi}_2(z_1,z_2;\rho_Z)\,dz_1\,dz_2 \tag{2.31}$$

given that $\mathcal{T}$ is monotonic and nonconstant almost everywhere. Expand the bivariate normal PDF in the Fourier Hermite series, as

$$\tilde{\phi}_2(z_1,z_2;\rho_Z) = \tilde{\phi}(z_1)\tilde{\phi}(z_2)\sum_{n=0}^{\infty}\frac{\rho_Z^n}{n!}H_n(z_1)H_n(z_2) \tag{2.32}$$

By substituting (2.24), (2.32) and the following recurrence relation of the Hermite polynomials

$$\frac{dH_n(z)}{dz} = nH_{n-1}(z) \tag{2.33}$$

into the right-hand side of (2.31), the latter becomes

$$\int_{-\infty}^{\infty}\int_{-\infty}^{\infty}\sum_{m=1}^{\infty} a_m \frac{mH_{m-1}(z_1)}{\sqrt{m!}}\sum_{n=1}^{\infty} a_n \frac{nH_{n-1}(z_2)}{\sqrt{n!}}\left[\tilde{\phi}(z_1)\tilde{\phi}(z_2)\sum_{k=0}^{\infty}\frac{\rho^k}{k!}H_k(z_1)H_k(z_2)\right]dz_1 dz_2 \tag{2.34}$$

With the pair orthogonality (2.28), it is straightforward to simplify (2.34) into

$$\sum_{n=1}^{\infty} n a_n^2 \rho_Z^{n-1} \tag{2.35}$$

which is identical to the derivative directly obtained from (2.29). Equation (2.31) is therefore established. □

This statement can be succinctly rephrased as follows:

An underlying Gaussian image is uniquely determined by a monotonic nonlinearity and the correlation coefficient of its non-Gaussian output.

With this result, it is straightforward to obtain Property *d*. Note that Properties *a*, *b*, and *c* are applicable to general point-wise nonlinearities, or zero-memory nonlinearities in the time domain, while *d* is restricted to monotonic point-wise nonlinearities only.

It is recognized that the expression (2.31) is also a special case of Price's theorem (Price, 1958). By using the series expansion, the exact Price's theorem can be recovered without the additional restrictions imposed on nonlinearity $\mathcal{T}$ (Price, 1958; Cambanis & Masry, 1978), which was also shown in Papoulis (1989) by using a different approach.

2.2.2 Algorithm to find the underlying Gaussian image

There are few cases of analytical solvability available between a non-Gaussian translation and its underlying Gaussian image. One example is the so-called Debye symmetric random medium (which is further discussed in Section 2.3) that has the volume fraction $c_1 = c_2 = 0.5$ and the correlation coefficient

$$\rho_Y(\boldsymbol{r}) = \exp\left(-\frac{|\boldsymbol{r}|}{\ell_c}\right) \tag{2.36}$$

The correlation coefficient of the underlying Gaussian image is analytically available as (Kotz, Balakrishnan, & Johnson, 2000, p. 263)

$$\rho_Z(\boldsymbol{r}) = \sin\left[\frac{\pi}{2}\exp\left(-\frac{|\boldsymbol{r}|}{\ell_c}\right)\right] \tag{2.37}$$

In general, given the prescribed PDF and the correlation coefficient of a non-Gaussian target, the Gaussian image need be solved numerically and a number of algorithms have been developed (e.g. Li and Hammond, 1975;

Yamazaki and Shinozuka, 1988; Grigoriu, 1998; Deodatis and Micaletti, 2001; Puig, Poirion, & Soize, 2002). A straightforward approach is to list a one-to-one mapping table obtained from Equation (2.29), with a_n calculated via (2.25). Alternatively, a numerically efficient iterative algorithm (Xu, 2005) is described as follows:

To solve the correlation coefficient ρ_Z, rewrite (2.29) as

$$\rho_Z(\boldsymbol{r}) = g\big(\rho_Z(\boldsymbol{r})\big) \tag{2.38a}$$

$$g\big(\rho_Z(\boldsymbol{r})\big) = \frac{1}{a_1^2}\left[\rho_Y(\boldsymbol{r}) - \sum_{n=2}^{\infty} a_n^2 \rho_Z^n(\boldsymbol{r})\right] \tag{2.38b}$$

Accordingly, a fixed-point iteration algorithm is established

$$\rho_Z^{(i)}(\boldsymbol{r}) = \rho_Y(\boldsymbol{r}); \quad i = 0 \tag{2.39a}$$

$$\rho_Z^{(i)}(\boldsymbol{r}) = \frac{\rho_Y(\boldsymbol{r}) - \rho_Y^{(i-1)}(\boldsymbol{r})}{a_1^2} + \rho_Z^{(i-1)}(\boldsymbol{r}); \quad i = 1, 2\cdots \tag{2.39b}$$

where

$$\rho_Y^{(i-1)}(\boldsymbol{r}) = \sum_{n=1}^{\infty} a_n^2 \left(\rho_Z^{(i-1)}(\boldsymbol{r})\right)^n \tag{2.40}$$

(2.38a,b) can be rewritten together in the following expansion:

$$\rho_Z(\boldsymbol{r}) = \frac{1}{a_1^2}\left[\rho_Y(\boldsymbol{r}) - \frac{a_2^2}{a_1^4}\rho_Y^2(\boldsymbol{r}) - \frac{a_3^2}{a_1^6}\rho_Y^3(\boldsymbol{r}) - \cdots\right] \tag{2.41}$$

In general, the convergence of this fixed-point iteration is fast since in the translation model the coefficient a_1 is close to 1 and the rest of coefficients a_n, $n = 2,3,\ldots$ decay to zero rapidly. An example of the coefficients is shown in Table 2.1 for three cases of the Beta probability distribution, of which the third is strongly non-Gaussian and bimodal.

2.2.3 Limitations of the translation model

As shown in Equation (2.29), given a homogeneous Gaussian input to the translation model, correspondingly there is a unique translated non-Gaussian output. An inverse question is, with any given homogeneous non-Gaussian output, whether there always exists a Gaussian image.

Table 2.1 Coefficients of the fourier hermite series

Coeff.	*Beta (4,2,x)*	*Beta (2,20,x)*	*Beta (0.5,0.5,x)*
a_1^2	0.975511	0.924597	0.899519
a_2^2	0.017435	0.075109	0
a_3^2	0.006415	0.000022	0.075210
a_4^2	0.000499	0.000270	0
a_5^2	0.000122	0	0.017100
a_6^2	0.000015	0.000002	0
a_7^2	0.000003	0	0.004468
a_8^2	0	0	0
a_9^2	0	0	0.001809
a_{10}^2	0	0	0
a_{11}^2	0	0	0.000514
a_{12}^2	0	0	0
a_{13}^2	0	0	0.000061
SUM	1.000000	1.000000	0.998681

Note: The calculations are based on Eq. (2.25) with an adaptive recursive Simpson's rule.

The answer is no, since the calculated correlation coefficient ρ_Z does not necessarily satisfy the requirement of positive semidefiniteness. In such a case, we say the CDF-based translation F_Y^{-1} and ρ_Y are incompatible.

As an example, let us apply the translation model to a fiber-reinforced two-phase composite as follows:

$$Y(\boldsymbol{r}) = F_Y^{-1}\left(\Phi\left(\tilde{Z}(\boldsymbol{r},\vartheta)\right)\right) = \begin{cases} 1 & \tilde{Z}(\boldsymbol{r}) > \Phi^{-1}(c_1) \quad \text{Fiber} \\ 0 & \quad\quad\quad\ \ \textit{else} \quad \text{Matrix} \end{cases} \tag{2.42}$$

where c_1 denotes the volume fraction of fibers. Assume the non-Gaussian random field is ergodic, with its autocorrelation calculated from a two-phase sample, shown in Figure 2.1a. The correlation coefficient of the underlying Gaussian image is then calculated by using the fixed-point iterative algorithm in 2.2.2, which in this case is negative-definite shown as the negative values of its power spectral density (Figure 2.2). By truncating the negative part of the PSD to zero, a Gaussian sample is generated from such a truncated PSD, as shown in Figure 2.1b. Such a generated sample presents a very different morphology from the target sample shown in Figure 2.1a.

As a concluding remark, the CDF-based translation is just one of many monotonic nonlinear operators that enable one-to-one mapping between an underlying Gaussian field and its non-Gaussian translation. A unique advantage of the CDF-based translation is that the model enables the most important statistical information, the marginal CDF or PDF, to be exactly

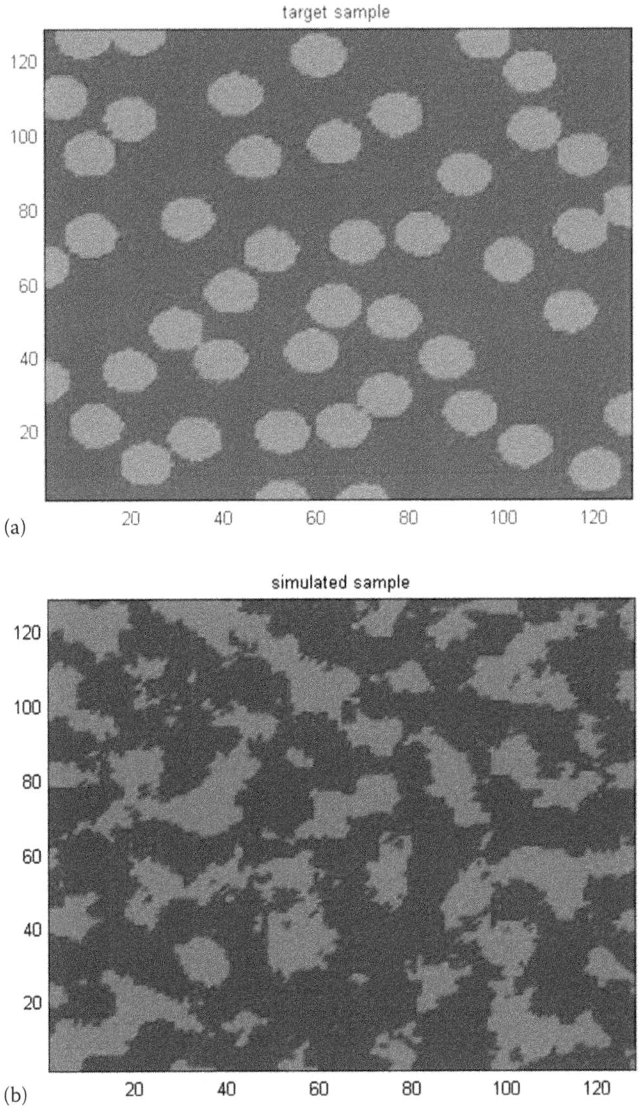

(a)

(b)

Figure 2.1 Inadequacy of the translation model on capturing of morphological features. (a) An underlying two-phase random composite sample or target upon which new random samples are generated; (b) a generated sample based on the translation model.

matched in the compatible cases. There are two major limitations of the translation model:

i. There is no promise to the existence of the underlying Gaussian image for a given non-Gaussian target; and

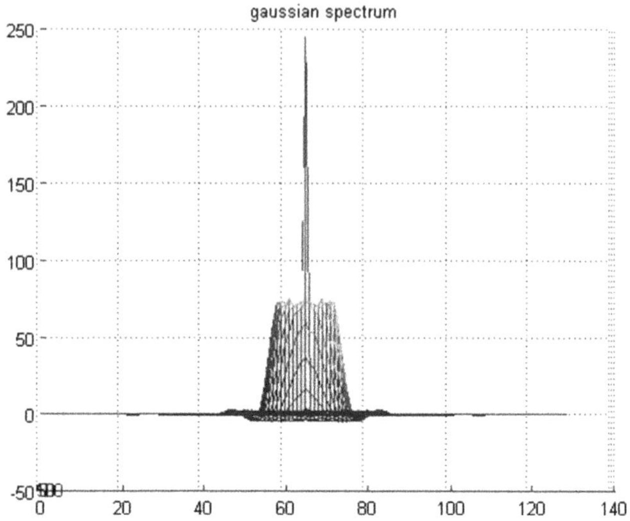

Figure 2.2 Power spectral density of a Gaussian image presenting negative values.

ii. The model can only handle up to the second-order statistics of a non-Gaussian target, while in real heterogeneous materials certain important morphological features are characterized by higher order statistics.

As illustrated in Figure 2.3, the CDF-based translation results in an unsymmetrical one-to-one mapping between the set of homogeneous Gaussian fields and the set of the translated fields; more specifically, the latter of which is a subset of homogeneous non-Gaussian fields.

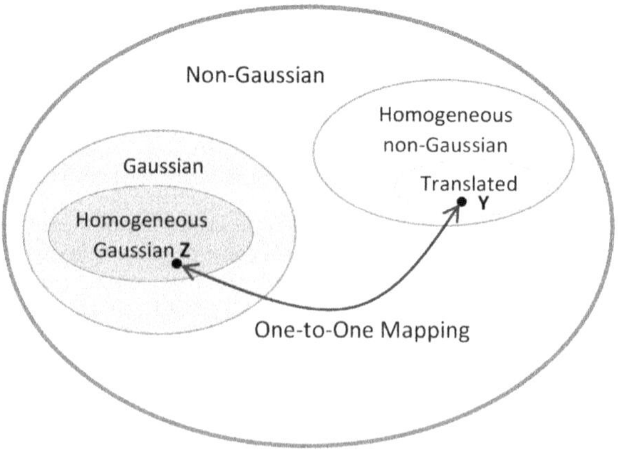

Figure 2.3 Unsymmetrical one-to-one mapping of the translation model.

2.3 NON-GAUSSIAN RANDOM FIELDS: THE CORRELATION MODEL

The relationship between the *n*th-order statistics and a random morphology was first investigated in vision research (Julesz, 1962). The gray-level cooccurrence, or correlation model, was then applied to texture analysis and image processing (Haralick & Shapiro, 1992) and random media (Hazlett, 1997; Yeong & Torquato, 1998). With regard to the task of characterization asserted at the beginning of this chapter, the correlation model effectively translates morphological information into a hierarchical order of correlation functions. The task of simulation, i.e., to generate a random sample based on prescribed correlation functions, is an interesting inverse or global optimization problem. In the case of multiphase composites, the optimization becomes specifically combinatorial. An optimization algorithm based on the full range of an autocorrelation conceptually leads to deterministic image reconstruction, a problem intensively studied in optics but irrelevant in simulation of random morphology.

In this section we introduce a short-range-correlation model on random morphology (Xu, 2005; Graham-Brady & Xu, 2008). The idea was first developed in texture synthesis (Gagalowicz & Ma, 1985) for the purpose of information reduction. It should be noted that the short-range dependence, corresponding to windowed morphology leaving the statistical measurement invariant, is closely related to the concept of representative volume element discussed in Chapters 1 and 7, and the correlation length in multidisciplinary fields such as stochastic process theory, neighborhood system in Markov random fields, textural resolution in texture analysis, and local roving window in vision research, etc.

2.3.1 The short-range-correlation model

Let $\mathcal{S}=\left\{i\middle|1\le i\le M_1\cdot M_2\right\}$ index a discrete set of sites for a morphological set X on a rectangular lattice $M_1\times M_2$. In Markov random field theory, a neighborhood set $\mathcal{N}$ means that only neighboring sites or a so-called clique have direct interactions with each other. The conditional probability of the value x_i is therefore conditioned only on its neighborhood set $\mathcal{N}\subseteq\mathcal{S}$

$$P(x_i|x_{\mathcal{S}\setminus\{i\}})=P(x_i|x_{\mathcal{N}\setminus\{i\}}) \tag{2.43}$$

where the subscript $\mathcal{S}\setminus\{i\}$ indicates the full lattice excluding site i. The equivalence of Markov and Gibbs random field models, known as Hammersley-Clifford theorem (Hammersley & Clifford, 1971), connects spatial statistics to statistical physics. The set X becomes a Gibbs random field on $\mathcal{S}$ and has Gibbs multivariate distribution

$$P(x)=\mathcal{Z}^{-1}e^{-\frac{1}{T}U(x)} \tag{2.44}$$

where the normalization constant $\mathcal{Z}$

$$\mathcal{Z} = \sum_{x \in X} e^{-\frac{1}{T}U(x)} \tag{2.45}$$

is the partition function, and the energy

$$U(x) = \sum_{n=1} V_n(x) \tag{2.46}$$

with $V_n(x)$ denoting the nth-order potential. With $V_n(x)$ given, a specific morphological configuration is generated by sampling the configuration space X according to the distribution (2.44). Note that the temperature T in (2.44) is a parameter related to the desired configuration. When the temperature $T \to +\infty$, X tends to be equally distributed, exhibiting a white noise behavior; conversely, when T approaches zero, X tends to concentrate at a specific point, which is one of energy minima as shown in the next subsection.

The n-point correlation function $R_n(\boldsymbol{r}_1,\ldots,\boldsymbol{r}_n)$ is defined as

$$R_n(\boldsymbol{r}_1,\ldots,\boldsymbol{r}_n) = \sum_{x \in X} x_1 x_2 \ldots x_n P(x) \tag{2.47}$$

where $\boldsymbol{r}_n$ denotes the spatial vector of the nth lattice site. Given an ergodic random field, R_n is calculated from a single configuration as

$$R_n(\boldsymbol{r}_2-\boldsymbol{r}_1,\ldots,\boldsymbol{r}_n-\boldsymbol{r}_1) = \frac{1}{M_1 M_2}\sum_{r} x(\boldsymbol{r})x(\boldsymbol{r}_2-\boldsymbol{r}_1+\boldsymbol{r})x(\boldsymbol{r}_3-\boldsymbol{r}_1+\boldsymbol{r})\ldots\ldots x(\boldsymbol{r}_n-\boldsymbol{r}_1+\boldsymbol{r}) \tag{2.48}$$

where the sum about $\boldsymbol{r}$ is made over all $M_1 M_2$ lattice sites.

When a site at which x is evaluated falls outside of the rectangular lattice $M_1 \times M_2$, an appropriate boundary condition such as a periodic one can be applied. The bias of the estimation (2.48) is considered to be negligible when the size $M_1 \times M_2$ is sufficiently large (e.g. 256×256). A closely related concept is the nth-order statistics $P_n\ (x_1,x_2,\cdots,x_n)$ or n-gram statistics, which in textural analysis is termed as *gray-level cooccurrence*. Note that the n-point correlation function is a subset of the nth-order statistics.

The characteristics of a Gibbs random field are specified by appropriate formulation of the nth-order potential $V_n(x)$. In the short-range-correlation (SRC) model, a *potential* is defined as the metric norm of a distance between two windowed correlation functions, such as a simulated $\tilde{R}_n$ and a target R_n. The first-order potential makes use of the first-order statistics, i.e. the full histogram f_X, not just the mean value only. In a multiphase composite, x takes values from a finite set $\mathcal{D}$ of grey levels, e.g. in a two-phase composite $\mathcal{D} = \{0,1\}$. The first three orders of the potentials are given here, with the rest following the pattern:

$$V_1(x) = \alpha_1 \left[\sum_{x_i \in \mathcal{D}} \left| f_X(x_i) - \tilde{f}_X(x_i) \right|^q \right]^{1/q} \tag{2.49}$$

$$V_2(x) = \alpha_2 \left[\sum_{i \in \mathcal{N}} \left| R_2(r_i) - \tilde{R}_2(r_i) \right|^q \right]^{1/q} \tag{2.50}$$

$$V_3(x) = \alpha_3 \left[\sum_{i,j \in \mathcal{N}} \left| R_3(r_i, r_j) - \tilde{R}_3(r_i, r_j) \right|^q \right]^{1/q} \tag{2.51}$$

where $1 \le q \le \infty$ and α_1, α_2, α_3 are coefficients used to manipulate the weights of the potentials.

Denote a function $d_q(\tilde{x}, x) = U(\tilde{x})$ the distance between any two configurations $x, \tilde{x} \in X$, i.e., $d_q : X \times X \rightarrow \Re$ taking pairs of morphological configurations into real numbers. By the definition of a metric, there are the following four sufficient and necessary conditions:

i. $d_q(x, \tilde{x}) \ge 0$ *for every* $x, \tilde{x} \in X$;
ii. $d_q(x, \tilde{x}) = 0$ *if and only if* $x = \tilde{x}$;
iii. $d_q(x, \tilde{x}) = d_q(\tilde{x}, x)$ *for every* $x, \tilde{x} \in X$; and
iv. $d_q(x, \tilde{x}) \le d_q(x, \tilde{x}') + d_q(\tilde{x}', \tilde{x})$ *for every* $x, \tilde{x}, \tilde{x}' \in X$.

Conditions i, ii, and iii are directly obtained from (2.49–2.51), and condition iv is derived by using Minkowski's Inequality.

Theorem

The short-range-correlation energy function $d_q : X \times X \rightarrow \Re$, *as defined in (2.46) and (2.49–2.51), is a metric on the morphological set X.*

Clearly for a finite set $\mathcal{D}$ a metric space (d_q, X) is the ℓ_q space, and as shown next we confine algorithms to the ℓ_2 space that is convenient for numerical operations.

2.3.2 Sampling algorithm

In Markov/Gibbs random field models, there are two well-established random sampling algorithms. The Metropolis sampler (Metropolis et al., 1953) uses a Monte Carlo procedure to generate a Markov chain of configurations, with acceptance of each configurational change based on (2.44). The Gibbs sampler (Geman & Geman, 1984) generates the next configuration using conditional probability (2.43) instead of energy change. The both sampling algorithms have the Gibbs distribution as equilibrium. Based on an initial configuration $x^{(0)}$, conveniently chosen to be white noise, the maximum probability is written as

$$\max_X P(x|x^{(0)}) = \max_X \frac{P(x^{(0)}|x)P(x)}{P(x^{(0)})} \tag{2.52}$$

Since a white noise $x^{(0)}$ can be generally treated as being independent from a desired configuration, i.e.

$$P(x^{(0)}|x) \approx P(x^{(0)}) \tag{2.53}$$

then (2.52) becomes

$$\max_X P(x|x^{(0)}) \approx \max_X P(x) \tag{2.54}$$

Therefore, generation of a desired configuration corresponds to minimization of the energy in (2.44) at a fixed temperature. When temperature is scheduled to cool down gradually till close to zero, the combined scheme is known as simulated annealing to escape from local minima into the global minimum, which requires significant computing time. In morphological simulation using the SRC model, the objective is to find many qualified local minima as desired configuration samples, and not the absolute global minimum that corresponds to deterministic reconstruction of the original configuration. Therefore, use of a fixed temperature results in a simple and fast Metropolis algorithm:

a. Given a target histogram f_X and correlation functions R_n, $n = 2, 3, \dots$;
b. Set SRC window size $L_1 \times L_2$;
c. Initialize a white noise $x^{(0)}$;
d. Calculate the initial SRC energy $U(x^{(0)})$ using (2.46) and (2.49–2.51);
e. Iteration step for the mth configuration
   i. Generate $x^{(m)}$ based on perturbation of $x^{(m-1)}$
   ii. Calculate the SRC energy $U(x^{(m)})$ using (2.46) and (2.49–2.51)
   iii. $\Delta U^{(m)} = U(x^{(m)}) - U(x^{(m-1)})$
   iv. If $\Delta U^{(m)} > 0$, then go to (i) to try another $x^{(m)}$
   Otherwise, $x^{(m)}$ is accepted and go to e) for next iteration $m + 1$
f. Until a prescribed criterion, i.e. the equilibrium, is reached.

There are two important algorithmic issues remaining in substep e(i). The Metropolis algorithm has two branches, spin-flip and spin-exchange. The spin-exchange approach, say in a two-phase medium, in each perturbation makes exchange of two sites in different phases, which always keeps a fixed volume fraction. The spin-flip approach, in another way, flips a single site individually to make a perturbation. The spin-flip approach has been shown to be advantageous over the spin-exchange approach in texture analysis (Copeland, Ravichandran, & Trivedi, 2001). In addition, the first-order marginal distribution or histogram in (2.49) is not deterministically constrained, which is consistent with the essential of stochastic simulation. Therefore the spin-flip approach fits the SRC model perfectly. There are three ways to select spin-flip sites, i.e. random scanning, periodic scanning, and raster scanning. In substep e(ii) the energy calculation is computationally the most demanding part of the model. The auto- and triple correlation functions of a morphological configuration x can be computed employing fast Fourier transform (e.g. Nikias & Petropulu, 1993):

$$R_2 = \mathrm{IFFT2}\left\{\left|\hat{x}(\xi)\right|^2\right\} \tag{2.55}$$

$$R_3 = \mathrm{IFFT3}\left\{\hat{x}(\xi_1)\hat{x}(\xi_2)\hat{x}^*(\xi_1+\xi_2)\right\} \tag{2.56}$$

where superscript * denotes complex conjugate, IFFT2 two-dimensional inverse FFT, and the Fourier image

$$\hat{x}(\xi) = \mathrm{FFT2}\left\{x(\boldsymbol{r})\right\} \tag{2.57}$$

When the number of spin-flips is large, this repetitive calculation of correlation functions becomes computationally intensive, which can be improved by using a more efficient updating scheme, detailed next.

2.3.3 Simulation of two-phase media

In a two-phase disordered medium, specifically, the information of the n-point correlation function is equivalent to the nth-order statistics (Frisch, 1965). There can be a variety of formulations corresponding to same nature of the nth-order SRC model. To separate the first-order statistics from the second order statistics, we choose a volume fraction c and the correlation coefficient

$$\rho(l_1,l_2) = \frac{R_2(l_1,l_2)-c^2}{c-c^2} \tag{2.58}$$

as parameters in the second-order SRC metric function

$$U_2 = \alpha_1 c + \alpha_2 \sqrt{\sum_{l_1=-L_1}^{L_1} \sum_{l_2=0}^{L_2} \left| \tilde{\rho}(l_1,l_2) - \rho(l_1,l_2) \right|^2} \tag{2.59}$$

where l_1 and l_2 refer to the coordinates in Figure 2.4. The short range window is such chosen in Figure 2.4 by taking advantage of statistical isotropy.

In the third-order SRC model, we simply let $\alpha_1 = \alpha_2 = 0$ and $\alpha_3 = 1$ in (2.49–2.51) since the triple correlation function contains the information of the first two orders, and the energy thus becomes

$$U_3 = \sqrt{\sum_{l_1=-L_1}^{L_1} \sum_{l_2=0}^{L_2} \sum_{l_1'=-L_1}^{L_1} \sum_{l_2'=0}^{L_2} \left| \tilde{R}_3(l_1,l_2;l_1',l_2') - R_3(l_1,l_2;l_1',l_2') \right|^2} \tag{2.60}$$

When there is a spin-flip at site $x^{(m)}(\boldsymbol{r})$ in an $M_1 \times M_2$ image of a two-phase medium with the value set $\mathcal{D} = \{0,1\}$, the autocorrelation function is updated as

$$R_2^{(m+1)}(\boldsymbol{s}) = R_2^{(m)}(\boldsymbol{s}) + \frac{1-2x^m(\boldsymbol{r})}{M_1 M_2} \cdot \left[x^{(m)}(\boldsymbol{r}+\boldsymbol{s}) + x^{(m)}(\boldsymbol{r}-\boldsymbol{s}) \right], s \neq 0 \tag{2.61}$$

$$R_2^{(m+1)}(0) = R_2^{(m)}(0) + \frac{1-2x^{(m)}(\boldsymbol{r})}{M_1 M_2} \tag{2.62}$$

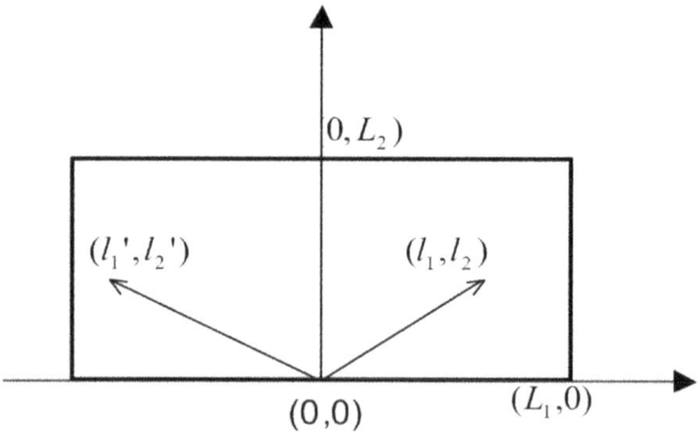

Figure 2.4 Short-range-correlation window. (Republished with permission of American Society of Mechanical Engineers (ASME), from Graham-Brady, L. and X.F. Xu, *J. Appl. Mech.*, 75: 061001, doi:10.1115/1.2957598, 2008. Permission conveyed through Copyright Clearance Center, Inc.)

Compared to the fast Fourier transform in (2.55) taking $O(M_1M_2 \log M_1M_2)$ multiplication operations, the updating scheme (2.61–2.62) requires only $O\left((2L_1+1)(L_2+1)\right)$ operations. In case of a 256 × 256 image with the SRC window size $L_1 = L_2 = 10$, the ratio of computing efficiency in terms of the number of multiplication operations is at the order of 10^4.

The updating scheme for the triple correlation function is given here:

$$R_3^{(m+1)}(\boldsymbol{s},\boldsymbol{s}') = R_3^{(m)}(s,s') + \frac{1-2x^{(m)}(\boldsymbol{r})}{M_1M_2}\Big[x^{(m)}(\boldsymbol{r}+\boldsymbol{s})x^{(m)}(\boldsymbol{r}+\boldsymbol{s}') + x^{(m)}(\boldsymbol{r}-\boldsymbol{s})$$

$$x^{(m)}(\boldsymbol{r}-\boldsymbol{s}+\boldsymbol{s}') + x^{(m)}(\boldsymbol{r}-\boldsymbol{s}')x^{(m)}(\boldsymbol{r}-\boldsymbol{s}'+\boldsymbol{s})\Big],$$

$$\boldsymbol{s} \neq 0, \boldsymbol{s}' \neq 0, \boldsymbol{s} \neq \boldsymbol{s}' \tag{2.63}$$

$$R_3^{(m+1)}(\boldsymbol{s},0) = R_3^{(m)}(\boldsymbol{s},0) + \frac{1-2x^{(m)}(\boldsymbol{r})}{M_1M_2}\Big[x^{(m)}(\boldsymbol{r}+\boldsymbol{s}) + x^{(m)}(\boldsymbol{r}-\boldsymbol{s})\Big], \boldsymbol{s} \neq 0 \tag{2.64a}$$

$$R_3^{(m+1)}(0,\boldsymbol{s}) = R_3^{(m)}(0,\boldsymbol{s}) + \frac{1-2x^{(m)}(\boldsymbol{r})}{M_1M_2}\Big[x^{(m)}(\boldsymbol{r}+\boldsymbol{s}) + x^{(m)}(\boldsymbol{r}-\boldsymbol{s})\Big], \boldsymbol{s} \neq 0 \tag{2.64b}$$

$$R_3^{(m+1)}(\boldsymbol{s},\boldsymbol{s}) = R_3^{(m)}(\boldsymbol{s},\boldsymbol{s}) + \frac{1-2x^{(m)}(\boldsymbol{r})}{M_1M_2}\Big[x^{(m)}(\boldsymbol{r}+\boldsymbol{s}) + x^{(m)}(\boldsymbol{r}-\boldsymbol{s})\Big], \boldsymbol{s} \neq 0 \tag{2.65}$$

$$R_3^{(m+1)}(0,0) = R_3^{(m)}(0,0) + \frac{1-2x^{(m)}(\boldsymbol{r})}{M_1M_2} \tag{2.66}$$

In each cycle of spin-flip throughout the whole image domain, the third-order simulation (2.63–2.66) requires $O\left(M_1M_2(2L_1+1)^2(L_2+1)^2\right)$ multiplication operations, $(2L_1 + 1)(L_2 + 1)$ times of that in the second-order model. The MATLAB® code of the third-order SRC simulation is attached in Appendix I.

Examples

In Figure 2.5, the second-order SRC simulation obtains the best quality samples with the window size $L_1 = L_2 = L$ being 10, 15, and 20, which is close to the correlation length of the target configuration. The interpretation is that too small a window size contains insufficient morphological information, while a window size too large beyond the correlation length includes unnecessary information overly constraining the numerical optimization. In the SRC model, the relation between the image size and the window size actually can be understood as Type-II scale separation mentioned in Chapter 1, which is further discussed at the end of this chapter.

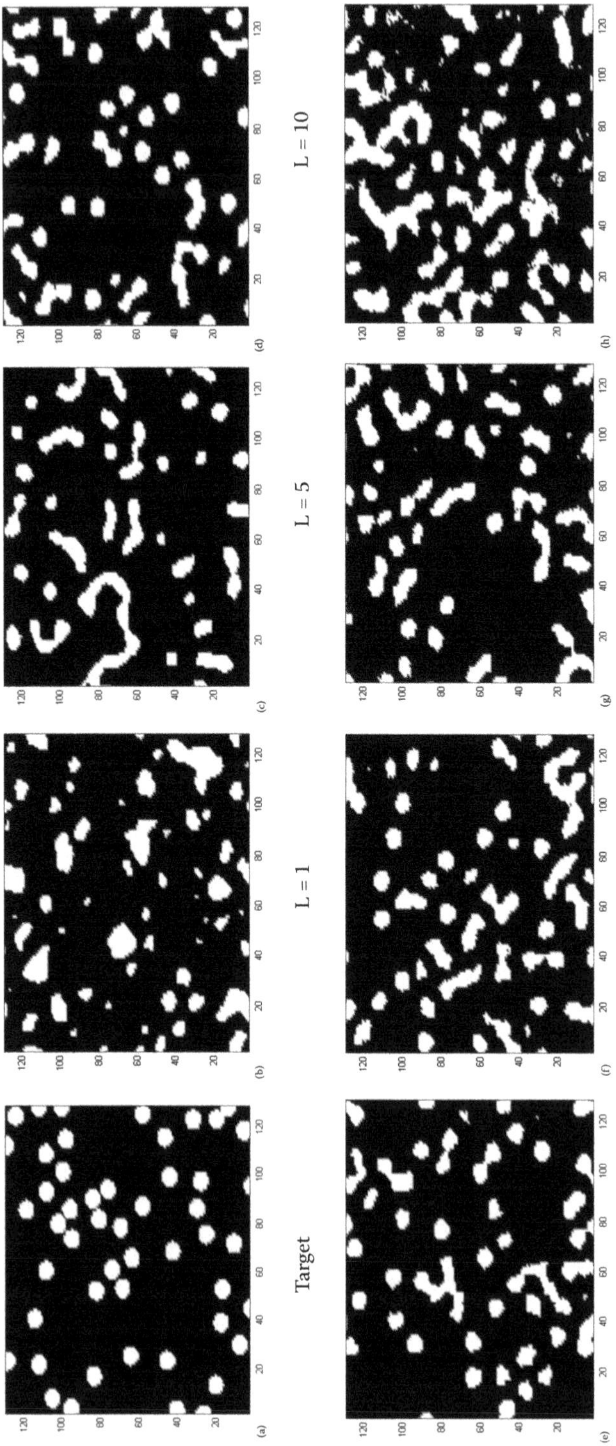

Figure 2.5 Second-order SRC simulation with different window sizes ($\alpha_1 = 0.7$, $\alpha_2 = 1.0$). (a) The target; (b) the window size L = 1; (c) L = 5; (d) L = 10; (e) L = 15; (f) L = 20; (g) L = 30; (h) L = 63. (Republished with permission of American Society of Mechanical Engineers (ASME), from Graham-Brady, L. and X.F. Xu, *J. Appl. Mech.*, 75: 061001, doi:10.1115/1.2957598, 2008. Permission conveyed through Copyright Clearance Center, Inc.)

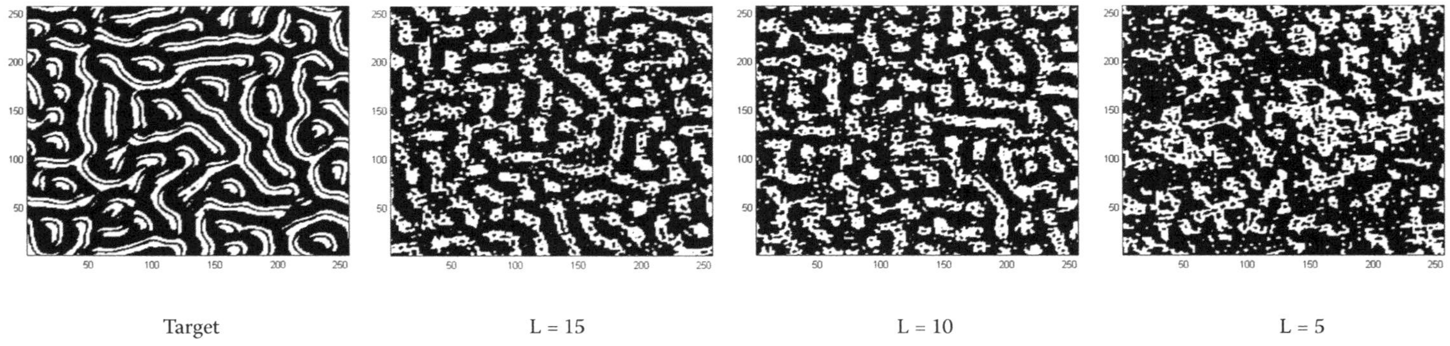

Figure 2.6 Second-order SRC simulation with different window sizes ($\alpha_1 = 0.3$, $\alpha_2 = 1.0$). (Republished with permission of American Society of Mechanical Engineers (ASME), from Graham-Brady, L. and X.F. Xu, *J. Appl. Mech.*, 75: 061001, doi:10.1115/1.2957598, 2008. Permission conveyed through Copyright Clearance Center, Inc.)

In Figure 2.6 the second-order simulation gives a more striking representation of the window size effect. Gagalowicz and Ma (1985) found that for a macroscopic texture, the window size should be larger than the extents of the basic pattern. The results here consistently show that to obtain a good simulation sample, the window size should be close to the correlation length. Compared with the second-order SRC simulation, the third-order SRC simulation in Figure 2.7 shows much improved sample quality, which confirms that morphological information missing in the second-order statistics can be retrieved from higher order statistics.

2.3.4 Classification of random morphologies

In vision research, efforts have been invested to seek counterexamples to the Julesz conjecture (Julesz et al., 1973), i.e. to find pairs of visually distinct configurations having same lower order statistics, and there were arguments with regard to ergodicity and local statistics (Gagalowicz, 1981). Choose the Debye symmetric random medium as an example, of which the third-order statistics are specifically determined by the second-order statistics (Frisch, 1965), i.e.

$$R_3(\boldsymbol{r},\boldsymbol{s}) = \frac{1}{2}\left[R_2(\boldsymbol{r}) + R_2(\boldsymbol{s}) + R_2(\boldsymbol{r}-\boldsymbol{s}) - \frac{1}{2}\right] \tag{2.67}$$

with the fourth- and higher order statistics left unspecified. The second-order SRC model is then used to simulate a 256 × 256 sample of the Debye medium matching the correlation coefficient (2.36) with $\ell_c = 2$, shown as one quadrant of Figure 2.8a. The other three quadrants each with resolution 256 × 256 are generated by translating a Gaussian sample simulated with the correlation coefficient (2.37), which are visually indiscriminate from the first one. Given (2.67), these samples are close to each other in the metric space up to the third-order statistics, with all the other higher order metrics remaining unconstrained. Therefore, the Debye medium is a morphological example in which the first two orders of statistics predominately control morphological appearance.

In the second example (Figure 2.8b), three quadrants are generated by using the same underlying Gaussian image with a different volume fraction $c = 0.25$. By using the autocorrelation measured from the three quadrants, the fourth quadrant is simulated with the translation model, which is again visually indistinguishable from the other three.

In the translation model, the morphological information of a non-Gaussian field is completely determined by the first two orders of statistics. As a result, a translated non-Gaussian morphological sample presents no visually structural features, which is demonstrated by the two examples in

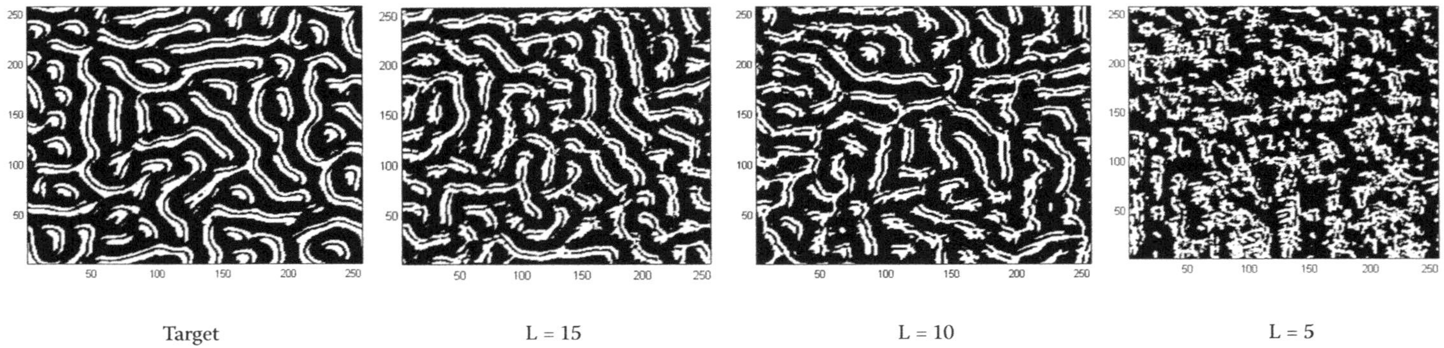

Figure 2.7 Third-order SRC simulation with different window sizes. (Republished with permission of American Society of Mechanical Engineers (ASME), from Graham-Brady, L. and X.F. Xu, *J. Appl. Mech.*, 75: 061001, doi:10.1115/1.2957598, 2008. Permission conveyed through Copyright Clearance Center, Inc.)

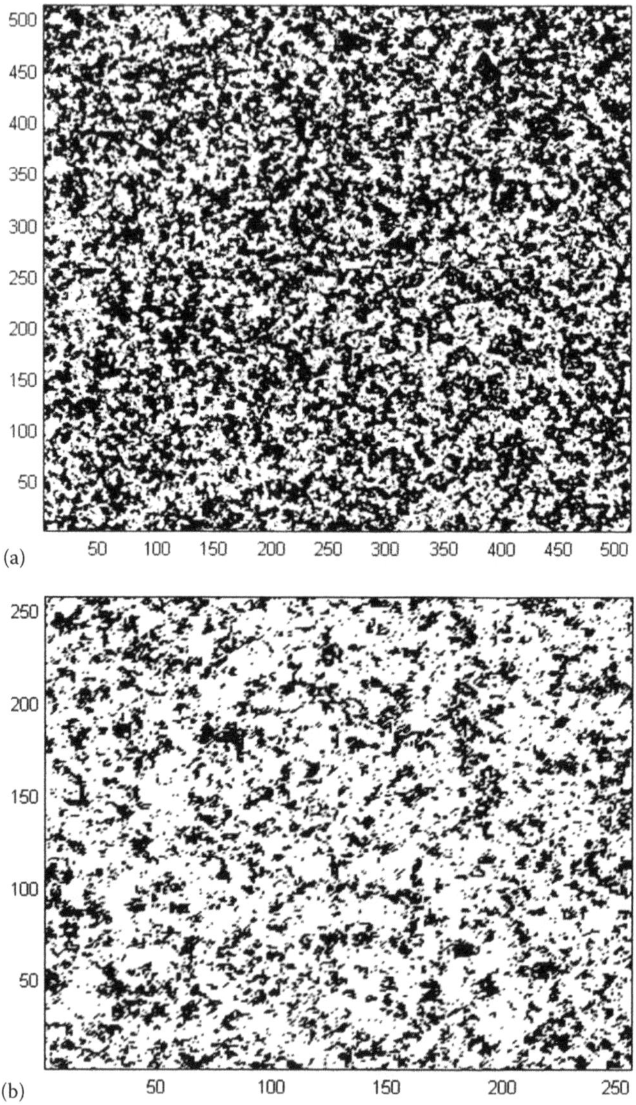

Figure 2.8 Morphology visually defined by the second-order statistics (a quadrant of each indistinguishable from the other three). (a) three quadrants are generated based on the translation model with c = 0.5, and one based on the SRC model; (b) same as (a) except for c = 0.25.

Figure 2.8. This class of morphologies that are predominately controlled by the second-order statistics is called *nonstructured morphologies*.

With regard to those morphologies mostly controlled by the second-order statistics but still presenting visually structural features, we categorize them

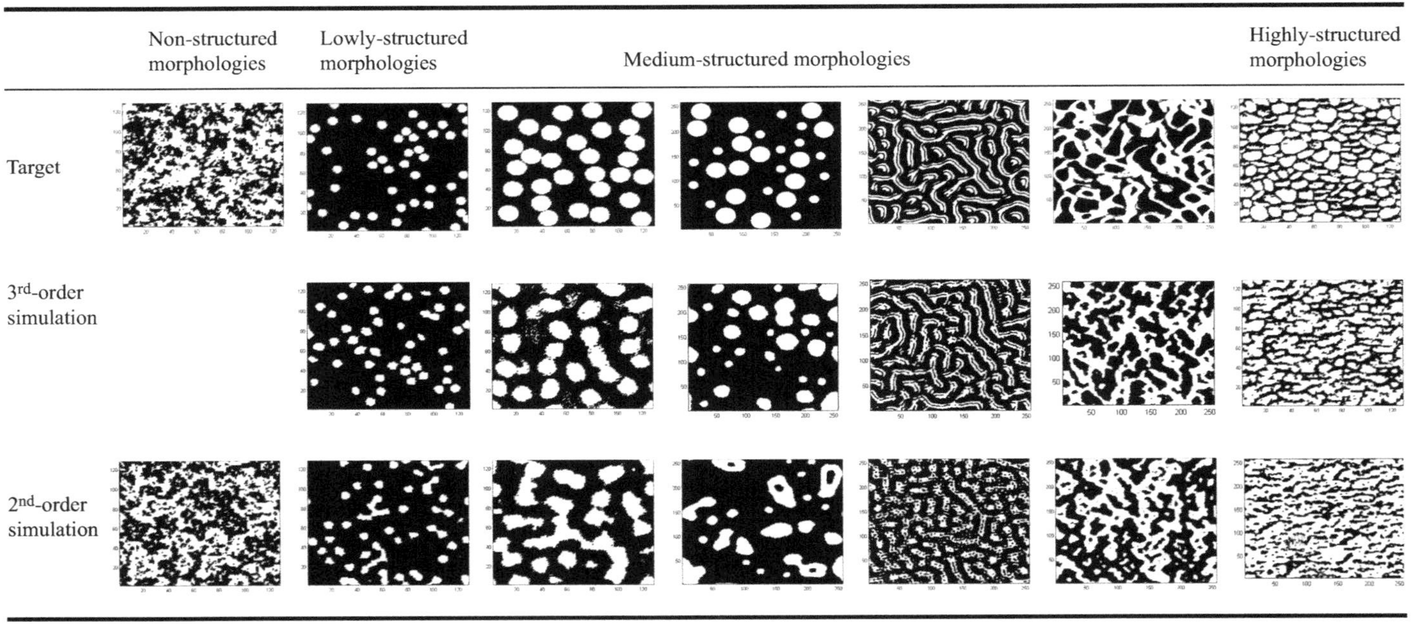

Figure 2.9 Classification of morphologies based on the SRC model. (Adapted with permission of American Society of Mechanical Engineers (ASME), from Graham-Brady, L. and X.F. Xu, *J. Appl. Mech.*, 75: 061001, doi:10.1115/1.2957598, 2008. Permission conveyed through Copyright Clearance Center, Inc.)

as *lowly structured morphologies*. An example is shown in Figure 2.9, where the shape of circles is not well defined due to the low resolution image.

In the case that the third-order statistics sufficiently capture essential morphological features, while the second-order statistics are inadequate, this class is named as *medium-structured morphologies*. It appears that most of morphologies in application belong to this class.

The class of *highly structured morphologies* refers to those that require the order of statistical information beyond the third. As shown in the example in Figure 2.9, the third-order SRC model has difficulty capturing connectivity and percolation that are considered to be of global behavior. One plausible reason is that the ergodicity hypothesis is no longer valid in such a morphology, which however is an underlying assumption implicitly made in the SRC model. Note that the SRC model is built upon a local window concept. When the correlation length of an image becomes comparable to the image size, the hypothesis of Type-II scale separation is violated. In such a scale-coupling case, it is hard for the SRC model based on a local window to capture global information beyond the correlation length, unless the window size increased to the image size, in which case the SRC model will lose its stochastic character and degenerate into deterministic reconstruction. The conflict between the SRC model and highly structured morphologies is a good reflection of the scale-coupling effect on morphological modeling and simulation.

Part I

Analytical homogenization of scale separation problems

Chapter 3

Green-function-based variational principles

Microstructure of a composite material, either by nature or engineering, is usually subject to morphological randomness and measurement uncertainty. Effective properties of a random composite therefore are not deterministically fixed unless the statistical information is completely known, while the latter of which hardly occurs, if not impossible. With limited statistical information known about a composite, such as volume fractions, spatial isotropy, etc., it can be theoretically proved, as shown in this chapter, that the effective elastic moduli or the effective conductivity of a composite belong to a bounded range. The lower and upper bounds are important in both theory and engineering practices. To find the two bounds, the variational method (e.g. Oden & Reddy, 1976) is considered to be the most effective approach. In the 1950s, Hill (1952) first applied the variational method to successfully recover Voigt bound and Reuss bound of composites, the first-order bounds involving the first-order statistics only (i.e. volume fractions). A decade later Hashin and Shtrikman (1962) discovered an important new variational principle named after them, and made a great breakthrough obtaining the second-order bounds. In their derivation, spatial isotropy of a composite, a main ingredient of the second-order statistics, is implicitly employed in addition to volume fractions. Shortly after that, the third-order bounds were derived by Beran (1965) on conductivity, and by Beran and Molyneux (1966) on the bulk modulus in elasticity, in which the third-order statistics are further involved. Dederichs and Zeller (1973) showed that invoking of the classical potential energy principles yields odd order bounds, and Kröner (1977) summarized the bounding theory by further recognizing that the Hashin-Shtrikman (HS) variational principle leads to even-order bounds.

In this chapter, the previously discussed fundamental variational principles of composites are formulated in detail, which are distinguished with the Green-function terms. In Section 3.1, the classical potential energy principles of elasticity are applied to a representative volume element (RVE) to obtain odd-order bounds of the potential energy. Next, in Section 3.2, the HS variational principle and the corresponding even-order energy bounds are derived. Finally, in Section 3.3, the principles derived in elasticity are adapted to conductivity and other similar transport properties of composites.

3.1 ODD-ORDER VARIATIONAL PRINCIPLE

3.1.1 Decomposition of an RVE problem

A Dirichlet RVE problem of a random elastic composite is considered in a unit volume domain Y with the unit volume |Y| = 1. The equilibrium equation of the RVE problem is written as

$$\nabla \cdot \boldsymbol{\sigma}(\boldsymbol{x}, \vartheta) = 0 \quad \text{in Y} \tag{3.1}$$

with a displacement boundary condition

$$\boldsymbol{u}(\boldsymbol{x}, \vartheta) = \tilde{\boldsymbol{u}}_0 \quad \text{on } \partial\text{Y} \tag{3.2}$$

that corresponds to a global strain $\boldsymbol{\varepsilon}_0$, i.e.

$$\int_{\text{Y}} \boldsymbol{\varepsilon}(\boldsymbol{x}, \vartheta)\, d\boldsymbol{x} = \boldsymbol{\varepsilon}_0 \tag{3.3}$$

The constitutive law and displacement-strain relation are given, respectively, as

$$\sigma(\boldsymbol{x}, \vartheta) = \boldsymbol{L}(\boldsymbol{x}, \vartheta)\boldsymbol{\varepsilon}(\boldsymbol{x}, \vartheta) \tag{3.4}$$

$$\boldsymbol{\varepsilon}(\boldsymbol{x}, \vartheta) = \nabla^s \boldsymbol{u}(\boldsymbol{x}, \vartheta) \tag{3.5}$$

where the superscript s denotes symmetric operation, or $\varepsilon_{ij} = \dfrac{1}{2}\left(\dfrac{\partial u_i}{\partial x_j} + \dfrac{\partial u_j}{\partial x_i}\right)$, $i,j = 1,2,3$. Note the stress $\boldsymbol{\sigma}(\boldsymbol{x}, \vartheta)$, strain $\boldsymbol{\varepsilon}(\boldsymbol{x}, \vartheta)$, and displacement $\boldsymbol{u}(\boldsymbol{x}, \vartheta)$, are all randomly fluctuating fields as a result of external forces or constraints acting upon random microstructure.

To clearly illustrate the concept of homogenization in modeling of heterogeneous materials, the RVE problem (3.1–3.5) is decomposed into a reference problem, or so-called comparison problem, and a fluctuation problem, as shown in Figure 3.1. The term *fluctuation*, instead of *perturbation*, is purposely used to distinguish the decomposition from that in the perturbation method, in that the fluctuation problem is free of the inherent constraint imposed on the perturbation method wherein a perturbation

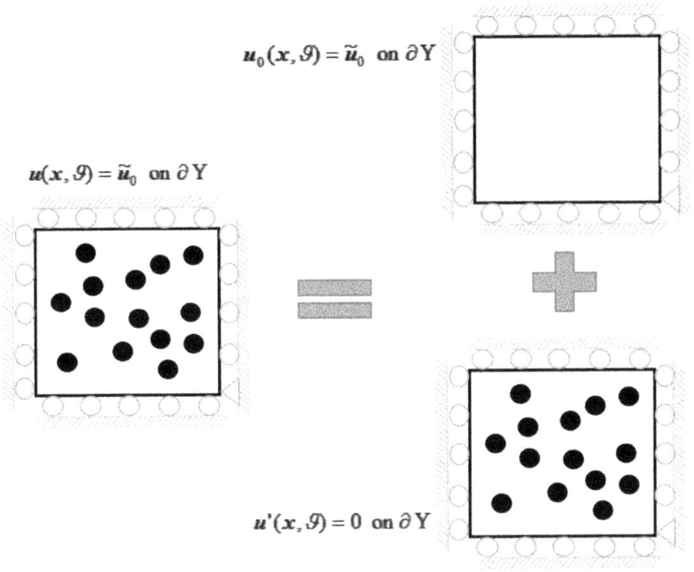

Figure 3.1 Schematic for decomposition of the Dirichlet RVE problem.

must be relatively small. Application of the perturbation method to random media is specifically discussed in Chapter 9. Relevant studies on the decomposition technique include Hashin and Shtrikman (1963), Ben-Amoz (1966), Mal and Knopoff (1967), Willis (1981) and Xu (2009), among others.

The reference problem corresponds to the original RVE problem characterized with a homogeneous medium, i.e.

$$\nabla \cdot \boldsymbol{\sigma}_0 = 0 \quad \text{in Y} \tag{3.6}$$

$$\boldsymbol{u}_0(\boldsymbol{x}, \vartheta) = \tilde{\boldsymbol{u}}_0 \quad \text{on } \partial\text{Y} \tag{3.7}$$

with

$$\boldsymbol{\sigma}_0 = \boldsymbol{L}_0 \boldsymbol{\varepsilon}_0 \tag{3.8}$$

$$\boldsymbol{\varepsilon}_0 = \nabla^s \boldsymbol{u}_0 \tag{3.9}$$

Given the reference moduli $\boldsymbol{L}_0$ and a global strain $\boldsymbol{\varepsilon}_0$ being uniformly constant, it is straightforward to find the stress $\boldsymbol{\sigma}_0$ and the displacement

$\boldsymbol{u}_0$ uniformly constant as well throughout the RVE domain. Following the superposition principle in linear elasticity, the difference between the original RVE problem and the reference problem results in the following fluctuation problem:

$$\nabla \cdot \boldsymbol{\sigma}'(\boldsymbol{x},\vartheta) + \nabla \cdot \boldsymbol{p}(\boldsymbol{x},\vartheta) = 0 \text{ in Y} \tag{3.10}$$

$$\boldsymbol{u}'(\boldsymbol{x},\vartheta) = 0 \text{ on } \partial\text{Y} \tag{3.11}$$

with

$$\sigma(\boldsymbol{x},\vartheta) = \boldsymbol{L}_0\boldsymbol{\varepsilon}'(\boldsymbol{x},\vartheta) \tag{3.12}$$

$$\boldsymbol{\varepsilon}'(\boldsymbol{x},\vartheta) = \nabla^s\boldsymbol{u}'(\boldsymbol{x},\vartheta) \tag{3.13}$$

where the variables of the fluctuation problem are denoted with the prime, and a newly introduced function $\boldsymbol{p}$ in (3.10) is called *stress polarization*.

In this decomposition, the displacement and strain fields of the reference and fluctuation problems are complementary to each other, i.e.

$$\boldsymbol{u}(\boldsymbol{x},\vartheta) = \boldsymbol{u}_0 + \boldsymbol{u}'(\boldsymbol{x},\vartheta) \tag{3.14}$$

$$\boldsymbol{\varepsilon}(\boldsymbol{x},\vartheta) = \boldsymbol{\varepsilon}_0 + \boldsymbol{\varepsilon}'(\boldsymbol{x},\vartheta) \tag{3.15}$$

while the total stress distinctively contains the stress polarization in addition

$$\boldsymbol{\sigma}(\boldsymbol{x},\vartheta) = \boldsymbol{\sigma}_0 + \boldsymbol{\sigma}'(\boldsymbol{x},\vartheta) + \boldsymbol{p}(\boldsymbol{x},\vartheta) \tag{3.16}$$

With (3.3) and (3.15), it is obvious the fluctuation strain $\boldsymbol{\varepsilon}'$ has a zero mean in the domain, i.e

$$\int_{\text{Y}} \boldsymbol{\varepsilon}'(\boldsymbol{x},\vartheta)\,d\boldsymbol{x} = 0 \tag{3.17}$$

The displacement solution of the fluctuation problem can be expressed in terms of the Green function as

$$\boldsymbol{u}'(\boldsymbol{x},\vartheta) = \int_{\text{Y}} \boldsymbol{G}(\boldsymbol{x},\boldsymbol{x}')\nabla \cdot \boldsymbol{p}(\boldsymbol{x}',\vartheta)\,d\boldsymbol{x}' \tag{3.18}$$

where the Green function $\boldsymbol{G}$ satisfies the boundary condition (3.11), i.e

$$\int_{\partial Y} \boldsymbol{G}(\boldsymbol{x},\boldsymbol{x}')\boldsymbol{t}(\boldsymbol{x}')d\boldsymbol{x}' = 0 \tag{3.19}$$

for any traction $\boldsymbol{t}$ acting upon the boundary. By applying Gauss' divergence theorem to (3.18) and taking the constraint (3.19) into account, it results in

$$u_i'(\boldsymbol{x}) = -\int_Y \frac{\partial G_{ik}(\boldsymbol{x},\boldsymbol{x}')}{\partial x'_l} p_{kl}(\boldsymbol{x}')d\boldsymbol{x}' \tag{3.20}$$

Application of (3.13) to (3.20) yields the fluctuation strain

$$\varepsilon_{ij}'(\boldsymbol{x}) = -\int_Y \Gamma_{ijkl}(\boldsymbol{x},\boldsymbol{x}')p_{kl}(\boldsymbol{x}')d\boldsymbol{x}' \tag{3.21}$$

with the fourth-rank modified Green function

$$\Gamma_{ijkl}(\boldsymbol{x},\boldsymbol{x}') = \frac{1}{2}\left(\frac{\partial^2 G_{ik}(\boldsymbol{x},\boldsymbol{x}')}{\partial x_j \partial x_l'} + \frac{\partial^2 G_{jk}(\boldsymbol{x},\boldsymbol{x}')}{\partial x_i \partial x_l'}\right)$$

that is symmetric between suffices i and j but not between k and l. Since the stress polarization is symmetric with $p_{kl} = p_{lk}$, the modified Green function can be further symmetrized as

$$\boxed{\Gamma_{ijkl}(\boldsymbol{x},\boldsymbol{x}') = \frac{1}{4}\left(\frac{\partial^2 G_{ik}(\boldsymbol{x},\boldsymbol{x}')}{\partial x_j \partial x_l'} + \frac{\partial^2 G_{jk}(\boldsymbol{x},\boldsymbol{x}')}{\partial x_i \partial x_l'} + \frac{\partial^2 G_{il}(\boldsymbol{x},\boldsymbol{x}')}{\partial x_j \partial x_k'} + \frac{\partial^2 G_{jl}(\boldsymbol{x},\boldsymbol{x}')}{\partial x_i \partial x_k'}\right)} \tag{3.22}$$

so that it holds all minor symmetries, i.e. $\Gamma_{ijkl} = \Gamma_{jikl} = \Gamma_{ijlk} = \Gamma_{jilk}$.

It should be noted that in this decomposition the stress polarization is taken as a basic unknown. Based on the equivalence between (3.1) and the superposition of (3.6) and (3.10), the exact stress polarization should satisfy the following compatibility condition:

$$\boldsymbol{p}(\boldsymbol{x},\vartheta) = \left(\boldsymbol{L}(\boldsymbol{x},\vartheta) - \boldsymbol{L}_0\right)\left(\boldsymbol{\varepsilon}_0 + \boldsymbol{\varepsilon}'(\boldsymbol{x},\vartheta)\right) \tag{3.23}$$

Given a global strain $\boldsymbol{\varepsilon}_0$, substitution of (3.23) into (3.21) leads to an iterative equation to compute the fluctuation strain $\boldsymbol{\varepsilon}'$, which is a fast numerical approach to deterministically solve problems of individual RVE samples (Moulinec & Suquet, 1998). The iterative equation is further incorporated into a stochastic solver on a lognormal random media problem (Xu & Graham-Brady, 2005, 2006), which however is inapplicable to random multiphase composites discussed here.

Noted as Type-II scale separation in Chapter 1, an RVE size must be sufficiently larger than the characteristic size of heterogeneities, or more precisely, the correlation length of random microstructure. With the Type-II scale separation condition satisfied, the infinite or free space Green function $\boldsymbol{G}^{\infty}$ and corresponding $\boldsymbol{\Gamma}^{\infty}$ replace the counterparts in the previous equations, where the RVE domain is treated as infinite space normalized with a unit volume. When the underlying reference tensor $\boldsymbol{L}_0$ is isotropic, the free space $\boldsymbol{G}^{\infty}$ and $\boldsymbol{\Gamma}^{\infty}$ are analytically available in both real space and Fourier space, as provided in Appendix II.

3.1.2 Principle of minimum potential energy

Based on the classical principle of minimum potential energy, the potential energy of the Dirichlet RVE problem (3.1–3.5) is bounded from above

$$\mathcal{U} \le \mathcal{U}^{+}(\boldsymbol{\varepsilon}) = \frac{1}{2}\langle \boldsymbol{\varepsilon}, \boldsymbol{L}\boldsymbol{\varepsilon} \rangle \tag{3.24}$$

with the exact solution

$$\mathcal{U} = \operatorname*{Min}_{\varepsilon \in \mathrm{E}} \left\{ \frac{1}{2}\langle \boldsymbol{\varepsilon}, \boldsymbol{L}\boldsymbol{\varepsilon} \rangle \right\} \tag{3.25}$$

where E denotes the strain function space containing all kinematically admissible strain functions satisfying the boundary condition (3.2), and the operation $\langle \cdot,\cdot \rangle$ denotes an inter product over the RVE domain.

By substituting (3.21) into (3.15) and taking the polarization $\boldsymbol{p}$ as an unknown, the total strain is given in terms of $\boldsymbol{p}$, as

$$\boldsymbol{\varepsilon} = \boldsymbol{\varepsilon}_0 - \boldsymbol{\Gamma}^{\infty}\boldsymbol{p} \tag{3.26}$$

where contracted notation is used. Substitution of (3.26) into the right-hand side of the inequality (3.24) yields the functional of the upper bound potential energy as

$$\boxed{\mathcal{U}^{+}(\boldsymbol{p}, \boldsymbol{L}_0) = \frac{1}{2}\langle \boldsymbol{\varepsilon}_0, \boldsymbol{L}\boldsymbol{\varepsilon}_0 \rangle + \frac{1}{2}\langle \boldsymbol{p}, \boldsymbol{\Gamma}^{\infty}\boldsymbol{p} \rangle + \frac{1}{2}\langle \boldsymbol{\Gamma}^{\infty}\boldsymbol{p}, (\boldsymbol{L}-\boldsymbol{L}_0)\boldsymbol{\Gamma}^{\infty}\boldsymbol{p} \rangle - \langle \boldsymbol{\Gamma}^{\infty}\boldsymbol{p}, (\boldsymbol{L}-\boldsymbol{L}_0)\boldsymbol{\varepsilon}_0 \rangle} \tag{3.27}$$

Note that in the upper bound functional (3.27), besides the elastic moduli $\boldsymbol{L}$, the argument $\boldsymbol{p}$ is also a random field tensor.

According to the variational method, the upper bound functional (3.27) is minimized by seeking the optimal values for the two arguments, and such optima are obtained by solving the equations resulting from vanishing of the functional derivatives, or so-called variational derivatives of (3.27). In this minimization process, it is important to construct an appropriate trial function space for $\boldsymbol{p}$, and the idea is obtained by observing the exact solution of the polarization in terms of $\boldsymbol{L}$ and $\boldsymbol{\varepsilon}_0$. By substituting (3.21) into the compatibility condition (3.23), it yields an infinite series solution for the polarization

$$\boldsymbol{p} = (\boldsymbol{L}-\boldsymbol{L}_0)\left(\boldsymbol{I}-\boldsymbol{\Gamma}^{\infty}(\boldsymbol{L}-\boldsymbol{L}_0)+(\boldsymbol{\Gamma}^{\infty}(\boldsymbol{L}-\boldsymbol{L}_0))^2-\cdots\right)\boldsymbol{\varepsilon}_0 \tag{3.28}$$

A trial polarization function is chosen by truncating this expansion at a certain order (e.g. the first-order truncation $\boldsymbol{p}^{(1)} \propto (\boldsymbol{L}-\boldsymbol{L}_0)$, the second-order truncation $\boldsymbol{p}^{(2)} \propto (\boldsymbol{L}-\boldsymbol{L}_0)\,\boldsymbol{\Gamma}^{\infty}\,(\boldsymbol{L}-\boldsymbol{L}_0)$, etc.).

It is observed that the third term on the right-hand side of the functional (3.27) involves the triple correlation as follows:

$$\int_Y\int_Y\int_Y \boldsymbol{\Gamma}^{\infty}(\boldsymbol{x}-\boldsymbol{y})\boldsymbol{\Gamma}^{\infty}(\boldsymbol{x}-\boldsymbol{z})\left(\boldsymbol{L}(\boldsymbol{x},\vartheta)-\boldsymbol{L}_0\right)\boldsymbol{p}(\boldsymbol{y},\vartheta)\boldsymbol{p}(\boldsymbol{z},\vartheta)\,d\boldsymbol{z}\,d\boldsymbol{y}\,d\boldsymbol{x}$$

i.e. the triple correlation among the elastic moduli $\boldsymbol{L}$ at point $\boldsymbol{x}$ and the stress polarization values at points $\boldsymbol{y}$ and $\boldsymbol{z}$. Given any trial polarization obtained from a truncation of the series expansion (3.28), it is clear the highest order correlation involved in the upper bound (3.27) is always of odd order about the underlying random field $\boldsymbol{L}(\boldsymbol{x}, \vartheta)$. The upper bound functional (3.27) is therefore called the *odd-order upper bound functional.*

By invoking the ergodicity hypothesis that is usually assumed in random composites, the upper bound potential energy (3.27) is alternatively expressed in ensemble averaging, as

$$\mathcal{U}^{+}(\boldsymbol{p},\boldsymbol{L}_0)=\frac{1}{2}\boldsymbol{\varepsilon}_0\bar{\boldsymbol{L}}\boldsymbol{\varepsilon}_0+\frac{1}{2}\overline{(\boldsymbol{p}\boldsymbol{\Gamma}^{\infty}\boldsymbol{p}}+\frac{1}{2}\overline{(\boldsymbol{\Gamma}^{\infty}\boldsymbol{p})(\boldsymbol{L}-\boldsymbol{L}_0\boldsymbol{\Gamma}^{\infty}\boldsymbol{p}}-\overline{(\boldsymbol{\Gamma}^{\infty}\boldsymbol{p})(\boldsymbol{L}-\boldsymbol{L}_0)\boldsymbol{\varepsilon}_0} \tag{3.29}$$

3.1.3 Principle of minimum complementary energy

Apply the classical principle of minimum complementary energy to the Dirichlet RVE problem (3.1–3.5), and we have the complementary energy bounded from above

$$\mathcal{U}^{c} \le \mathcal{U}^{c+} = \frac{1}{2}\langle \boldsymbol{\sigma},\boldsymbol{M}\boldsymbol{\sigma}\rangle - (\tilde{\boldsymbol{u}}_0,\boldsymbol{t}) \tag{3.30}$$

where the superscript c denotes the complementary potential energy, the plus sign the upper bound, the compliance tensor $\boldsymbol{M} = \boldsymbol{L}^{-1}$, and the operation $(\cdot,\cdot)$ an inter product of a displacement and a traction over the part of the RVE boundary imposed with a displacement or traction condition. In this case, the surface integration is conducted over the whole RVE boundary ∂Y. The exact complementary potential energy in (3.30) is given as the minimum

$$\mathcal{U}^c = \underset{\sigma\in\Sigma}{\mathrm{Min}}\left\{\frac{1}{2}\langle\boldsymbol{\sigma},\boldsymbol{M}\boldsymbol{\sigma}\rangle-(\tilde{\boldsymbol{u}}_0,\boldsymbol{t})\right\} \tag{3.31}$$

where Σ denotes the stress function space containing all statically admissible stress functions satisfying the equilibrium equation and any traction boundary condition.

By applying Gauss' divergence theorem, we obtain the following relations

$$\langle\boldsymbol{\sigma},\boldsymbol{\varepsilon}_0\rangle = (\tilde{\boldsymbol{u}}_0,\boldsymbol{t}) \tag{3.32}$$

$$\langle\boldsymbol{\sigma}',\boldsymbol{\varepsilon}'\rangle = -\langle\boldsymbol{p},\boldsymbol{\varepsilon}'\rangle \tag{3.33}$$

Substitution of (3.32) into (3.30) yields the upper bound complementary energy

$$\mathcal{U}^{c+}(\boldsymbol{\sigma}) = \frac{1}{2}\langle\boldsymbol{\sigma},\boldsymbol{M}\boldsymbol{\sigma}\rangle-\langle\boldsymbol{\sigma},\boldsymbol{\varepsilon}_0\rangle \tag{3.34}$$

By employing (3.16), (3.21), (3.12) and (3.8), a trial stress function is accordingly constructed as

$$\boldsymbol{\sigma} = \boldsymbol{L}_0\boldsymbol{\varepsilon}_0 - \boldsymbol{L}_0\boldsymbol{\Gamma}^\infty\boldsymbol{p}+\boldsymbol{p} \tag{3.35}$$

which is compatible to the trial strain function (3.26).

By substituting the trial function (3.35) into (3.34), the upper bound complementary energy is expanded in terms of the stress polarization as

$$\begin{aligned}\mathcal{U}^{c+}(\boldsymbol{p},\boldsymbol{L}_0) = &-\frac{1}{2}\langle\boldsymbol{\varepsilon}_0,(\boldsymbol{L}_0-\boldsymbol{L}_0(\boldsymbol{M}-\boldsymbol{M}_0)\boldsymbol{L}_0)\boldsymbol{\varepsilon}_0\rangle+\frac{1}{2}\langle\boldsymbol{p},\boldsymbol{M}\boldsymbol{p}\rangle+\langle\boldsymbol{L}_0\boldsymbol{\varepsilon}_0,(\boldsymbol{M}-\boldsymbol{M}_0)\boldsymbol{p}\rangle\\ &+\frac{1}{2}\langle\boldsymbol{\Gamma}^\infty\boldsymbol{p},\boldsymbol{L}_0\boldsymbol{M}\boldsymbol{L}_0\boldsymbol{\Gamma}^\infty\boldsymbol{p}\rangle-\langle\boldsymbol{\varepsilon}_0,\boldsymbol{L}_0(\boldsymbol{M}-\boldsymbol{M}_0)\boldsymbol{L}_0\boldsymbol{\Gamma}^\infty\boldsymbol{p}\rangle-\langle\boldsymbol{L}_0\boldsymbol{M}\boldsymbol{p},\boldsymbol{\Gamma}^\infty\boldsymbol{p}\rangle\end{aligned} \tag{3.36}$$

where the reference compliance tensor $\boldsymbol{M}_0 = \boldsymbol{L}_0^{-1}$. By taking (3.21) into account, the relation (3.33) becomes

$$\left\langle \boldsymbol{\Gamma}^{\infty}\boldsymbol{p}, \boldsymbol{L}_0\boldsymbol{\Gamma}^{\infty}\boldsymbol{p}\right\rangle = \left\langle \boldsymbol{p}, \boldsymbol{\Gamma}^{\infty}\boldsymbol{p}\right\rangle \tag{3.37}$$

and the upper bound (3.36) is accordingly rewritten as

$$\begin{aligned}\mathcal{U}^{c+}(\boldsymbol{p},\boldsymbol{L}_0) = &-\frac{1}{2}\left\langle \boldsymbol{\varepsilon}_0, \left(\boldsymbol{L}_0 - \boldsymbol{L}_0(\boldsymbol{M}-\boldsymbol{M}_0)\boldsymbol{L}_0\right)\boldsymbol{\varepsilon}_0\right\rangle \\ &+\frac{1}{2}\left\langle \boldsymbol{p}, \boldsymbol{M}\boldsymbol{p}\right\rangle + \left\langle \boldsymbol{L}_0\boldsymbol{\varepsilon}_0, (\boldsymbol{M}-\boldsymbol{M}_0)\boldsymbol{p}\right\rangle + \frac{1}{2}\left\langle \boldsymbol{\Gamma}^{\infty}\boldsymbol{p}, \boldsymbol{L}_0(\boldsymbol{M}-\boldsymbol{M}_0)\boldsymbol{L}_0\boldsymbol{\Gamma}^{\infty}\boldsymbol{p}\right\rangle \\ &+\frac{1}{2}\left\langle \boldsymbol{p}, \boldsymbol{\Gamma}^{\infty}\boldsymbol{p}\right\rangle - \left\langle \boldsymbol{\varepsilon}_0, \boldsymbol{L}_0(\boldsymbol{M}-\boldsymbol{M}_0)\boldsymbol{L}_0\boldsymbol{\Gamma}^{\infty}\boldsymbol{p}\right\rangle - \left\langle \boldsymbol{L}_0\boldsymbol{M}\boldsymbol{p}, \boldsymbol{\Gamma}^{\infty}\boldsymbol{p}\right\rangle\end{aligned} \tag{3.38}$$

Owning to the complementary relation

$$\mathcal{U} = -\mathcal{U}^{c} \tag{3.39}$$

the upper bound complementary potential energy is translated into the lower bound potential energy as

$$\mathcal{U} = -\mathcal{U}^{c} \geq -\mathcal{U}^{c+} = \mathcal{U}^{-} \tag{3.40}$$

where the superscripted negative sign denotes the lower bound. The lower bound potential energy is accordingly obtained as the negative of (3.38), i.e.

$$\boxed{\begin{aligned}\mathcal{U}^{-}(\boldsymbol{p},\boldsymbol{L}_0) = &\frac{1}{2}\left\langle \boldsymbol{\varepsilon}_0, \left(\boldsymbol{L}_0 - \boldsymbol{L}_0(\boldsymbol{M}-\boldsymbol{M}_0)\boldsymbol{L}_0\right)\boldsymbol{\varepsilon}_0\right\rangle \\ &-\frac{1}{2}\left\langle \boldsymbol{p}, \boldsymbol{M}\boldsymbol{p}\right\rangle - \left\langle \boldsymbol{L}_0\boldsymbol{\varepsilon}_0, (\boldsymbol{M}-\boldsymbol{M}_0)\boldsymbol{p}\right\rangle - \frac{1}{2}\left\langle \boldsymbol{\Gamma}^{\infty}\boldsymbol{p}, \boldsymbol{L}_0(\boldsymbol{M}-\boldsymbol{M}_0)\boldsymbol{L}_0\boldsymbol{\Gamma}^{\infty}\boldsymbol{p}\right\rangle \\ &-\frac{1}{2}\left\langle \boldsymbol{p}, \boldsymbol{\Gamma}^{\infty}\boldsymbol{p}\right\rangle + \left\langle \boldsymbol{\varepsilon}_0, \boldsymbol{L}_0(\boldsymbol{M}-\boldsymbol{M}_0)\boldsymbol{L}_0\boldsymbol{\Gamma}^{\infty}\boldsymbol{p}\right\rangle + \left\langle \boldsymbol{L}_0\boldsymbol{M}\boldsymbol{p}, \boldsymbol{\Gamma}^{\infty}\boldsymbol{p}\right\rangle\end{aligned}} \tag{3.41}$$

Similar to the observation made on the upper bound potential energy (3.27), the highest order correlation involved in the functional (3.41) is always of odd order about the underlying random field $\boldsymbol{L}$, and therefore (3.41) is called the *odd-order lower bound functional*. With the ergodicity assumption, the lower bound potential energy is alternatively expressed as

$$\begin{aligned}\mathcal{U}^{-}(\boldsymbol{p},\boldsymbol{L}_0) = &\frac{1}{2}\boldsymbol{\varepsilon}_0\overline{(\boldsymbol{L}_0 - \boldsymbol{L}_0(\boldsymbol{M}-\boldsymbol{M}_0)\boldsymbol{L}_0)}\boldsymbol{\varepsilon}_0 - \frac{1}{2}\overline{\boldsymbol{p}\boldsymbol{M}\boldsymbol{p}} - (\boldsymbol{L}_0\boldsymbol{\varepsilon}_0)\overline{(\boldsymbol{M}-\boldsymbol{M}_0)\boldsymbol{p}} \\ -\frac{1}{2}&\overline{(\boldsymbol{\Gamma}^{\infty}\boldsymbol{p})\boldsymbol{L}_0(\boldsymbol{M}-\boldsymbol{M}_0)\boldsymbol{L}_0\boldsymbol{\Gamma}^{\infty}\boldsymbol{p}} - \frac{1}{2}\overline{\boldsymbol{p}\boldsymbol{\Gamma}^{\infty}\boldsymbol{p}} + \boldsymbol{\varepsilon}_0\overline{\boldsymbol{L}_0(\boldsymbol{M}-\boldsymbol{M}_0)\boldsymbol{L}_0\boldsymbol{\Gamma}^{\infty}\boldsymbol{p}} + \overline{\boldsymbol{L}_0\boldsymbol{M}\boldsymbol{p}\boldsymbol{\Gamma}^{\infty}\boldsymbol{p}}\end{aligned} \tag{3.42}$$

These two bounds of the potential energy, (3.27) and (3.41), can be alternatively derived by choosing a Neumann RVE instead and applying the two classical minimum energy principles. The detail is left to the reader as an exercise.

According to the variational method, extremization of the two odd-order functionals (3.27) and (3.41) about the two arguments yields, respectively, the lowest upper bound and the highest lower bound. The corresponding variational principle is therefore called the *odd-order variational principle*. By further specifying the linear relation between the potential energy and the effective elastic moduli, two bounds of the effective elastic moduli are explicitly obtained, to which Chapter 4 is devoted.

3.2 EVEN-ORDER HASHIN-SHTRIKMAN PRINCIPLE

In this section we show that by imposing a condition of positive or negative semidefiniteness to $\boldsymbol{L} - \boldsymbol{L}_0$, the odd-order variational principle derived in the preceding section can be transformed into the Hashin-Shtrikman variational principle that is of even order.

Reorganize the terms in the upper and lower bounds (3.27) and (3.41) into the following expressions, respectively:

$$\mathcal{U}^+(\boldsymbol{p},\boldsymbol{L}_0)=\frac{1}{2}\langle\boldsymbol{\varepsilon}_0,\boldsymbol{L}_0\boldsymbol{\varepsilon}_0\rangle-\frac{1}{2}\langle\boldsymbol{p},\boldsymbol{\Gamma}^{\infty}\boldsymbol{p}\rangle-\frac{1}{2}\langle\boldsymbol{p},(\Delta\boldsymbol{L})^{-1}\boldsymbol{p}\rangle+\langle\boldsymbol{\varepsilon}_0,\boldsymbol{p}\rangle+\frac{1}{2}\langle\Delta\boldsymbol{\varepsilon},(\boldsymbol{L}-\boldsymbol{L}_0)\Delta\boldsymbol{\varepsilon}\rangle \tag{3.43}$$

$$\begin{aligned}\mathcal{U}^-(\boldsymbol{p},\boldsymbol{L}_0)=\frac{1}{2}\langle\boldsymbol{\varepsilon}_0,\boldsymbol{L}_0\boldsymbol{\varepsilon}_0\rangle-\frac{1}{2}\langle\boldsymbol{p},\boldsymbol{\Gamma}^{\infty}\boldsymbol{p}\rangle-\frac{1}{2}\langle\boldsymbol{p},(\boldsymbol{L}-\boldsymbol{L}_0)^{-1}\boldsymbol{p}\rangle+\langle\boldsymbol{\varepsilon}_0,\boldsymbol{p}\rangle\\-\frac{1}{2}\langle\Delta\boldsymbol{\varepsilon},\boldsymbol{L}_0(\boldsymbol{M}-\boldsymbol{M}_0)\boldsymbol{L}_0\Delta\boldsymbol{\varepsilon}\rangle\end{aligned} \tag{3.44}$$

where

$$\Delta\boldsymbol{\varepsilon}=\left((\boldsymbol{L}-\boldsymbol{L}_0)^{-1}+\boldsymbol{\Gamma}^{\infty}\right)\boldsymbol{p}-\boldsymbol{\varepsilon}_0 \tag{3.45}$$

When the trial polarization is taken to be exact that satisfies the compatibility condition (3.21), we immediately have

$$\Delta\boldsymbol{\varepsilon}=0 \tag{3.46}$$

Consequently, the last terms on the right-hand side of (3.43) and (3.44) identically vanish. The last term of (3.43) becomes nonnegative or nonpositive when the tensor ($\boldsymbol{L} - \boldsymbol{L}_0$), or equivalently $-(\boldsymbol{M} - \boldsymbol{M}_0)$, is positive or negative semidefinite, respectively. Conversely, the last term of (3.44) is nonpositive or nonnegative when $(\boldsymbol{M} - \boldsymbol{M}_0)$, or equivalently $-(\boldsymbol{L} - \boldsymbol{L}_0)$, is positive or negative semidefinite, respectively. It should be noted that the condition $\Delta\boldsymbol{\varepsilon} = 0$ is not the necessary condition for the last terms of the bounds (3.43) and (3.44) to vanish. In special cases when $\boldsymbol{L} - \boldsymbol{L}_0$ is neither positive nor negative semidefinite, the volume integration in the two last terms can still result in a zero value. An important case is found when reducing the third-order bounds to the second-order Hashin-Shtrikman bounds, as shown in Section 4.3 of Chapter 4.

By defining the Hashin-Shtrikman functional as

$$\boxed{\mathcal{H}(\boldsymbol{p},\boldsymbol{L}_0) = \frac{1}{2}\langle \boldsymbol{\varepsilon}_0, \boldsymbol{L}_0\boldsymbol{\varepsilon}_0\rangle - \frac{1}{2}\langle \boldsymbol{p}, \boldsymbol{\Gamma}^{\infty}\boldsymbol{p}\rangle - \frac{1}{2}\langle \boldsymbol{p}, (\boldsymbol{L} - \boldsymbol{L}_0)^{-1}\boldsymbol{p}\rangle + \langle \boldsymbol{\varepsilon}_0, \boldsymbol{p}\rangle} \tag{3.47}$$

we obtain two bounds wider than the previous two odd order bounds, i.e.

$$\mathcal{H}^-(\boldsymbol{p},\boldsymbol{L}_0) \le \mathcal{U}^-(\boldsymbol{p},\boldsymbol{L}_0) \le \mathcal{U} \le \mathcal{U}^+(\boldsymbol{p},\boldsymbol{L}_0) \le \mathcal{H}^+(\boldsymbol{p},\boldsymbol{L}_0) \tag{3.48}$$

where the HS upper bound $\mathcal{H}^+$ and the HS lower bound $\mathcal{H}^-$ correspond to the expression (3.47) with $(\boldsymbol{L} - \boldsymbol{L}_0)$ being negative and positive semidefinite, respectively. With regard to an ergodic random composite, (3.47) is alternatively expressed as

$$\mathcal{H}(\boldsymbol{p},\boldsymbol{L}_0) = \frac{1}{2}\boldsymbol{\varepsilon}_0\boldsymbol{L}_0\boldsymbol{\varepsilon}_0 - \frac{1}{2}\overline{\boldsymbol{p}\boldsymbol{\Gamma}^{\infty}\boldsymbol{p}} - \frac{1}{2}\overline{\boldsymbol{p}(\boldsymbol{L} - \boldsymbol{L}_0)^{-1}\boldsymbol{p}} + \boldsymbol{\varepsilon}_0\boldsymbol{p} \tag{3.49}$$

It is observed that the second term on the right-hand side of (3.47) involves the highest order correlation (i.e. the correlation between the polarization values at two points). Given any trial polarization obtained from a truncation of the infinite series (3.28), such a correlation is always of even order about the underlying random field $\boldsymbol{L}$. Therefore the HS energy bounds and the HS principle are also called *even-order energy bounds*, and *even-order variational principle*, respectively.

The inequalities in (3.48) show that, by using same reference moduli and a same trial polarization, the resulting odd-order energy bounds are always narrower than the corresponding even-order bounds. This outcome indeed makes sense since the $(2n+1)$th odd-order bounds take into account one more order of correlation information than the $(2n)$th even-order bounds.

It is further noted that to extremize the HS functional (3.47) the reference moduli $\boldsymbol{L}_0$ should be optimally chosen subject to negative or positive semidefiniteness of $(\boldsymbol{L} - \boldsymbol{L}_0)$. In a well-ordered composite with $(\boldsymbol{L}_m - \boldsymbol{L}_n)$

being either positive or negative semidefinite for any phases $m \neq n$, a common choice for the reference moduli is to simply have

$$\boldsymbol{L}_0 = \underset{n}{\mathrm{Min}}\{\boldsymbol{L}_n\} \text{ or } \underset{n}{\mathrm{Max}}\{\boldsymbol{L}_n\} \tag{3.50}$$

When such a phase with the maximum or minimum moduli does not exist, a mixed choice can be made for the reference moduli (e.g. a combination of the maxima of the bulk and shear moduli). Further discussion is given in Subsection 4.2 of Chapter 4.

3.3 ODD- AND EVEN-ORDER VARIATIONAL PRINCIPLES ON CONDUCTIVITY

Owning to the common underlying Laplace equation about the scalar potential, the bounds on thermal or electrical conductivity of a composite are identically applicable to a variety of other physical properties, e.g. electrical permittivity, magnetic and fluid permeability, diffusivity, etc. Denote $K(\boldsymbol{x}, \vartheta)$ the isotropic conductivity of a random medium, $\boldsymbol{q}$ the heat flux vector, and u the temperature. Given the Fourier law of heat conduction

$$\boldsymbol{q} = -K\nabla u \tag{3.51}$$

the Dirichlet RVE problem (3.1–3.2) is accordingly rewritten as

$$\nabla \cdot \boldsymbol{q}(\boldsymbol{x}, \vartheta) = 0 \text{ in Y} \tag{3.52}$$

with a temperature boundary condition

$$u(\boldsymbol{x}, \vartheta) = \tilde{u}_0 \text{ on } \partial\mathrm{Y} \tag{3.53}$$

that corresponds to a global temperature gradient ∇u_0, i.e.

$$\int_{\mathrm{Y}} \nabla u(\boldsymbol{x}, \vartheta)\, d\boldsymbol{x} = \nabla u_0 \tag{3.54}$$

The Dirichlet problem is then decomposed into a reference problem and a fluctuation problem, exactly like those described in (3.6–3.13). Following the same derivation in Sections 3.1 and 3.2, the upper and lower odd-order bounds are obtained, respectively, as

$$\mathcal{U}^{+}(\boldsymbol{p},K_0)=\frac{1}{2}\langle\nabla u_0,K\nabla u_0\rangle+\frac{1}{2}\langle\boldsymbol{p},\boldsymbol{\Gamma}^{\infty}\boldsymbol{p}\rangle$$
$$+\frac{1}{2}\langle\boldsymbol{\Gamma}^{\infty}\boldsymbol{p},(K-K_0)\boldsymbol{\Gamma}^{\infty}\boldsymbol{p}\rangle-\langle\nabla u_0,(K-K_0)\boldsymbol{\Gamma}^{\infty}\boldsymbol{p}\rangle \tag{3.55}$$

$$\mathcal{U}^{-}(\boldsymbol{p},K_0)=\frac{1}{2}\left\langle\nabla u_0,\left(2-\frac{K_0}{K}\right)K_0\nabla u_0\right\rangle-\frac{1}{2}\langle\boldsymbol{p},K^{-1}\boldsymbol{p}\rangle-\left\langle\nabla u_0,\left(\frac{K_0}{K}-1\right)\boldsymbol{p}\right\rangle$$
$$-\frac{1}{2}\left\langle\boldsymbol{\Gamma}^{\infty}\boldsymbol{p},K_0\left(\frac{K_0}{K}-1\right)\boldsymbol{\Gamma}^{\infty}\boldsymbol{p}\right\rangle$$
$$-\frac{1}{2}\langle\boldsymbol{p},\boldsymbol{\Gamma}^{\infty}\boldsymbol{p}\rangle+\left\langle\nabla u_0,\left(\frac{K_0}{K}-1\right)K_0\boldsymbol{\Gamma}^{\infty}\boldsymbol{p}\right\rangle+\left\langle\left(\frac{K_0}{K}-1\right)\boldsymbol{p},\boldsymbol{\Gamma}^{\infty}\boldsymbol{p}\right\rangle \tag{3.56}$$

while the even order HS bounds are given as

$$\mathcal{H}(\boldsymbol{p},K_0)=\frac{1}{2}\langle\nabla u_0,K_0\nabla u_0\rangle-\frac{1}{2}\langle\boldsymbol{p},\boldsymbol{\Gamma}^{\infty}\boldsymbol{p}\rangle-\frac{1}{2}\langle\boldsymbol{p},(K-K_0)^{-1}\boldsymbol{p}\rangle+\langle\nabla u_0,\boldsymbol{p}\rangle \tag{3.57}$$

with $(K-K_0)<0$ and $(K-K_0)>0$ in the case of the upper bound and lower bound, respectively.

As a concluding remark, the formulation presented in this section can be directly extended to anisotropic conductivity by changing the scalar K to a rank-two tensor, which is left to the reader as an exercise.

Chapter 4

Nth-order variational bounds

The first derived variational bounds are Voigt-Reuss bounds (Voigt, 1887; Reuss, 1929) on elasticity or Wiener bounds (Wiener, 1912) on transport properties of composites, which correspond to the simple rule of mixture or the first-order bounds since only the first-order statistics, i.e the volume fractions, are employed. The second-order bounds were derived by Hashin and Shtrikman on magnetic permeability (Hashin & Shtrikman, 1962) and elastic moduli (Hashin & Shtrikman, 1963) of composites. The third-order bounds were derived by Beran (1965) on permittivity, and Beran and Molyneux (1966) on the bulk modulus of composites, which contain a pair of geometric parameters of the bulk modulus resulting from integration of the triple correlation. The third-order bounds of the effective shear modulus were derived by Milton and Phan-Thien (1982) and finalized in a symmetric format (Xu, 2011a), which further involve a pair of geometric parameters of the shear modulus. The fourth-order bounds of the effective conductivity were first derived in Fourier space (Milton & Phan-Thien, 1982), and in this chapter are derived in real space and simplified with a fourth-order geometric parameter. The fourth-order bounds derived here also exclude a range that is physically unattainable.

Besides the previously discussed variational bounds derived based on correlation functions of microstructure, other types of bounds mainly consist of ellipsoidal bounds introduced in Chapter 5, and analytical bounds (e.g. optimal bounds using phase-exchange inequality; Milton, 2002). Variational bounds of nonlinear composites as an ongoing research topic are not covered in this book. Readers interested in the topic are referred to Suquet (1993), Talbot and Willis (1997), Ponte Castañeda (2002), and Xu and Jie (2014a), etc.

In this chapter, we apply the odd- and even-order variational principles of Chapter 3 to multiphase composites to derive the first-, second-, third-, and fourth-order bounds in the following four sections, respectively. Based on these bounds, in Section 4.5 a phase transition is particularly discussed from a mechanical point of view, which predicts a peculiar phenomenon of fluid–antifluid annihilation in terms of elastic rigidities.

4.1 FIRST-ORDER VOIGT-REUSS BOUNDS

4.1.1 Elastic moduli

The simplest trial polarization is to take $\boldsymbol{p} = 0$, and by applying the odd-order variational principle (3.27), it yields the Voigt-Reuss bounds for the effective elastic moduli of composites as

$$\boxed{\langle \boldsymbol{L}^{-1} \rangle^{-1} = \boldsymbol{L}_-^{(1)} < \boldsymbol{L}^e < \boldsymbol{L}_+^{(1)} = \langle \boldsymbol{L} \rangle} \tag{4.1}$$

where the superscript (1) denotes the order of the bounds, and the bracket volume averaging. Note the first-order upper bound or the Voigt bound obtained from (3.27) is independent of the choice for the reference moduli $\boldsymbol{L}_0$ since the latter is eventually cancelled out in (3.27). By substituting the trial polarization $\boldsymbol{p} = 0$ into the lower bound energy (3.41) or (3.42), it follows

$$\mathcal{U}^-(\boldsymbol{p} = 0, \boldsymbol{L}_0) = \frac{1}{2}\boldsymbol{\varepsilon}_0\left(2\boldsymbol{L}_0 - \boldsymbol{L}_0\langle M \rangle \boldsymbol{L}_0\right)\boldsymbol{\varepsilon}_0 \tag{4.2}$$

To extremize (4.2), solving of the following equation

$$\begin{aligned} \frac{d\mathcal{U}^-(\boldsymbol{p} = 0, \boldsymbol{L}_0)}{d\boldsymbol{L}_0} &= \boldsymbol{\varepsilon}_0(\boldsymbol{I} - \langle \boldsymbol{M} \rangle \boldsymbol{L}_0)\boldsymbol{\varepsilon}_0 \\ &= 0 \end{aligned} \tag{4.3}$$

yields the optimal reference moduli $\boldsymbol{L}_0 = \langle \boldsymbol{M} \rangle^{-1}$, substitution of which into (4.2) leads to the Reuss bound $\langle \boldsymbol{L}^{-1} \rangle^{-1}$. Note that in (4.1) strict inequalities are imposed because the Voigt-Reuss bounds are not simultaneously attainable for all components of the effective elastic moduli. A composite material can be optimally designed to realize the Voigt-Reuss bounds for individual components of the effective elastic moduli only and not all of them at the same time. For example, the effective Young's modulus of an anisotropic laminate-type composite reaches the Voigt bound and the Reuss bound, respectively, in the directions parallel and transverse to the layers. Suppose an N-phase statistically isotropic composite has the elastic moduli of each phase being isotropic as well; according to (4.1) the Voigt-Reuss bounds of the effective bulk modulus κ^e and shear modulus μ^e are expressed, respectively, as

$$\boxed{\left(\sum_{n=1}^{N} c_n \kappa_n^{-1}\right)^{-1} = \kappa_-^{(1)} < \kappa^e < \kappa_+^{(1)} = \sum_{n=1}^{N} c_n \kappa_n} \tag{4.4}$$

$$\left(\sum_{n=1}^{N} c_n \mu_n^{-1}\right)^{-1} = \mu_-^{(1)} < \mu^e < \mu_+^{(1)} = \sum_{n=1}^{N} c_n \mu_n \tag{4.5}$$

where c_n denotes the volume fraction of Phase-n.

4.1.2 Conductivity

Following the same justification previously discussed, substitution of the trial function $\boldsymbol{p} = 0$ into (3.55) and (3.56) yields the Voigt-Reuss bounds for the effective conductivity of an N-phase composite as

$$\left(\sum_{n=1}^{N} c_n K_n^{-1}\right)^{-1} = K_-^{(1)} < K^e < K_+^{(1)} = \sum_{n=1}^{N} c_n K_n \tag{4.6}$$

4.2 SECOND-ORDER HASHIN-SHTRIKMAN BOUNDS

4.2.1 General results

Consider an N-phase random composite in an RVE domain. The random field for the elastic moduli of the composite is expressed as

$$\boldsymbol{L}(\boldsymbol{x},\vartheta) = \sum_{n=1}^{N} \chi_n(\boldsymbol{x},\vartheta)\boldsymbol{L}_n \tag{4.7}$$

where the indicator function

$$\chi_n(\boldsymbol{x},\vartheta) = \begin{cases} 1 & \boldsymbol{x} \in D_n \\ 0 & \boldsymbol{x} \notin D_n \end{cases} \tag{4.8}$$

with D_n the domain of Phase-n. It is clear the RVE domain $Y = \bigcup_{n=1}^{N} D_n$ and $D_m \bigcap D_n = \varnothing$ when $m \neq n$. In an ergodic random composite, the volume fraction of Phase-n is simply ensemble or volume averaging of the corresponding indicator field

$$\begin{aligned} c_n &= \overline{\chi_n(\boldsymbol{x},\vartheta)} \\ &= \langle \chi_n(\boldsymbol{x},\vartheta) \rangle \end{aligned}$$

with the bracket denoting the volume integration over the RVE domain. The autocorrelation between a point in Phase-m and another in Phase-n is defined as

$$\begin{aligned} c_{mn}(\boldsymbol{x}-\boldsymbol{x}') &= \overline{\chi_m(\boldsymbol{x},\vartheta)\chi_n(\boldsymbol{x}',\vartheta)} \\ &= \langle \chi_m(\boldsymbol{x},\vartheta)\chi_n(\boldsymbol{x}',\vartheta) \rangle \end{aligned} \tag{4.9}$$

with the corresponding correlation coefficient given as

$$\begin{aligned} \rho_{mn}(\boldsymbol{x}-\boldsymbol{x}') &= \frac{c_{mn}(\boldsymbol{x}-\boldsymbol{x}')-c_m c_n}{c_{mn}(0)-c_m c_n} \\ &= \frac{c_{mn}(\boldsymbol{x}-\boldsymbol{x}')-c_m c_n}{c_n\left(\delta_{mn}-c_m\right)} \end{aligned} \tag{4.10}$$

where δ_{mn} denotes Kronecker delta.

The simplest non-zero trial function for the stress polarization is the first-order truncation of the series expansion (3.28), i.e. $\boldsymbol{p}^{(1)} \propto \boldsymbol{L}-\boldsymbol{L}_0$. Accordingly, the trial polarization is constructed as a piece-wise constant field

$$\boldsymbol{p}(\boldsymbol{x},\vartheta) = \sum_{n=1}^{N} \chi_n(\boldsymbol{x},\vartheta)\boldsymbol{p}_n \tag{4.11}$$

where $\boldsymbol{p}_n$ represents the constant stress polarization in Phase-n. By substituting the trial polarization (4.11) into the HS functional (3.47) and equating its derivative about $\boldsymbol{p}_n$ to zero, it yields the following N equations

$$c_n(\boldsymbol{L}_n-\boldsymbol{L}_0)^{-1}\boldsymbol{p}_n + \boldsymbol{p}_m \sum_{m=1}^{N} \int_Y \boldsymbol{\Gamma}^{\infty}(\boldsymbol{x}-\boldsymbol{x}')\left(c_n\,\rho_{mn}(\boldsymbol{x}-\boldsymbol{x}')(\delta_{mn}-c_m)+c_m c_n\right)d\boldsymbol{x}' = c_n\,\boldsymbol{\varepsilon}_0 \tag{4.12}$$

which reduces to

$$(\boldsymbol{L}_n-\boldsymbol{L}_0)^{-1}\boldsymbol{p}_n + \boldsymbol{p}_m \sum_{m=1}^{N} \int_Y \boldsymbol{\Gamma}^{\infty}(\boldsymbol{x})\left(\rho_{mn}(\boldsymbol{x})(\delta_{mn}-c_m)+c_m\right)d\boldsymbol{x} = \boldsymbol{\varepsilon}_0 \tag{4.13}$$

with $n = 1,2, \ldots,N$. The reference moduli tensor $\boldsymbol{L}_0$ is normally chosen to be isotropic. Assume that the spatial distribution of all the N phases is statistically isotropic, i.e. all the correlation coefficient functions $\rho_{mn}(\boldsymbol{x})$

are isotropic, denoted as $\rho_0(\boldsymbol{x})$. To solve $\boldsymbol{p}_n$ from (4.13), the following two properties are needed:

Property

Given that the correlation coefficient $\rho_0(\boldsymbol{x})$ and the reference moduli $\boldsymbol{L}_0$ are both isotropic, it follows that

$$\int_Y \boldsymbol{\Gamma}^{\infty}(\boldsymbol{x})\rho_0(\boldsymbol{x})\,d\boldsymbol{x} = \mathbf{S}_0\boldsymbol{L}_0^{-1} \tag{4.14}$$

where $\mathbf{S}_0$ is the spherical Eshelby tensor (A3.2).

Proof: Expand the left-hand side of (4.14) in Fourier space with spherical coordinates (r, θ, φ)

$$\begin{aligned}\int_Y \boldsymbol{\Gamma}^{\infty}(\mathrm{x})\rho_0(\mathrm{x})\,d\mathrm{x} &= \frac{1}{(2\pi)^3}\int_{-\infty}^{+\infty} \hat{\boldsymbol{\Gamma}}^{\infty}(\boldsymbol{\xi})\hat{\rho}_0(\boldsymbol{\xi})\,d\boldsymbol{\xi} \\ &= \frac{1}{(2\pi)^3}\int_0^{\pi}\int_0^{2\pi} \hat{\boldsymbol{\Gamma}}^{\infty}(\theta,\varphi)\sin\varphi\;\, d\theta\, d\varphi \int_0^{\infty} \hat{\rho}_0(r) r^2\, dr\end{aligned} \tag{4.15}$$

where the RVE domain is infinitely large. Since

$$\int_{-\infty}^{+\infty} \hat{\rho}(\boldsymbol{\xi})\,d\boldsymbol{\xi} = 4\pi\int_0^{\infty} \hat{\rho}_0(r) r^2\, dr \tag{4.16}$$

and

$$\begin{aligned}\frac{1}{(2\pi)^3}\int_{-\infty}^{+\infty} \hat{\rho}(\boldsymbol{\xi})\,d\boldsymbol{\xi} &= \rho_0(0) \\ &= 1\end{aligned} \tag{4.17}$$

(4.15) becomes

$$\frac{1}{4\pi}\int_0^{\pi}\int_0^{2\pi} \hat{\boldsymbol{\Gamma}}^{\infty}(\theta,\varphi)\sin\varphi\;\, d\theta\, d\varphi \tag{4.18}$$

By substituting (A2.6) into (4.18), and considering

$$\begin{cases} \xi_1 = |\boldsymbol{\xi}|\sin\varphi\cos\theta \\ \xi_2 = |\boldsymbol{\xi}|\sin\varphi\sin\theta \\ \xi_3 = |\boldsymbol{\xi}|\cos\varphi \end{cases} \tag{4.19a, b, c}$$

(4.18) becomes $S_0 L_0^{-1}$, with S_0 given in (A3.2). □

Property

$$\int_Y \boldsymbol{\Gamma}^{\infty}(\boldsymbol{x}-\boldsymbol{x}')d\boldsymbol{x}' = 0 \tag{4.20}$$

Proof: The left-hand side of (4.20) corresponds to a uniform polarization field $\boldsymbol{p} = -1$ in (3.21) as follows:

$$\boldsymbol{\varepsilon}'(\boldsymbol{x},\vartheta) = -\int_Y \boldsymbol{\Gamma}(\boldsymbol{x},\boldsymbol{x}')\boldsymbol{p}(\boldsymbol{x}',\vartheta)d\boldsymbol{x}'$$

Given such a polarization field for the fluctuation problem (3.10–3.11), $\nabla\cdot\boldsymbol{p}(\boldsymbol{x}, \vartheta)$ the "body force" in (3.10) is zero. With a zero displacement boundary condition (3.11), and no any external or internal force acting upon the RVE, it results in a zero displacement and a zero strain throughout the RVE domain. □

Note that Property (4.14) can be alternatively obtained via the approach presented in Property (5.11) of Chapter 5. With Properties (4.14) and (4.20), (4.13) becomes

$$(\boldsymbol{L}_n - \boldsymbol{L}_0)^{-1}\boldsymbol{p}_n + \boldsymbol{S}_0\boldsymbol{L}_0^{-1}\sum_{m=1}^{N}(\delta_{mn} - c_m)\boldsymbol{p}_m = \boldsymbol{\varepsilon}_0 \tag{4.21}$$

with $n = 1, 2, \ldots, N$. The stress polarization in each phase is solved as follows

$$\boldsymbol{p}_n = (\boldsymbol{L}_n - \boldsymbol{L}_0)\tilde{\boldsymbol{A}}_n\boldsymbol{\varepsilon}_0 \quad n = 1,2,\ldots,N \tag{4.22}$$

with

$$\tilde{\boldsymbol{A}}_n = \boldsymbol{A}_n\left(\sum_{m=1}^{N} c_m\boldsymbol{A}_m\right)^{-1} \tag{4.23a}$$

$$\boldsymbol{A}_n = \left[\boldsymbol{I} + \boldsymbol{S}_0(\boldsymbol{L}_0^{-1}\boldsymbol{L}_n - \boldsymbol{I})\right]^{-1} \tag{4.23b}$$

By comparing (3.23) with (4.22), the strain in Phase-n is given in terms of the reference strain, as

$$\boldsymbol{\varepsilon}_n = \tilde{\boldsymbol{A}}_n\boldsymbol{\varepsilon}_0 \quad n = 1,2,\ldots,N \tag{4.24}$$

The tensors $\boldsymbol{A}_n$ and $\tilde{\boldsymbol{A}}_n$ are therefore called *concentration tensor* and *normalized concentration tensor*, respectively. Equation (4.23a) immediately yields

$$\sum_{n=1}^{N} c_n \tilde{\boldsymbol{A}}_n = \boldsymbol{I}$$

which, by taking (4.24) into account, leads to

$$\sum_{n=1}^{N} c_n \boldsymbol{\varepsilon}_n = \boldsymbol{\varepsilon}_0 \tag{4.25}$$

since $\boldsymbol{\varepsilon}_0$ is uniformly constant throughout the RVE domain. Note that (4.25) becomes invalid when the reference strain $\boldsymbol{\varepsilon}_0$ varies over a spatial domain, or the polarization is nonconstant in each phase (e.g. the case detailed in Section 8.2).

By substituting the solution (4.22) into the HS functional (3.47) and applying the HS principle, the second-order bound, or the so-called Hashin-Shtrikman bound, is obtained as a weighted average of the elastic moduli tensors of all the phases

$$\boldsymbol{L}^{(2)} = \sum_{n=1}^{N} c_n \boldsymbol{L}_n \tilde{\boldsymbol{A}}_n \tag{4.26}$$

which serves as the upper bound $\boldsymbol{L}_+^{(2)}$ with $\boldsymbol{L}$-$\boldsymbol{L}_0$ being negative semidefinite, or the lower bound $\boldsymbol{L}_-^{(2)}$ with $\boldsymbol{L}$-$\boldsymbol{L}_0$ positive semidefinite. When there is an extreme phase holding the maximum or minimum moduli, e.g. a well-ordered composite with $\boldsymbol{L}_m$ - $\boldsymbol{L}_n$ being either positive or negative semidefinite for any phases $m \neq n$, the optimal reference moduli is simply taken as (3.50), i.e. $\boldsymbol{L}_0 = \underset{n=1,2,\ldots N}{\text{Min}} \{\boldsymbol{L}_n\}$ or $\underset{n=1,2,\ldots N}{\text{Max}} \{\boldsymbol{L}_n\}$. Note that when the reference moduli tensor $\boldsymbol{L}_0$ is such chosen that $\boldsymbol{L}$-$\boldsymbol{L}_0$ satisfies neither negative nor positive semidefiniteness, the result (4.26) degenerates to a certain approximation without any bounding property. The self-consistent schemes in micromechanics (Hill, 1965), or the so-called effective medium approximations in conductivity (e.g. Tuck, 1999), are well-known examples of such an approximation by choosing the effective elastic moduli tensor or the effective conductivity as the elastic moduli or conductivity of the reference medium, i.e. $\boldsymbol{L}_0 = \boldsymbol{L}^{(2)}$.

4.2.2 HS bounds for phases characterized with the isotropic elastic moduli

When the elastic moduli of all the N phases are isotropic, the "bulk modulus" and "shear modulus" of the concentration tensor $\boldsymbol{A}_n \sim (3\kappa_A, 2\mu_A)$

are obtained from (4.23) by using Hill's symbolic notation and the following operation(Hill, 1964):

$$3\kappa_A = \left[1 + \frac{3\kappa_S}{3\kappa_0}(3\kappa_n - 3\kappa_0)\right]^{-1} = \frac{3\kappa_0 + 4\mu_0}{3\kappa_n + 4\mu_0} \tag{4.27}$$

$$2\mu_A = \left[1 + \frac{2\mu_S}{2\mu_0}(2\mu_n - 2\mu_0)\right]^{-1} = \frac{5\mu_0(3\kappa_0 + 4\mu_0)}{\mu_0(9\kappa_0 + 8\mu_0) + 6\mu_n(\kappa_0 + 2\mu_0)} \tag{4.28}$$

where the two "moduli" of the spherical Eshelby tensor $\boldsymbol{S}_0$ (A3.2) are available as

$$3\kappa_S = \frac{3\kappa_0}{3\kappa_0 + 4\mu_0} \tag{4.29}$$

$$2\mu_S = \frac{6(\kappa_0 + 2\mu_0)}{5(3\kappa_0 + 4\mu_0)} \tag{4.30}$$

With (4.27–4.30), by applying the symbolic operation onto (4.26), the HS bounds of the effective bulk and shear moduli are expressed as functions of the reference moduli, respectively, as

$$\kappa^{(2)}(\mu_0) = \left(\sum_{m=1}^{N} \frac{c_m}{\frac{4\mu_0}{3} + \kappa_m}\right)^{-1} \sum_{n=1}^{N} \frac{c_n\kappa_n}{\frac{4\mu_0}{3} + \kappa_n} \tag{4.31}$$

$$\mu^{(2)}(\kappa_0, \mu_0) = \left(\sum_{m=1}^{N} \frac{c_m}{\frac{(9\kappa_0 + 8\mu_0)\mu_0}{6(\kappa_0 + 2\mu_0)} + \mu_m}\right)^{-1} \sum_{n=1}^{N} \frac{c_n\mu_n}{\frac{(9\kappa_0 + 8\mu_0)\mu_0}{6(\kappa_0 + 2\mu_0)} + \mu_n} \tag{4.32}$$

When the reference shear modulus is chosen to be the maximum $\mu_0 = \max_n\{\mu_n\}$ and the minimum $\min_n\{\mu_n\}$, respectively, the optimal multiphase HS bounds of the effective bulk modulus are finally obtained as

$$\boxed{\kappa_+^{(2)} = \left(\sum_{m=1}^{N} \frac{c_m}{\frac{4}{3}\max_j\{\mu_j\} + \kappa_m} \right)^{-1} \sum_{n=1}^{N} \frac{c_n \kappa_n}{\frac{4}{3}\max_j\{\mu_j\} + \kappa_n}} \tag{4.33}$$

$$\boxed{\kappa_-^{(2)} = \left(\sum_{m=1}^{N} \frac{c_m}{\frac{4}{3}\min_j\{\mu_j\} + \kappa_m} \right)^{-1} \sum_{n=1}^{N} \frac{c_n \kappa_n}{\frac{4}{3}\min_j\{\mu_j\} + \kappa_n}} \tag{4.34}$$

When there exists such a phase that its bulk and shear moduli are simultaneously the maxima or minima among all the phases in a composite, the optimal multiphase HS bounds of the effective shear modulus are explicitly obtained from (4.32) as

$$\boxed{\mu_+^{(2)} = \left(\sum_{m=1}^{N} \frac{c_m}{\max_j\left\{\frac{(9\kappa_j + 8\mu_j)\mu_j}{6(\kappa_j + 2\mu_j)}\right\} + \mu_m} \right)^{-1} \sum_{n=1}^{N} \frac{c_n \mu_n}{\max_j\left\{\frac{(9\kappa_j + 8\mu_j)\mu_j}{6(\kappa_j + 2\mu_j)}\right\} + \mu_n}} \tag{4.35}$$

$$\boxed{\mu_-^{(2)} = \left(\sum_{m=1}^{N} \frac{c_m}{\min_j\left\{\frac{(9\kappa_j + 8\mu_j)\mu_j}{6(\kappa_j + 2\mu_j)}\right\} + \mu_m} \right)^{-1} \sum_{n=1}^{N} \frac{c_n \mu_n}{\min_j\left\{\frac{(9\kappa_j + 8\mu_j)\mu_j}{6(\kappa_j + 2\mu_j)}\right\} + \mu_n}} \tag{4.36}$$

The HS bounds (4.33–4.36) were first derived by Hashin and Shtrikman (1963). A well-ordered composite holding the inequality $(\kappa_n - \kappa_m)(\mu_n - \mu_m) > 0$ for any phases $m \neq n$ is an example that satisfies the previously discussed existence condition of a hardest or a softest phase.

When the condition is not satisfied, the multiphase HS bounds of the effective shear modulus are obtained by choosing $\kappa_0 = \max_n\{\kappa_n\}$,

$\mu_0 = \max\{\mu_n\}$ and $\kappa_0 = \min_n\{\kappa_n\}$, $\mu_0 = \min_n\{\mu_n\}$ (Walpole, 1966), respectively, as

$$\mu_+^{(2)} = \left(\sum_{m=1}^{N} \frac{c_m}{\max_{(i,j)}\left\{ \frac{(9\kappa_i + 8\mu_j)\mu_j}{6(\kappa_i + 2\mu_j)} \right\} + \mu_m} \right)^{-1} \sum_{n=1}^{N} \frac{c_n \mu_n}{\max_{(i,j)}\left\{ \frac{(9\kappa_i + 8\mu_j)\mu_j}{6(\kappa_i + 2\mu_j)} \right\} + \mu_n} \tag{4.37}$$

$$\mu_-^{(2)} = \left(\sum_{m=1}^{N} \frac{c_m}{\min_{(i,j)}\left\{ \frac{(9\kappa_i + 8\mu_j)\mu_j}{6(\kappa_i + 2\mu_j)} \right\} + \mu_m} \right)^{-1} \sum_{n=1}^{N} \frac{c_n \mu_n}{\min_{(i,j)}\left\{ \frac{(9\kappa_i + 8\mu_j)\mu_j}{6(\kappa_i + 2\mu_j)} \right\} + \mu_n} \tag{4.38}$$

Since the reference phase with the such combined moduli does not physically exist in an original composite, the bounds (4.37–38) should not be the narrowest bounds. In other words, they are physically unattainable, and the following two-phase case in (4.41–4.42) confirms the statement.

In fact, physical existence of the reference phase in an original composite is not a sufficient condition to realize physical attainability. There are other conditions to match as well. Let us use Hashin coated sphere assemblage (Hashin,1962; Milton, 2002) to reason these conditions of attainability. A coated sphere is embedded into an infinite medium within which a uniform current flows from infinity. Given the isotropic conductivities of the core and the shell, and the ratio between the radiuses of the core and the sphere, there is such a corresponding value of the isotropic conductivity chosen for the infinite medium that the coated sphere presents no disturbance to the uniform current. In other words, the medium effectively does not feel any difference with insertion of such a coated sphere, and the insertion can be continuously applied until the medium is completely filled with such coated spheres of various sizes (Figure 4.1). This chosen value of the conductivity for the medium is exactly the HS bound of the effective conductivity that is realized with the Hashin coated sphere assemblage. When such an assemblage idea is extended to multicoated spheres, clearly there are certain constraints imposed on the thicknesses of shells and the conductivities of multiphases, so that the insertion of no disturbance can be realized.

Therefore, the multiphase HS bounds (4.33–4.36) are not always the narrowest bounds. Certain parameter constraints or regimes were provided by Milton (1981) to attain the multiphase HS bounds of the effective bulk modulus (4.33–4.34), and by Gibiansky and Sigmund (2000) on the shear modulus (4.35–4.36).

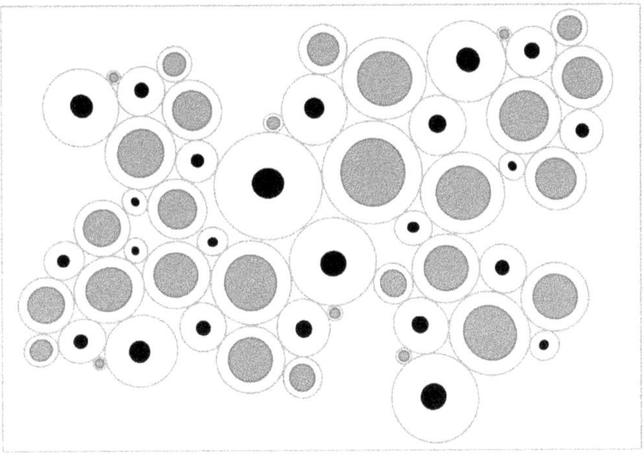

Figure 4.1 Schematic of a three-phase Hashin coated sphere assemblage consisting of two types of coated spheres (white = matrix phase, gray or dark = inclusion phases).

When specialized to two-phase composites, the multiphase HS bounds (4.33–4.36) reduce to

$$\kappa_{+}^{(2)} = (c_1\kappa_1 + c_2\kappa_2) - \frac{c_1c_2(\kappa_1 - \kappa_2)^2}{c_1\kappa_2 + c_2\kappa_1 + \frac{4}{3}\max\{\mu_1, \mu_2\}} \tag{4.39}$$

$$\kappa_{-}^{(2)} = (c_1\kappa_1 + c_2\kappa_2) - \frac{c_1c_2(\kappa_1 - \kappa_2)^2}{c_1\kappa_2 + c_2\kappa_1 + \frac{4}{3}\min\{\mu_1, \mu_2\}} \tag{4.40}$$

$$\mu_{+}^{(2)} = (c_1\mu_1 + c_2\mu_2) - \frac{c_1c_2(\mu_1 - \mu_2)^2}{c_1\mu_2 + c_2\mu_1 + \max\limits_{n}\left\{\frac{\mu_n(9\kappa_n + 8\mu_n)}{6(\kappa_n + 2\mu_n)}\right\}} \tag{4.41}$$

$$\mu_{-}^{(2)} = (c_1\mu_1 + c_2\mu_2) - \frac{c_1c_2(\mu_1 - \mu_2)^2}{c_1\mu_2 + c_2\mu_1 + \min\limits_{n}\left\{\frac{\mu_n(9\kappa_n + 8\mu_n)}{6(\kappa_n + 2\mu_n)}\right\}} \tag{4.42}$$

In a non-well-ordered case, by reducing the third-order bounds (4.133) and (4.146) to the second-order bounds, the loose bounds (4.37–4.38) with $N = 2$ are tightened to (4.41–4.42), which confirms unattainability of (4.37–38).

The two-phase bounds of the effective bulk modulus (4.39–4.40) are physically realizable with the Hashin coated sphere assemblage (two-phase version of Figure 4.1). The bounds of the effective shear modulus (4.41–4.42) are attainable for well-ordered composites with several types of microgeometry (e.g. finite-rank laminates [Francfort & Murat, 1986], reiterated cell structures [Lukkassen, 1999]).

It is finally noted that the bound (4.26) is applicable to the elastic moduli of phases being anisotropic as well. Examples of such obtained optimal HS bounds are left to the reader as an exercise, e.g. in a fiber-reinforced composite the elastic moduli of fibers being transversely isotropic.

4.2.3 HS bounds in 2D elasticity

Elasticity of fiber-reinforced composites typically involves a two-dimensional (2D) plane strain model. A thin plate containing holes under in-plane loading is a 2D plane stress example. To find the HS bounds for such a 2D problem we need the 2D counterparts of Properties (4.14) and (4.20).

Property

Given that the correlation coefficient $\rho_0(\boldsymbol{x})$ and the reference moduli $\mathbf{L}_0$ are both isotropic in the plane strain case, it follows that

$$\int_Y \boldsymbol{\Gamma}^{\infty}(\boldsymbol{x})\rho_0(\boldsymbol{x})d\boldsymbol{x} = \mathbf{S}_0\boldsymbol{L}_0^{-1} \tag{4.43}$$

where $\mathbf{S}_0$ the circular Eshelby tensor is given by taking $a_1 = a_2$ in (A3.3) i.e.

$$\mathbf{S}_0 = \begin{bmatrix} \dfrac{5-4\nu_0}{8(1-\nu_0)} & \dfrac{4\nu_0-1}{8(1-\nu_0)} & 0 \\ \dfrac{4\nu_0-1}{8(1-\nu_0)} & \dfrac{5-4\nu_0}{8(1-\nu_0)} & 0 \\ 0 & 0 & \dfrac{3-4\nu_0}{4(1-\nu_0)} \end{bmatrix} \tag{4.44}$$

and the plane strain elastic moduli

$$\boldsymbol{L}_0 = \begin{bmatrix} \kappa_0^{\varepsilon}+\mu_0 & \kappa_0^{\varepsilon}-\mu_0 & 0 \\ \kappa_0^{\varepsilon}-\mu_0 & \kappa_0^{\varepsilon}+\mu_0 & 0 \\ 0 & 0 & \mu_0 \end{bmatrix} \tag{4.45}$$

with the in-plane bulk modulus

$$\kappa_0^{\varepsilon} = \kappa_0 + \frac{\mu_0}{3} \tag{4.46}$$

Proof: Expand the left-hand side of (4.43) in Fourier space with polar coordinates (r, θ)

$$\int_Y \boldsymbol{\Gamma}^{\infty}(\boldsymbol{x})\rho_0(\boldsymbol{x})d\boldsymbol{x} = \frac{1}{(2\pi)^2}\int_0^{2\pi} \hat{\boldsymbol{\Gamma}}^{\infty}(\theta)\ d\theta \int_0^{\infty} \hat{\rho}_0(r) r\,dr \tag{4.47}$$

which, given that

$$\int_{-\infty}^{\infty} \hat{\rho}_0(\boldsymbol{\xi})d\boldsymbol{\xi} = 2\pi \int_0^{\infty} \hat{\rho}_0(r) r\,dr \tag{4.48}$$

and

$$\frac{1}{(2\pi)^2}\int_{-\infty}^{\infty} \hat{\rho}_0(\boldsymbol{\xi})d\boldsymbol{\xi} = \rho_0(0) = 1 \tag{4.49}$$

becomes

$$\frac{1}{2\pi}\int_0^{2\pi} \hat{\boldsymbol{\Gamma}}^{\infty}(\theta)d\theta \tag{4.50}$$

Since in the plane strain case the wave vector reduces to

$$\boldsymbol{\xi} = \{\xi_1, \xi_2\}^T \tag{4.51}$$

$$\begin{cases} \xi_1 = |\boldsymbol{\xi}|\cos\theta \\ \xi_2 = |\boldsymbol{\xi}|\sin\theta \end{cases} \tag{4.52}$$

substitution of (A2.6) and (4.52) into (4.50) yields (4.43). □

Similar to the proof of Property (4.43), it follows in two-dimensional elasticity

Property

$$\int_{Y} \boldsymbol{\Gamma}^{\infty}(\boldsymbol{x}-\boldsymbol{x}')d\boldsymbol{x}' = 0 \tag{4.53}$$

With Properties (4.43) and (4.53), from (4.21) we similarly obtain the plane strain version of (4.22–4.26). When the elastic moduli of all the N phases are isotropic, the 2D "bulk modulus" and "shear modulus" of $\tilde{\boldsymbol{A}}_n \sim (2\kappa_A, 2\mu_A)$ is obtained from (4.23) by using the following 2D symbolic operation

$$\begin{aligned} 2\kappa_A &= \left[1 + \frac{2\kappa_S}{2\kappa_0^{\varepsilon}}(2\kappa_n^{\varepsilon} - 2\kappa_0^{\varepsilon})\right]^{-1} \\ &= \frac{\kappa_0^{\varepsilon} + \mu_0}{\kappa_n^{\varepsilon} + \mu_0} \end{aligned} \tag{4.54}$$

$$\begin{aligned} 2\mu_A &= \left[1 + \frac{2\mu_S}{2\mu_0}(2\mu_n - 2\mu_0)\right]^{-1} \\ &= \frac{2\mu_0(\kappa_0^{\varepsilon} + \mu_0)}{\mu_n(\kappa_0^{\varepsilon} + 2\mu_0) + \kappa_0^{\varepsilon}\mu_0} \end{aligned} \tag{4.55}$$

given the "moduli" of the circular Eshelby tensor $\boldsymbol{S}_0$ as

$$2\kappa_S = \frac{\kappa_0^{\varepsilon}}{\kappa_0^{\varepsilon} + \mu_0} \tag{4.56}$$

$$2\mu_S = \frac{\kappa_0^{\varepsilon} + 2\mu_0}{2(\kappa_0^{\varepsilon} + \mu_0)} \tag{4.57}$$

With (4.54–4.57), by applying the symbolic operation onto (4.26), the HS bounds are expressed in terms of the reference moduli as

$$\kappa^{(2)}(\mu_0) = \left(\sum_{m=1}^{N} \frac{c_m}{\mu_0 + \kappa_m^{\varepsilon}}\right)^{-1} \sum_{n=1}^{N} \frac{c_n \kappa_n^{\varepsilon}}{\mu_0 + \kappa_n^{\varepsilon}} \tag{4.58}$$

$$\mu^{(2)}(\kappa_0^\varepsilon,\mu_0)=\left(\sum_{m=1}^{N}\frac{c_m}{\dfrac{\kappa_0^\varepsilon\mu_0}{\kappa_0^\varepsilon+2\mu_0}+\mu_m}\right)^{-1}\sum_{n=1}^{N}\frac{c_n\mu_n}{\dfrac{\kappa_0^\varepsilon\mu_0}{\kappa_0^\varepsilon+2\mu_0}+\mu_n} \tag{4.59}$$

Similar to the HS bounds (4.33–4.36), the optimal HS bounds in the plane strain case are therefore given as

$$\boxed{\kappa_+^{(2)}=\left(\sum_{m=1}^{N}\frac{c_m}{\max\limits_j\{\mu_j\}+\kappa_m^\varepsilon}\right)^{-1}\sum_{n=1}^{N}\frac{c_n\kappa_n^\varepsilon}{\max\limits_j\{\mu_j\}+\kappa_n^\varepsilon}} \tag{4.60}$$

$$\boxed{\kappa_-^{(2)}=\left(\sum_{m=1}^{N}\frac{c_m}{\min\limits_j\{\mu_j\}+\kappa_m^\varepsilon}\right)^{-1}\sum_{n=1}^{N}\frac{c_n\kappa_n^\varepsilon}{\min\limits_j\{\mu_j\}+\kappa_n^\varepsilon}} \tag{4.61}$$

$$\boxed{\mu_+^{(2)}=\left(\sum_{m=1}^{N}\frac{c_m}{\max\limits_j\left\{(2\kappa_j^{\varepsilon-1}+\mu_j^{-1})^{-1}\right\}+\mu_m}\right)^{-1}\sum_{n=1}^{N}\frac{c_n\mu_n}{\max\limits_j\left\{\left(2\kappa_j^{\varepsilon-1}+\mu_j^{-1}\right)^{-1}\right\}+\mu_n}} \tag{4.62}$$

$$\boxed{\mu_-^{(2)}=\left(\sum_{m=1}^{N}\frac{c_m}{\min\limits_j\left\{\left(2\kappa_j^{\varepsilon-1}+\mu_j^{-1}\right)^{-1}\right\}+\mu_m}\right)^{-1}\sum_{n=1}^{N}\frac{c_n\mu_n}{\min\limits_j\left\{\left(2\kappa_j^{\varepsilon-1}+\mu_j^{-1}\right)^{-1}\right\}+\mu_n}} \tag{4.63}$$

Equations (4.62–4.63) are applicable when there is such a phase that its bulk and shear moduli are simultaneously the maxima or minima among all the phases in a composite. When the condition is not satisfied, the HS bounds are loosely given as

$$\mu_+^{(2)}=\left(\sum_{m=1}^{N}\frac{c_m}{\max\limits_{(i,j)}\left\{\left(2\kappa_i^{\varepsilon-1}+\mu_j^{-1}\right)^{-1}\right\}+\mu_m}\right)^{-1}\sum_{n=1}^{N}\frac{c_n\mu_n}{\max\limits_{(i,j)}\left\{\left(2\kappa_i^{\varepsilon-1}+\mu_j^{-1}\right)^{-1}\right\}+\mu_n} \tag{4.64}$$

$$\mu_-^{(2)} = \left(\sum_{m=1}^{N} \frac{c_m}{\min\limits_{(i,j)} \left\{ \left(2\kappa_i^{\varepsilon-1} + \mu_j^{-1} \right)^{-1} \right\} + \mu_m} \right)^{-1} \sum_{n=1}^{N} \frac{c_n \mu_n}{\min\limits_{(i,j)} \left\{ \left(2\kappa_i^{\varepsilon-1} + \mu_j^{-1} \right)^{-1} \right\} + \mu_n} \tag{4.65}$$

which are not physically attainable.

Specialized to two-phase composites, the optimal HS bounds (4.60–4.63) in the plane strain case reduce to

$$\boxed{\kappa_+^{(2)} = (c_1 \kappa_1^{\varepsilon} + c_2 \kappa_2^{\varepsilon}) - \frac{c_1 c_2 (\kappa_1^{\varepsilon} - \kappa_2^{\varepsilon})^2}{c_1 \kappa_2^{\varepsilon} + c_2 \kappa_1^{\varepsilon} + \max\{\mu_1, \mu_2\}}} \tag{4.66}$$

$$\boxed{\kappa_-^{(2)} = (c_1 \kappa_1^{\varepsilon} + c_2 \kappa_2^{\varepsilon}) - \frac{c_1 c_2 (\kappa_1^{\varepsilon} - \kappa_2^{\varepsilon})^2}{c_1 \kappa_2^{\varepsilon} + c_2 \kappa_1^{\varepsilon} + \min\{\mu_1, \mu_2\}}} \tag{4.67}$$

$$\boxed{\mu_+^{(2)} = (c_1 \mu_1 + c_2 \mu_2) - \frac{c_1 c_2 (\mu_1 - \mu_2)^2}{c_1 \mu_2 + c_2 \mu_1 + \max\limits_{n} \left\{ \left(2\kappa_n^{\varepsilon-1} + \mu_n^{-1} \right)^{-1} \right\}}} \tag{4.68}$$

$$\boxed{\mu_-^{(2)} = (c_1 \mu_1 + c_2 \mu_2) - \frac{c_1 c_2 (\mu_1 - \mu_2)^2}{c_1 \mu_2 + c_2 \mu_1 + \min\limits_{n} \left\{ (2\kappa_n^{\varepsilon-1} + \mu_n^{-1})^{-1} \right\}}} \tag{4.69}$$

The plane strain HS bounds (4.66–4.67) and (4.68–4.69) are given by Hill (1964) and Hashin (1965), respectively, with (4.68–4.69) subjected to the well-ordered condition

$$(\kappa_1^{\varepsilon} - \kappa_2^{\varepsilon})(\mu_1 - \mu_2) > 0 \tag{4.70}$$

As shown later in Section 4.3, in a non-well-ordered case, the bounds of the effective shear modulus (4.64–4.65) when $N = 2$ actually can also be tightened to (4.68–4.69), which indicates the condition (4.70) can be waived on (4.68–4.69).

In the plane stress case, Properties (4.43) and (4.53) of the plane strain case can be similarly derived, with $\boldsymbol{S}_0$ and $\boldsymbol{L}_0$ obtained from (4.44–4.45) by replacing ν_0 with $\nu_0/(1 + \nu_0)$ and keeping μ_0 unchanged. The multiphase HS bounds in the plane stress case are found to be identical to (4.60–4.65) except for replacing the plane strain in-plane bulk modulus (4.46) with the plane stress in-plane bulk modulus

$$\kappa_0^\sigma = \frac{9\kappa_0\mu_0}{3\kappa_0 + 4\mu_0} \tag{4.71}$$

Attainability of the HS bounds in two-dimensional elasticity is similar to that in three-dimension discussed in Subsection 4.2.2.

4.2.4 HS bounds of the effective conductivity

The HS bounds of the effective conductivity are similarly derived by substituting the trial polarization (4.11) into the HS functional (3.57). In the isotropic case, a simple way is to take advantage of the derived HS bounds of the effective bulk modulus by adopting a zero value for the Poisson's ratio. With the Poisson's ratio being zero, it follows that in 3D

$$\mu_j = \frac{3}{2}\kappa_j \tag{4.72}$$

and in 2D

$$\mu_j = \kappa_j^\varepsilon \quad \text{or} \quad \kappa_j^\sigma \tag{4.73}$$

By substituting (4.72–4.73) into the HS bounds of the effective bulk modulus (4.33–4.34) and (4.60–4.61), the multiphase HS bounds of the effective conductivity are obtained as

$$\boxed{K_+^{(2)} = \left(\sum_{m=1}^{N} \frac{c_m}{(d-1)\max_j\{K_j\} + K_m}\right)^{-1} \sum_{n=1}^{N} \frac{c_n K_n}{(d-1)\max_j\{K_j\} + K_n}} \tag{4.74}$$

$$\boxed{K_-^{(2)} = \left(\sum_{m=1}^{N} \frac{c_m}{(d-1)\min_j\{K_j\} + K_m}\right)^{-1} \sum_{n=1}^{N} \frac{c_n K_n}{(d-1)\min_j\{K_j\} + K_n}} \tag{4.75}$$

where $d = 2$ or 3 indicates dimensionality. In the case of two-phase composites, the bounds (4.74–4.75) are specialized to

$$\boxed{K_+^{(2)} = (c_1K_1 + c_2K_2) - \frac{c_1c_2(K_1 - K_2)^2}{c_1K_2 + c_2K_1 + (d-1)\max\{K_1, K_2\}}} \tag{4.76}$$

$$\boxed{K_-^{(2)} = (c_1K_1 + c_2K_2) - \frac{c_1c_2(K_1 - K_2)^2}{c_1K_2 + c_2K_1 + (d-1)\min\{K_1, K_2\}}} \tag{4.77}$$

Attainability of HS bounds on conductivity is similar to that on the bulk modulus discussed in Subsection 4.2.2.

4.3 THIRD-ORDER BOUNDS

4.3.1 Third-order bounds of the effective bulk modulus

Similar to the derivation of the second order HS bounds, by substituting the trial polarization (4.11) with the number of phases $N = 2$ into the upper bound potential energy functional (3.27) and equating its derivative about $\Delta \boldsymbol{p} = \boldsymbol{p}_1 - \boldsymbol{p}_2$ to zero, it yields

$$\begin{aligned}
&\left[\int_Y \boldsymbol{\Gamma}^\infty(\boldsymbol{x}')(\boldsymbol{L}_1 - \boldsymbol{L}_2)\int_Y \boldsymbol{\Gamma}^\infty(\boldsymbol{x}'')c_{111}(,\boldsymbol{x}'\boldsymbol{x}'')\,d\boldsymbol{x}''\,d\boldsymbol{x}'\right.\\
&\left.+\int_Y \boldsymbol{\Gamma}^\infty(\boldsymbol{x}')(\boldsymbol{L}_2 - \boldsymbol{L}_0)\int_Y \boldsymbol{\Gamma}^\infty(\boldsymbol{x}'')c_{11}(\boldsymbol{x}'' - \boldsymbol{x}')\,d\boldsymbol{x}''\,d\boldsymbol{x}' + \int_Y \boldsymbol{\Gamma}^\infty(\boldsymbol{x}')c_{11}(\boldsymbol{x}')\,d\boldsymbol{x}'\right]\Delta\boldsymbol{p}\\
&= \int_Y \boldsymbol{\Gamma}^\infty(\boldsymbol{x}')c_{11}(\boldsymbol{x}')\,d\boldsymbol{x}'\,(\boldsymbol{L}_1 - \boldsymbol{L}_2)\boldsymbol{\varepsilon}_0
\end{aligned} \tag{4.78}$$

where the triple-correlation function

$$c_{111}(\boldsymbol{x}', \boldsymbol{x}'') = \overline{\chi_1(\boldsymbol{x}, \vartheta)\chi_1(\boldsymbol{x} + \boldsymbol{x}', \vartheta)\chi_1(\boldsymbol{x} + \boldsymbol{x}'', \vartheta)} \tag{4.79}$$

Given that the spatial distribution of two phases is statistically isotropic, by using (4.10) and Property (4.14), (4.78) is rewritten with concise notation as

$$\left(\boldsymbol{\Gamma}^\infty(\boldsymbol{L}_1 - \boldsymbol{L}_2)\boldsymbol{\Gamma}^\infty c_{111} + \boldsymbol{\Gamma}^\infty(\boldsymbol{L}_2 - \boldsymbol{L}_0)\boldsymbol{\Gamma}^\infty c_{11} + c_1c_2\boldsymbol{S}_0\boldsymbol{L}_0^{-1}\right)\Delta\boldsymbol{p} = c_1c_2\boldsymbol{S}_0\boldsymbol{L}_0^{-1}(\boldsymbol{L}_1 - \boldsymbol{L}_2)\boldsymbol{\varepsilon}_0 \tag{4.80}$$

Suppose the elastic moduli $\boldsymbol{L}_0$, $\boldsymbol{L}_1$, $\boldsymbol{L}_2$ are all isotropic. Given the reference strain as a hydrostatic strain

$$\varepsilon_{0,ij} = \delta_{ij}e_0/3 \tag{4.81}$$

the right-hand side of (4.80) results in a hydrostatic strain as well, and thereby the non-zero entries of the tensor $\Delta p_{ij} = \Delta p\delta_{ij}$ are diagonal ones only, with all the shear components being zero. Accordingly, the trial polarization (4.11) specialized to a two-phase composite is written as

$$p_{ij}(\boldsymbol{x},\vartheta) = \Delta p \delta_{ij} \chi_1(\boldsymbol{x},\vartheta) + p_{2,ij} \tag{4.82}$$

Similarly, the elastic moduli (4.7) and the compliance tensor in the two-phase case are written, respectively, as

$$\boldsymbol{L}(\boldsymbol{x},\vartheta) = (\boldsymbol{L}_1 - \boldsymbol{L}_2)\chi_1(\boldsymbol{x},\vartheta) + \boldsymbol{L}_2 \tag{4.83}$$

$$\boldsymbol{M}(\boldsymbol{x},\vartheta) = (\boldsymbol{M}_1 - \boldsymbol{M}_2)\chi_1(\boldsymbol{x},\vartheta) + \boldsymbol{M}_2 \tag{4.84}$$

By substituting (4.82–4.84) into the upper bound energy functional (3.27) and applying Properties (4.14) and (4.20), the first, second, and fourth terms of the upper bound (3.27) are directly obtained, respectively, as

$$\frac{1}{2}\langle \boldsymbol{\varepsilon}_0, \boldsymbol{L}\boldsymbol{\varepsilon}_0 \rangle = \frac{1}{2}(c_1\kappa_1 + c_2\kappa_2)e_0^2 \tag{4.85}$$

$$\begin{aligned} \frac{1}{2}\langle \boldsymbol{p}, \boldsymbol{\Gamma}^\infty \boldsymbol{p} \rangle &= \frac{1}{2}\Delta \boldsymbol{p}\left(c_1c_2\mathbf{S}_0\boldsymbol{L}_0^{-1}\right)\Delta \boldsymbol{p} \\ &= \frac{1}{2}c_1c_2\frac{(\Delta p)^2}{\kappa_0 + 4\mu_0/3} \end{aligned} \tag{4.86}$$

$$\begin{aligned} -\langle \boldsymbol{\Gamma}^\infty \boldsymbol{p}, (\boldsymbol{L} - \boldsymbol{L}_0)\boldsymbol{\varepsilon}_0 \rangle &= -\boldsymbol{\varepsilon}_0(\boldsymbol{L}_1 - \boldsymbol{L}_2)\left(c_1c_2\mathbf{S}_0\boldsymbol{L}_0^{-1}\right)\Delta \boldsymbol{p} \\ &= -c_1c_2\frac{(\kappa_1 - \kappa_2)}{\kappa_0 + 4\mu_0/3}\Delta p e_0 \end{aligned} \tag{4.87}$$

The third term is expanded into

$$\begin{aligned} &\frac{1}{2}\langle \boldsymbol{\Gamma}^\infty \boldsymbol{p}, (\boldsymbol{L} - \boldsymbol{L}_0)\boldsymbol{\Gamma}^\infty \boldsymbol{p} \rangle \\ &= \frac{1}{2}\Delta \boldsymbol{p}\left[\int_Y \boldsymbol{\Gamma}^\infty(\boldsymbol{x}')(\boldsymbol{L}_2 - \boldsymbol{L}_0)\int_Y \boldsymbol{\Gamma}^\infty(\boldsymbol{x}'')c_{11}(\boldsymbol{x}'' - \boldsymbol{x}')\,d\boldsymbol{x}''\,d\boldsymbol{x}'\right. \\ &\quad \left. + \int_Y \boldsymbol{\Gamma}^\infty(\boldsymbol{x}')(\boldsymbol{L}_1 - \boldsymbol{L}_2)\int_Y \boldsymbol{\Gamma}^\infty(\boldsymbol{x}'')c_{111}(\boldsymbol{x}'', \boldsymbol{x}')\,d\boldsymbol{x}''\,d\boldsymbol{x}'\right]\Delta \boldsymbol{p} \end{aligned} \tag{4.88}$$

To simplify the right hand side of (4.88), we need to look closely at the operator $\boldsymbol{\Gamma}^{\infty}\boldsymbol{L}\boldsymbol{\Gamma}^{\infty}$ that is subjected to a pressure of unity. According to the definition of strain energy, the operator $\boldsymbol{\Gamma}^{\infty}\boldsymbol{L}\boldsymbol{\Gamma}^{\infty}$ consists of the following three modes, i.e. normal, shear, and bulk

$$\begin{aligned}\boldsymbol{\Gamma}^{\infty}\boldsymbol{L}\boldsymbol{\Gamma}^{\infty} &= 2\left(\Gamma^{\infty}_{11nn}\mu\Gamma^{\infty}_{11nn} + \Gamma^{\infty}_{22nn}\mu\Gamma^{\infty}_{22nn} + \Gamma^{\infty}_{33nn}\mu\Gamma^{\infty}_{33nn}\right) \\ &\quad + 4\left(\Gamma^{\infty}_{12nn}\mu\Gamma^{\infty}_{12nn} + \Gamma^{\infty}_{13nn}\mu\Gamma^{\infty}_{13nn} + \Gamma^{\infty}_{23nn}\mu\Gamma^{\infty}_{23nn}\right) \\ &\quad + \left(\kappa - \frac{2}{3}\mu\right)\left(\Gamma^{\infty}_{11nn} + \Gamma^{\infty}_{22nn} + \Gamma^{\infty}_{33nn}\right)^2\end{aligned} \tag{4.89}$$

with

$$\Gamma^{\infty}_{11nn} = \Gamma^{\infty}_{1111} + \Gamma^{\infty}_{1122} + \Gamma^{\infty}_{1133} \tag{4.90}$$

By using (A2.6), it is straightforward to find that the bulk mode energy in (4.89) is zero, i.e.

Property

$$\Gamma^{\infty}_{11nn} + \Gamma^{\infty}_{22nn} + \Gamma^{\infty}_{33nn} = 0 \tag{4.91}$$

The first term on the right-hand side of (4.88) is written in Fourier space as

$$\frac{1}{2}\Delta\boldsymbol{p}\left(\boldsymbol{\Gamma}^{\infty}(\boldsymbol{L}_2 - \boldsymbol{L}_0)\boldsymbol{\Gamma}^{\infty}c_{11}\right)\Delta\boldsymbol{p} = \frac{c_1c_2}{2}\frac{1}{(2\pi)^3}\Delta\boldsymbol{p}\left(\hat{\boldsymbol{\Gamma}}^{\infty}(\boldsymbol{L}_2 - \boldsymbol{L}_0)\hat{\boldsymbol{\Gamma}}^{\infty}\hat{\rho}_0\right)\Delta\boldsymbol{p} \tag{4.92}$$

which, by using (4.89–4.91), (4.16–4.17), and (A2.6), is expanded and calculated as

$$\begin{aligned}&\frac{c_1c_2}{2}2(\mu_2 - \mu_0)(\Delta p)^2\frac{1}{4\pi}\int_0^{\pi}\int_0^{2\pi}\sum_{i=1,j=1}^{3}\left(\hat{\Gamma}^{\infty}_{ijnn}(\theta,\varphi)\right)^2\sin\varphi\, d\theta d\varphi \\ &= c_1c_2\frac{(\mu_2 - \mu_0)}{(\kappa_0 + 4\mu_0/3)^2}(\Delta p)^2\end{aligned} \tag{4.93}$$

The second term on the right-hand side of (4.88) is similarly expanded into

$$\begin{aligned}&\frac{1}{2}\Delta\boldsymbol{p}\left(\boldsymbol{\Gamma}^{\infty}(\boldsymbol{L}_1 - \boldsymbol{L}_2)\boldsymbol{\Gamma}^{\infty}c_{111}\right)\Delta\boldsymbol{p} \\ &= \frac{1}{2}2(\mu_1 - \mu_2)(\Delta p)^2\int_{Y}\int_{Y}\sum_{i=1,j=1}^{3}\Gamma^{\infty}_{ijnn}(\boldsymbol{x})\Gamma^{\infty}_{ijnn}(\boldsymbol{x}')c_{111}(\boldsymbol{x},\boldsymbol{x}')d\boldsymbol{x}d\boldsymbol{x}'\end{aligned} \tag{4.94}$$

which becomes

$$\frac{1}{2}2(\mu_1-\mu_2)(\Delta p)^2\int_Y\int_Y\sum_{i=1,j=1}^{3}\Gamma^{\infty}_{ijnn}(\boldsymbol{x})\Gamma^{\infty}_{ijnn}(\boldsymbol{x}')\left(c_{111}(\boldsymbol{x},\boldsymbol{x}')-\frac{c_{11}(\boldsymbol{x})c_{11}(\boldsymbol{x}')}{c_1}\right)d\boldsymbol{x}d\boldsymbol{x}'$$

$$+\frac{1}{2}c_1c_2^2\frac{\kappa_1-\kappa_2}{(\kappa_0+4\mu_0/3)^2}(\Delta p)^2 \tag{4.95}$$

by recognizing that

$$\begin{aligned}\frac{1}{2}\Delta\boldsymbol{p}\left(\boldsymbol{\Gamma}^{\infty}c_{11}(\boldsymbol{L}_1-\boldsymbol{L}_2)\boldsymbol{\Gamma}^{\infty}c_{11}\right)\Delta\boldsymbol{p}&=\frac{1}{2}c_1^2c_2^2\Delta\boldsymbol{p}\left(\mathbf{S}_0\boldsymbol{L}_0^{-1}(\boldsymbol{L}_1-\boldsymbol{L}_2)\mathbf{S}_0\boldsymbol{L}_0^{-1}\right)\Delta\boldsymbol{p}\\&=\frac{1}{2}c_1^2c_2^2\frac{\kappa_1-\kappa_2}{(\kappa_0+4\mu_0/3)^2}(\Delta p)^2\end{aligned} \tag{4.96}$$

Based on the spherical coordinates of two points $\boldsymbol{x}$ and $\boldsymbol{x}'$ shown in Figure 4.2, the relationship between the rectangular coordinates and the spherical coordinates is established as follows (Corson, 1974):

$$\begin{cases}x_1=|\boldsymbol{x}|\sin\varphi\cos\theta\\x_2=|\boldsymbol{x}|\sin\varphi\sin\theta\\x_3=|\boldsymbol{x}|\cos\varphi\end{cases} \tag{4.97a, b, c}$$

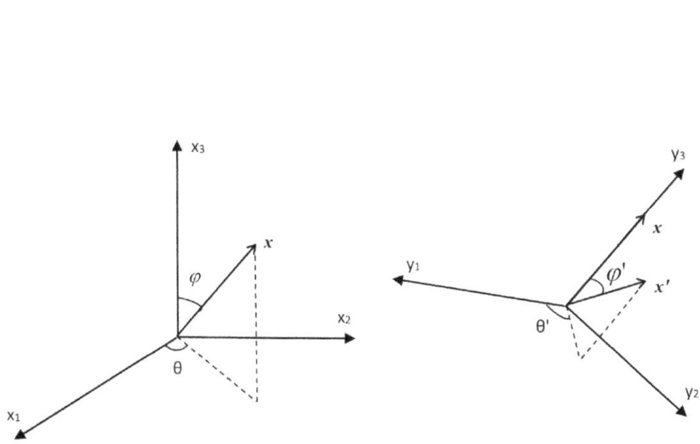

Figure 4.2 Rectangular and spherical coordinates of two points $\boldsymbol{x}$ and $\boldsymbol{x}'$.

$$\begin{cases} x_1' = |\boldsymbol{x}'|(\cos\varphi\sin\varphi\cos\theta + \sin\varphi'\sin\theta'\cos\varphi\cos\theta + \sin\varphi'\cos\theta'\sin\theta) \\ x_2' = |\boldsymbol{x}'|(\cos\varphi'\sin\varphi\sin\theta + \sin\varphi'\sin\theta'\cos\varphi\sin\theta - \sin\varphi'\cos\theta'\cos\theta) \\ x_3' = |\boldsymbol{x}'|(\cos\varphi'\cos\varphi - \sin\varphi'\sin\theta'\sin\varphi) \end{cases} \tag{4.98a, b, c}$$

Therefore the integral in (4.95) is expanded in the spherical coordinates as

$$\begin{aligned} &\int_Y\int_Y \sum_{i=1,j=1}^{3} \Gamma^{\infty}_{ijnn}(\boldsymbol{x})\Gamma^{\infty}_{ijnn}(\boldsymbol{x}')\left(c_{111}(\boldsymbol{x},\boldsymbol{x}') - \frac{c_{11}(\boldsymbol{x})c_{11}(\boldsymbol{x}')}{c_1}\right)d\boldsymbol{x}\,d\boldsymbol{x}' \\ &= \int_0^{2\pi}\int_0^{2\pi}\int_0^{\pi} \sum_{i=1,j=1}^{3} \Gamma^{\infty}_{ijnn}(r,\theta,\varphi)\Gamma^{\infty}_{ijnn}(s,\theta,\varphi,\theta',\varphi')\sin\varphi d\varphi d\theta d\theta' \\ &\int_0^{\infty}\int_0^{\infty}\int_0^{\pi}\left(c_{111}(r,s,\varphi') - \frac{c_{11}(r)c_{11}(s)}{c_1}\right)\sin\varphi'\, d\varphi'\, dr\, ds \end{aligned} \tag{4.99}$$

where $|\boldsymbol{x}| = r$ and $|\boldsymbol{x}'| = s$. With the free space modified Green function (A2.6), the first triple integral on the right-hand side of (4.99) is calculated as

$$\begin{aligned} &\int_0^{2\pi}\int_0^{2\pi}\int_0^{\pi} \sum_{i=1,j=1}^{3} \Gamma^{\infty}_{ijnn}(r,\theta,\varphi)\Gamma^{\infty}_{ijnn}(s,\theta,\varphi,\theta',\varphi')\sin\varphi d\varphi d\theta d\theta' \\ &= \frac{3}{2(\kappa_0 + 4\mu_0/3)^2}\frac{1+3\cos 2\varphi'}{rs} \end{aligned} \tag{4.100}$$

With the result (4.100), (4.95) is simplified into

$$\frac{2(\mu_1 - \mu_2)(\Delta p)^2}{3(\kappa_0 + 4\mu_0/3)^2}c_1c_2\zeta_1 + \frac{1}{2}c_1c_2^2\frac{\kappa_1 - \kappa_2}{(\kappa_0 + 4\mu_0/3)^2}(\Delta p)^2 \tag{4.101}$$

where ζ_1 the geometric parameter of the bulk modulus for Phase-1, or simply the bulk parameter of Phase-1, is defined as

Definition

$$\boxed{\zeta_1 = \frac{9}{2c_1c_2}\int_0^{\infty}\frac{dr}{r}\int_0^{\infty}\frac{ds}{s}\int_{-1}^{1}du\left(c_{111}(r,s,u) - \frac{c_{11}(r)c_{11}(s)}{c_1}\right)P_2(u)} \tag{4.102}$$

with the second Legendre polynomial $P_2(u) = \dfrac{3u^2 - 1}{2}$, $u = \cos(\varphi')$. The triple correlation $c_{111}(r,s,u)$ represents the probability of a triangle, with two sides of lengths r and s at angle $\cos^{-1}(u)$, having all three vertices lie in Phase-1 when placed randomly in the composite.

The sum of (4.85–4.87), (4.93), and (4.101) yields the upper bound energy

$$\begin{aligned} \mathcal{U}^{+}(\Delta p,\kappa_0,\mu_0) &= \frac{1}{2}(c_1\kappa_1 + c_2\kappa_2)e_0^2 + \frac{1}{2}c_1c_2\frac{(\Delta p)^2}{\kappa_0 + 4\mu_0/3} \\ &+ c_1c_2\frac{(\mu_2-\mu_0)(\Delta p)^2}{(\kappa_0+4\mu_0/3)^2} + \frac{2(\mu_1-\mu_2)(\Delta p)^2}{3(\kappa_0+4\mu_0/3)^2}c_1c_2\zeta_1 \\ &+ \frac{1}{2}c_1c_2^2\frac{(\kappa_1-\kappa_2)(\Delta p)^2}{(\kappa_0+4\mu_0/3)^2} - c_1c_2\frac{(\kappa_1-\kappa_2)\Delta p}{\kappa_0+4\mu_0/3}e_0 \end{aligned} \tag{4.103}$$

which equals

$$\frac{1}{2}\kappa_+^{(3)}e_0^2 \tag{4.104}$$

By extremizing (4.103) with respect to Δp, it results in the upper bound

$$\boxed{\kappa_+^{(3)} = (c_1\kappa_1 + c_2\kappa_2) - \frac{c_1c_2(\kappa_1-\kappa_2)^2}{c_1\kappa_2 + c_2\kappa_1 + \tilde{\kappa}_+}} \tag{4.105}$$

where

$$\tilde{\kappa}_+ = \frac{4}{3}(\mu_1\zeta_1 + \mu_2\zeta_2) + (\kappa_0 - \kappa_2) + \frac{2}{3}(\mu_2 - \mu_0) \tag{4.106}$$

with $\zeta_2 = 1 - \zeta_1$. By taking the reference moduli $\kappa_0 = \kappa_2$ and $\mu_0 = \mu_2$, physically meaning that Phase-2 serves as the matrix, it follows

$$\boxed{\tilde{\kappa}_+ = \frac{4}{3}\langle\mu\rangle_\zeta} \tag{4.107}$$

with $\langle\mu\rangle_\zeta = \mu_1\zeta_1 + \mu_2\zeta_2$. By comparing the third-order bound (4.105) with the second-order HS bounds (4.39–4.40), it yields that

$$0 \le \zeta_1 \le 1, \quad 0 \le \zeta_2 \le 1 \tag{4.108}$$

By exchanging Phase-1 and Phase-2, the bulk parameter ζ_2 for Phase-2 is found to be

$$\boxed{\zeta_2 = \frac{9}{2c_1c_2}\int_0^\infty \frac{dr}{r}\int_0^\infty \frac{ds}{s}\int_{-1}^{1} du\left(c_{222}(r,s,u) - \frac{c_{22}(r)c_{22}(s)}{c_2}\right)P_2(u)} \tag{4.109}$$

Similarly, by substituting (4.82–4.84) into the lower bound energy functional (3.41), it is found that the non-zero entries of $\boldsymbol{p}_1$ and $\boldsymbol{p}_2$ are diagonal only, i.e., $p_{n,ij} = p_n\delta_{ij}$, $n = 1, 2$. Accordingly, all the terms of the lower bound energy functional (3.41) except for the fourth term are directly obtained, respectively, as

$$\frac{1}{2}\left\langle \boldsymbol{\varepsilon}_0, \left(\boldsymbol{L}_0 - \boldsymbol{L}_0(\boldsymbol{M}-\boldsymbol{M}_0)\boldsymbol{L}_0\right)\boldsymbol{\varepsilon}_0\right\rangle = \frac{1}{2}\left(\kappa_0 - \kappa_0^2\left(\frac{c_1}{\kappa_1} + \frac{c_2}{\kappa_1} - \frac{1}{\kappa_0}\right)\right)e_0^2 \tag{4.110}$$

$$-\frac{1}{2}\left\langle \boldsymbol{p}, \boldsymbol{M}\boldsymbol{p}\right\rangle = -\frac{1}{2}\left(\frac{c_1p_1^2}{\kappa_1} + \frac{c_2p_2^2}{\kappa_1}\right) \tag{4.111}$$

$$-\left\langle \boldsymbol{L}_0\boldsymbol{\varepsilon}_0, (\boldsymbol{M}-\boldsymbol{M}_0)\boldsymbol{p}\right\rangle = -\kappa_0\left(c_1p_1\left(\frac{1}{\kappa_1} - \frac{1}{\kappa_0}\right) + c_2p_2\left(\frac{1}{\kappa_1} - \frac{1}{\kappa_0}\right)\right)e_0 \tag{4.112}$$

$$\begin{aligned} -\frac{1}{2}\left\langle \boldsymbol{p}, \boldsymbol{\Gamma}^\infty \boldsymbol{p}\right\rangle &= -\frac{1}{2}\Delta\boldsymbol{p}(c_1c_2\boldsymbol{S}_0\boldsymbol{L}_0^{-1})\Delta\boldsymbol{p} \\ &= -\frac{1}{2}c_1c_2\frac{(p_1-p_2)^2}{\kappa_0 + 4\mu_0/3} \end{aligned} \tag{4.113}$$

$$\begin{aligned} \left\langle \boldsymbol{\varepsilon}_0, \boldsymbol{L}_0(\boldsymbol{M}-\boldsymbol{M}_0)\boldsymbol{L}_0\Gamma^\infty \boldsymbol{p}\right\rangle &= \boldsymbol{\varepsilon}_0\boldsymbol{L}_0(\boldsymbol{M}_1 - \boldsymbol{M}_2)\boldsymbol{L}_0\left(c_1c_2\boldsymbol{S}_0\boldsymbol{L}_0^{-1}\right)\Delta\boldsymbol{p} \\ &= c_1c_2\frac{(1/\kappa_1 - 1/\kappa_2)\kappa_0^2}{\kappa_0 + 4\mu_0/3}(p_1 - p_2)e_0 \end{aligned} \tag{4.114}$$

$$\begin{aligned} \left\langle \boldsymbol{L}_0\boldsymbol{M}\boldsymbol{p}, \boldsymbol{\Gamma}^\infty \boldsymbol{p}\right\rangle &= \boldsymbol{L}_0\left(\boldsymbol{M}_1\boldsymbol{p}_1 - \boldsymbol{M}_2\boldsymbol{p}_2\right)\left(c_1c_2\boldsymbol{S}_0\boldsymbol{L}_0^{-1}(\boldsymbol{p}_1 - \boldsymbol{p}_2\right) \\ &= c_1c_2\frac{\kappa_0}{\kappa_0 + 4\mu_0/3}\left(p_1/\kappa_1 - p_2/\kappa_2\right)(p_1 - p_2) \end{aligned} \tag{4.115}$$

Note that, unlike the upper bound case, the two pressure variables p_1 and p_2 are not always reducible to Δp, and thus both p_1 and p_2 explicitly present in all the results. By replacing $(\boldsymbol{L}-\boldsymbol{L}_0)$ with $\boldsymbol{L}_0(\boldsymbol{M}-\boldsymbol{M}_0)\boldsymbol{L}_0$, the fourth term of (3.41) becomes identical to (4.88), and the result of the fourth term is thus obtained from (4.101) as

$$-\frac{1}{2}\left\langle \boldsymbol{\Gamma}^{\infty}\boldsymbol{p}, \boldsymbol{L}_0(\boldsymbol{M}-\boldsymbol{M}_0)\boldsymbol{L}_0\boldsymbol{\Gamma}^{\infty}\boldsymbol{p}\right\rangle$$
$$=-\frac{2(1/\mu_1-1/\mu_2)\,\mu_0^2(p_1-p_2)^2}{3(\kappa_0+4\mu_0/3)^2}c_1c_2\zeta_1-\frac{1}{2}c_1c_2^2\frac{(1/\kappa_1-1/\kappa_2)\kappa_0^2}{(\kappa_0+4\mu_0/3)^2}(p_1-p_2)^2 \tag{4.116}$$

The sum of all the terms yields the lower bound energy in terms of the four arguments as follows:

$$\begin{aligned}
&\mathcal{U}^-(p_1,p_2,\kappa_0,\mu_0)\\
&=\frac{1}{2}\left(\kappa_0-\kappa_0^2\left(\frac{c_1}{\kappa_1}+\frac{c_2}{\kappa_2}-\frac{1}{\kappa_0}\right)\right)e_0^2\\
&-\frac{1}{2}\left(\frac{c_1p_1^2}{\kappa_1}+\frac{c_2p_2^2}{\kappa_2}\right)-\kappa_0\left(c_1p_1\left(\frac{1}{\kappa_1}-\frac{1}{\kappa_0}\right)+c_2p_2\left(\frac{1}{\kappa_2}-\frac{1}{\kappa_0}\right)\right)e_0\\
&-\frac{1}{2}c_1c_2\frac{(p_1-p_2)^2}{\kappa_0+4\mu_0/3}+c_1c_2\frac{(1/\kappa_1-1/\kappa_2)\kappa_0^2}{\kappa_0+4\mu_0/3}(p_1-p_2)e_0\\
&+c_1c_2\frac{\kappa_0\,(p_1/\kappa_1-p_2/\kappa_2)(p_1-p_2)}{\kappa_0+4\mu_0/3}-\frac{2(1/\mu_1-1/\mu_2)\mu_0^2(p_1-p_2)^2}{3(\kappa_0+4\mu_0/3)^2}c_1c_2\zeta_1\\
&-\frac{1}{2}c_1c_2^2\frac{(1/\kappa_1-1/\kappa_2)\kappa_0^2}{(\kappa_0+4\mu_0/3)^2}(p_1-p_2)^2
\end{aligned} \tag{4.117}$$

which equals $\frac{1}{2}\kappa_-^{(3)}\boldsymbol{e}_0^2$. Extremization of (4.117) with respect to p_1 and p_2 yields the low bound of the effective bulk modulus as

$$\boxed{\kappa_-^{(3)}=(c_1\kappa_1+c_2\kappa_2)-\frac{c_1c_2(\kappa_1-\kappa_2)^2}{c_1\kappa_2+c_2\kappa_1+\tilde{\kappa}_-}} \tag{4.118}$$

with

$$\tilde{\kappa}_{-} = \frac{4}{3}\left(\frac{\zeta_1}{\mu_1} + \frac{\zeta_2}{\mu_2} + \frac{\mu_2 - \mu_0}{\mu_2 \mu_0} + \frac{3\kappa_0(\kappa_2 - \kappa_0)}{4\kappa_2 \mu_0^2}\right)^{-1} \tag{4.119}$$

By taking the reference moduli $\kappa_0 = \kappa_2$ and $\mu_0 = \mu_2$, it follows that

$$\boxed{\tilde{\kappa}_{-} = \frac{4}{3}\left\langle \mu^{-1} \right\rangle_{\zeta}^{-1}} \tag{4.120}$$

which is symmetric to the upper bound case (4.107).

When the bulk parameters take the values $\zeta_1 = 1$, $\zeta_2 = 0$ and $\zeta_1 = 0$, $\zeta_2 = 1$, respectively, the upper bound (4.107) and the lower bound (4.120) become identical, corresponding to the upper and lower HS bounds (4.39–4.40).

4.3.2 Third-order bounds of the effective shear modulus

Given the reference strain as a pure shear strain, say, of 12 and 21 components only, i.e.

$$\varepsilon_{0,ij} = (\delta_{i1}\delta_{j2} + \delta_{i2}\delta_{j1})\gamma_0/2 \tag{4.121}$$

with γ_0 denoting the engineering shear strain, the right-hand side of (4.80) results in a pure shear strain as well, and thereby the only non-zero components of $\Delta \boldsymbol{p}$ are $\Delta p_{12} = \Delta p_{21} = \Delta p$. The first, second, and fourth terms of the upper bound functional (3.27) are directly obtained, respectively, as

$$\frac{1}{2}\left\langle \boldsymbol{\varepsilon}_0, \boldsymbol{L}\boldsymbol{\varepsilon}_0 \right\rangle = \frac{1}{2}(c_1\mu_1 + c_2\mu_2)\gamma_0^2 \tag{4.122}$$

$$\begin{aligned} \frac{1}{2}\left\langle \boldsymbol{p}, \boldsymbol{\Gamma}^{\infty} \boldsymbol{p} \right\rangle &= \frac{1}{2}\Delta \boldsymbol{p}(c_1 c_2 \boldsymbol{S}_0 \boldsymbol{L}_0^{-1})\Delta \boldsymbol{p} \\ &= \frac{1}{2} c_1 c_2 \frac{6(\kappa_0 + 2\mu_0)}{5(3\kappa_0 + 4\mu_0)\mu_0}(\Delta p)^2 \end{aligned} \tag{4.123}$$

$$-\left\langle \boldsymbol{\Gamma}^{\infty}\boldsymbol{p},(\boldsymbol{L}-\boldsymbol{L}_0)\boldsymbol{\varepsilon}_0\right\rangle = -\boldsymbol{\varepsilon}_0(\boldsymbol{L}_1-\boldsymbol{L}_2)\left(c_1c_2\mathbf{S}_0\boldsymbol{L}_0^{-1}\right)\Delta\boldsymbol{p}$$
$$= -c_1c_2(\mu_1-\mu_2)\frac{6(\kappa_0+2\mu_0)}{5(3\kappa_0+4\mu_0)\mu_0}\Delta p\gamma_0 \tag{4.124}$$

Similar to the derivation (4.88–4.101) in the bulk modulus case, the third term of (3.27) is decomposed into the following three terms

$$\begin{aligned}
&\frac{1}{2}\left\langle \boldsymbol{\Gamma}^{\infty}\boldsymbol{p},(\boldsymbol{L}-\boldsymbol{L}_0)\boldsymbol{\Gamma}^{\infty}\boldsymbol{p}\right\rangle \\
&=\frac{1}{2}\Delta\boldsymbol{p}\left(\boldsymbol{\Gamma}^{\infty}(\boldsymbol{L}_2-\boldsymbol{L}_0)\boldsymbol{\Gamma}^{\infty}c_{11}\right)\Delta\boldsymbol{p}+\frac{1}{2}\Delta\boldsymbol{p}\left(\boldsymbol{\Gamma}^{\infty}(\boldsymbol{L}_1-\boldsymbol{L}_2)\boldsymbol{\Gamma}^{\infty}c_{111}\right)\Delta\boldsymbol{p} \\
&=\frac{1}{2}\Delta\boldsymbol{p}\left(\boldsymbol{\Gamma}^{\infty}(\boldsymbol{L}_2-\boldsymbol{L}_0)\boldsymbol{\Gamma}^{\infty}c_{11}\right)\Delta\boldsymbol{p}+\frac{1}{2}\Delta\boldsymbol{p}\left(\boldsymbol{\Gamma}^{\infty}(\boldsymbol{L}_1-\boldsymbol{L}_2)\boldsymbol{\Gamma}^{\infty}(c_{111}-c_{11}c_{11}/c_1)\right)\Delta\boldsymbol{p} \\
&+\frac{1}{2}\frac{1}{c_1}\Delta\boldsymbol{p}\left(\boldsymbol{\Gamma}^{\infty}c_{11}(\boldsymbol{L}_1-\boldsymbol{L}_2)\boldsymbol{\Gamma}^{\infty}c_{11}\right)\Delta\boldsymbol{p}
\end{aligned} \tag{4.125}$$

which are expanded and calculated, respectively, as

$$\begin{aligned}
&\frac{1}{2}\Delta\boldsymbol{p}\left(\boldsymbol{\Gamma}^{\infty}(\boldsymbol{L}_2-\boldsymbol{L}_0)\boldsymbol{\Gamma}^{\infty}c_{11}\right)\Delta\boldsymbol{p} \\
&=\frac{c_1c_2}{2}(\Delta p)^2\frac{1}{4\pi}\int_0^{\pi}\int_0^{2\pi}\sin\varphi\; d\theta d\varphi\left[2(\mu_2-\mu_0)\sum_{i=1,j=1}^{3}\left(\hat{\Gamma}^{\infty}_{ij12}(\theta,\varphi)+\hat{\Gamma}^{\infty}_{ij21}(\theta,\varphi)\right)^2\right. \\
&\left.+(\kappa_2-\frac{2}{3}\mu_2-\kappa_0+\frac{2}{3}\mu_0)\left(\sum_{k=1}^{3}\hat{\Gamma}^{\infty}_{kk12}(\theta,\varphi)+\hat{\Gamma}^{\infty}_{kk21}(\theta,\varphi)\right)^2\right] \\
&=\frac{3c_1c_2}{5(3\kappa_0+4\mu_0)^2}\left((\mu_2-\mu_0)\frac{3\kappa_0^2+8\kappa_0\mu_0+8\mu_0^2}{\mu_0^2}+2(\kappa_2-\kappa_0)\right)(\Delta p)^2
\end{aligned} \tag{4.126}$$

$$\frac{1}{2}\Delta\boldsymbol{p}\left(\boldsymbol{\Gamma}^{\infty}(\boldsymbol{L}_1-\boldsymbol{L}_2)\boldsymbol{\Gamma}^{\infty}\left(c_{111}-c_{11}c_{11}/c_1\right)\right)\Delta\boldsymbol{p}=\frac{1}{2}(\Delta p)^2\int_Y\int_Y\left(c_{111}(\boldsymbol{x},\boldsymbol{x}')-\frac{c_{11}(\boldsymbol{x})c_{11}(\boldsymbol{x}')}{c_1}\right)$$

$$d\boldsymbol{x}d\boldsymbol{x}'\left[2(\mu_1-\mu_2)\sum_{i=1,j=1}^{3}\left(\Gamma^{\infty}_{ij12}(\boldsymbol{x})+\Gamma^{\infty}_{ij21}(\boldsymbol{x})\right)\left(\Gamma^{\infty}_{ij12}(\boldsymbol{x}')+\Gamma^{\infty}_{ij21}(\boldsymbol{x}')\right)\right.$$

$$\left.+(\kappa_1-\frac{2}{3}\mu_1-\kappa_2+\frac{2}{3}\mu_2)\sum_{m=1}^{3}\left(\Gamma^{\infty}_{mm12}(\boldsymbol{x})+\Gamma^{\infty}_{mm21}(\boldsymbol{x})\right)\sum_{k=1}^{3}\left(\Gamma^{\infty}_{kk12}(\boldsymbol{x}')+\Gamma^{\infty}_{kk21}(\boldsymbol{x}')\right)\right]$$

$$=\frac{1}{2}\frac{9\Delta p^2}{10\mu_0^2(3\kappa_0+4\mu_0)^2}\int_0^{\infty}\frac{dr}{r}\int_0^{\infty}\frac{ds}{s}\int_{-1}^{1}du\left[6(-1+3u^2)(\kappa_1-\kappa_2)\mu_0^2\right.$$

$$+9(2-21u^2+25u^4)(\mu_1-\mu_2)\kappa_0^2+6(1-18u^2+25u^4)(\mu_1-\mu_2)\kappa_0\mu_0$$

$$\left.+(-7+6u^2+25u^4)(\mu_1-\mu_2)\mu_0^2\right]\left(c_{111}(r,s,u)-\frac{c_{11}(r)c_{11}(s)}{c_1}\right)$$

$$=c_1c_2(\Delta p)^2\frac{5\left(42(\kappa_1-\kappa_2)\mu_0^2+(\mu_1-\mu_2)(3\kappa_0+8\mu_0)^2\right)\zeta_1+16(\mu_1-\mu_2)(3\kappa_0+\mu_0)^2\eta_1}{175\mu_0^2(3\kappa_0+4\mu_0)^2} \tag{4.127}$$

$$\frac{1}{2}\frac{1}{c_1}\Delta\boldsymbol{p}\left(\boldsymbol{\Gamma}^{\infty}c_{11}(\boldsymbol{L}_1-\boldsymbol{L}_2)\boldsymbol{\Gamma}^{\infty}c_{11}\right)\Delta\boldsymbol{p}=\frac{1}{2}c_1c_2^2(\mu_1-\mu_2)\left(\frac{6(\kappa_0+2\mu_0)}{5(3\kappa_0+4\mu_0)\mu_0}\right)^2(\Delta p)^2 \tag{4.128}$$

Definition
In (4.127) the geometric parameter of the shear modulus for Phase-1 is defined as

$$\boxed{\eta_1=\frac{225}{8c_1c_2}\int_0^{\infty}\frac{dr}{r}\int_0^{\infty}\frac{ds}{s}\int_{-1}^{1}du\left(c_{111}(r,s,u)-\frac{c_{11}(r)c_{11}(s)}{c_1}\right)P_4(u)} \tag{4.129}$$

with $P_4(u)=\frac{1}{8}(35u^4-30u^2+3)$ the fourth Legendre polynomial.

The sum of (4.122–4.124) and (4.126–4.128) yields a quadratic function of the upper bound energy about two arguments B_1 and B_2, as

$$\begin{aligned}\mathcal{U}^{+}(B_1,B_2) &= \frac{1}{2}(c_1\mu_1 + c_2\mu_2)\gamma_0^2 \\ &+ \frac{c_1c_2}{90}B_1^2\left(10c_2(\mu_1-\mu_2)+6\zeta_1(\kappa_1-\kappa_2)+6\kappa_2+17\mu_2\right) \\ &- \frac{c_1c_2}{45}B_1\left[(\mu_1-\mu_2)(15+4\zeta_1B_2+4c_2B_2)\right. \\ &\left.+2B_2(3\zeta_1(\kappa_1-\kappa_2)+3\kappa_2+4\mu_2)\right] \\ &+ \frac{c_1c_2}{1575}B_2\left[210(\mu_1-\mu_2)+4B_2(\mu_1-\mu_2)(7c_2+10\zeta_1+18\eta_1)\right. \\ &\left.+35B_2(3\zeta_1(\kappa_1-\kappa_2)+3\kappa_2+4\mu_2)\right]\end{aligned} \tag{4.130}$$

with

$$B_1 = \frac{\Delta p}{\mu_0} \tag{4.131}$$

$$B_2 = \frac{(3\kappa_0+\mu_0)}{\mu_0(3\kappa_0+4\mu_0)}\Delta p \tag{4.132}$$

Extremization of the upper bound energy (4.130) with respect to B_1 and B_2 leads to the upper bound of the shear modulus, as

$$\boxed{\mu_+^{(3)} = (c_1\mu_1+c_2\mu_2) - \frac{c_1c_2(\mu_1-\mu_2)^2}{c_1\mu_2+c_2\mu_1+\tilde{\mu}_+}} \tag{4.133a}$$

$$\boxed{\tilde{\mu}_+ = \frac{\mu_+^*\left(9\kappa_+^*+8\mu_+^*\right)}{6\left(\langle\kappa\rangle_\zeta+2\mu_+^{**}\right)}} \tag{4.133b}$$

with

$$\begin{aligned}\mu_+^* &= \sqrt{\langle\mu\rangle_\zeta\langle\mu\rangle_\eta} \\ \kappa_+^* &= \left(\frac{5}{21}\sqrt{\frac{\langle\mu\rangle_\zeta}{\langle\mu\rangle_\eta}}+\frac{16}{21}\sqrt{\frac{\langle\mu\rangle_\eta}{\langle\mu\rangle_\zeta}}\right)\langle\kappa\rangle_\zeta \\ \mu_+^{**} &= \frac{1}{21}\langle\mu\rangle_\zeta+\frac{20}{21}\langle\mu\rangle_\eta\end{aligned} \tag{4.133c, d, e}$$

when B_1 and B_2 take the following optimal values

$$B_1 = \frac{5(\mu_1-\mu_2)\left(21\langle\kappa\rangle_\zeta + 4\langle\mu\rangle_\zeta + 24\langle\mu\rangle_\eta\right)}{56\langle\mu\rangle_\zeta\langle\mu\rangle_\eta + 4\langle\mu\rangle\left(\langle\mu\rangle_\zeta + 20\langle\mu\rangle_\eta\right) + 3\langle\kappa\rangle_\zeta\left(14\langle\mu\rangle + 5\langle\mu\rangle_\zeta + 16\langle\mu\rangle_\eta\right)}\gamma_0$$

$$B_2 = \frac{35(\mu_1-\mu_2)\left(3\langle\kappa\rangle_\zeta + \langle\mu\rangle_\zeta\right)}{56\langle\mu\rangle_\zeta\langle\mu\rangle_\eta + 4\langle\mu\rangle\left(\langle\mu\rangle_\zeta + 20\langle\mu\rangle_\eta\right) + 3\langle\kappa\rangle_\zeta\left(14\langle\mu\rangle + 5\langle\mu\rangle_\zeta + 16\langle\mu\rangle_\eta\right)}\gamma_0$$

with $\eta_2 = 1 - \eta_1$ and $\langle\mu\rangle = c_1\mu_1 + c_2\mu_2$. By comparing (4.133) with the second-order HS bounds (4.41–4.42), it follows that

$$0 \le \eta_1 \le 1, \quad 0 \le \eta_2 \le 1 \tag{4.134}$$

By exchanging Phase-1 and Phase-2, it is found that

$$\boxed{\eta_2 = \frac{225}{8c_1c_2}\int_0^\infty \frac{dr}{r}\int_0^\infty \frac{ds}{s}\int_{-1}^{1} du\left(c_{222}(r,s,u) - \frac{c_{11}(r)c_{11}(s)}{c_1}\right)P_4(u)} \tag{4.135}$$

It is remarked that, when $\zeta_1 \neq \eta_1$, the Poisson's ratio of the optimal reference medium

$$\begin{aligned} \nu_0 &= 1 - \frac{B_1}{2B_2} \\ &= \frac{21\langle\kappa\rangle_\zeta + 10\langle\mu\rangle_\zeta - 24\langle\mu\rangle_\eta}{14\left(3\langle\kappa\rangle_\zeta + \langle\mu\rangle_\zeta\right)} \end{aligned} \tag{4.136}$$

differs from either ν_1 or ν_2, which indicates the bounds are not attainable. When $\zeta_1 = \eta_1$, $\nu_0 = \dfrac{3\langle\kappa\rangle_\zeta - 2\langle\mu\rangle_\zeta}{2\left(3\langle\kappa\rangle_\zeta + \langle\mu\rangle_\zeta\right)}$ reduces exactly to that of the medium characterized with $\langle\kappa\rangle_\zeta$ and $\langle\mu\rangle_\zeta$. In the latter case, only when $\zeta_1 = \eta_1 = 0$ or 1 the reference Poisson's ratio becomes ν_1 or ν_2, indicating physical existence of the reference medium and thereby realization of the upper bound. In fact, when $\zeta_1 = \eta_1 = 0$ or 1, (4.133) reduces to the

HS bound (4.41) or (4.42) covering both the well-ordered and the non-well-ordered cases, which tightens (4.37–4.38) in the two-phase case to (4.41–4.42).

By substituting the reference strain (4.121) into the lower bound functional (3.41), it is found that the non-zero entries of $\boldsymbol{p}_1$ and $\boldsymbol{p}_2$ are shear components $p_{1,12} = p_{1,21}$ and $p_{2,12} = p_{2,21}$ only, which are simply denoted as p_1 and p_2, respectively. All the terms of the lower bound energy functional (3.41) except for the fourth term are directly obtained, respectively, as

$$\frac{1}{2}\left\langle \boldsymbol{\varepsilon}_0, (\boldsymbol{L}_0 - \boldsymbol{L}_0(\boldsymbol{M} - \boldsymbol{M}_0)\boldsymbol{L}_0)\boldsymbol{\varepsilon}_0 \right\rangle = \frac{1}{2}\left(\mu_0 - \mu_0^2\left(\frac{c_1}{\mu_1} + \frac{c_2}{\mu_2} - \frac{1}{\mu_0}\right)\right)\gamma_0^2 \tag{4.137}$$

$$-\frac{1}{2}\left\langle \boldsymbol{p}, \boldsymbol{M}\boldsymbol{p} \right\rangle = -\frac{1}{2}\left(\frac{c_1 p_1^2}{\mu_1} + \frac{c_2 p_2^2}{\mu_2}\right) \tag{4.138}$$

$$-\left\langle \boldsymbol{L}_0\boldsymbol{\varepsilon}_0, (\boldsymbol{M} - \boldsymbol{M}_0)\boldsymbol{p} \right\rangle = -\mu_0\left(c_1 p_1\left(\frac{1}{\mu_1} - \frac{1}{\mu_0}\right) + c_2 p_2\left(\frac{1}{\mu_2} - \frac{1}{\mu_0}\right)\right)\gamma_0 \tag{4.139}$$

$$\begin{aligned} -\frac{1}{2}\left\langle \boldsymbol{p}, \boldsymbol{\Gamma}^{\infty}\boldsymbol{p} \right\rangle &= -\frac{1}{2}\Delta\boldsymbol{p}(c_1 c_2 \mathbf{S}_0 \boldsymbol{L}_0^{-1})\Delta\boldsymbol{p} \\ &= -\frac{1}{2}c_1 c_2 \frac{6(\kappa_0 + 2\mu_0)}{5(3\kappa_0 + 4\mu_0)\mu_0}(p_1 - p_2)^2 \end{aligned} \tag{4.140}$$

$$\begin{aligned} \left\langle \boldsymbol{\varepsilon}_0, \boldsymbol{L}_0(\boldsymbol{M} - \boldsymbol{M}_0)\boldsymbol{L}_0\Gamma^{\infty}\boldsymbol{p} \right\rangle &= \boldsymbol{\varepsilon}_0\boldsymbol{L}_0(\boldsymbol{M}_1 - \boldsymbol{M}_2)\boldsymbol{L}_0\left(c_1 c_2 \mathbf{S}_0 \boldsymbol{L}_0^{-1}\right)\Delta\boldsymbol{p} \\ &= c_1 c_2 \mu_0^2\left(\frac{1}{\mu_1} - \frac{1}{\mu_2}\right)\frac{6(\kappa_0 + 2\mu_0)}{5(3\kappa_0 + 4\mu_0)\mu_0}(p_1 - p_2)\gamma_0 \end{aligned} \tag{4.141}$$

$$\begin{aligned} \left\langle \boldsymbol{L}_0\boldsymbol{M}\boldsymbol{p}, \boldsymbol{\Gamma}^{\infty}\boldsymbol{p} \right\rangle &= \boldsymbol{L}_0(\boldsymbol{M}_1\boldsymbol{p}_1 - \boldsymbol{M}_2\boldsymbol{p}_2)\left(c_1 c_2 \mathbf{S}_0 \boldsymbol{L}_0^{-1}(\boldsymbol{p}_1 - \boldsymbol{p}_2)\right) \\ &= c_1 c_2 \mu_0\left(\frac{p_1}{\mu_1} - \frac{p_2}{\mu_2}\right)\frac{6(\kappa_0 + 2\mu_0)}{5(3\kappa_0 + 4\mu_0)\mu_0}(p_1 - p_2) \end{aligned} \tag{4.142}$$

Similar to (4.125–4.128), the fourth term is expanded and calculated as

$$
\begin{aligned}
&-\frac{1}{2}\left\langle \boldsymbol{\Gamma}^{\infty}\boldsymbol{p}, \boldsymbol{L}_0(\boldsymbol{M}-\boldsymbol{M}_0)\boldsymbol{L}_0\boldsymbol{\Gamma}^{\infty}\boldsymbol{p}\right\rangle \\
&= -\frac{1}{2}\Delta\boldsymbol{p}\left(\boldsymbol{\Gamma}^{\infty}\boldsymbol{L}_0(\boldsymbol{M}_2-\boldsymbol{M}_0)\boldsymbol{L}_0\boldsymbol{\Gamma}^{\infty}c_{11}\right)\Delta\boldsymbol{p} \\
&\quad -\frac{1}{2}\Delta\boldsymbol{p}\left(\boldsymbol{\Gamma}^{\infty}\boldsymbol{L}_0(\boldsymbol{M}_1-\boldsymbol{M}_2)\boldsymbol{L}_0\boldsymbol{\Gamma}^{\infty}(c_{111}-c_{11}c_{11}/c_1)\right)\Delta\boldsymbol{p} \\
&\quad -\frac{1}{2}\frac{1}{c_1}\Delta\boldsymbol{p}\left(\boldsymbol{\Gamma}^{\infty}c_{11}\boldsymbol{L}_0(\boldsymbol{M}_1-\boldsymbol{M}_2)\boldsymbol{L}_0\boldsymbol{\Gamma}^{\infty}c_{11}\right)\Delta\boldsymbol{p} \\
&= -\frac{1}{2}c_1c_2\left[\left(\frac{1}{\mu_2}-\frac{1}{\mu_0}\right)\frac{6(3\kappa_0^2+8\kappa_0\mu_0+8\mu_0^2)}{5\mu_0^2(3\kappa_0+4\mu_0)^2}\right. \\
&\quad \left.+\left(\frac{1}{\kappa_2}-\frac{1}{\kappa_0}\right)\frac{12}{5(3\kappa_0+4\mu_0)^2}\right](p_1-p_2)^2 \\
&\quad -\frac{1}{2}\frac{2c_1c_2\left(p_1-p_2\right)^2}{175\mu_1\mu_2\kappa_1\kappa_2(3\kappa_0+4\mu_0)^2}(\mu_1-\mu_2)\left[16\eta_1\kappa_1\kappa_2(3\kappa_0+4\mu_0)^2\right. \\
&\quad \left.+5\zeta_1\left(48\mu_0\kappa_0\kappa_1\kappa_2+64\mu_0^2\kappa_1\kappa_2+\kappa_0^2\left(9\kappa_1\kappa_2+42\mu_1\mu_2\frac{\kappa_1-\kappa_2}{\mu_1-\mu_2}\right)\right)\right] \\
&\quad -\frac{1}{2}c_1c_2^2\left(\frac{6(\kappa_0+2\mu_0)}{5(3\kappa_0+4\mu_0)}\right)^2\left(\frac{1}{\mu_1}-\frac{1}{\mu_2}\right)\left(p_1-p_2\right)^2
\end{aligned}
\tag{4.143}
$$

Take γ_0 as a value of unity, and the sum of these eight terms results in a quadratic function of the energy about four arguments p_1, p_2, B_3, and μ_0, i.e.

$$
\begin{aligned}
\mathcal{U}^{-}(\mu_0,p_1,p_2,B_3) &= \frac{1}{2}\mu_0\left(2-\frac{c_1\mu_0}{\mu_1}-\frac{c_2\mu_0}{\mu_2}\right)-\frac{c_1p_1^2}{2\mu_1}-\frac{c_2p_2^2}{2\mu_2} \\
&\quad +\frac{18B_3^2c_1c_2^2(p_1-p_2)^2(\mu_1-\mu_2)}{25\mu_1\mu_2}-\frac{6c_1c_2(p_1-p_2)\mu_0B_3(\mu_1-\mu_2)}{5\mu_1\mu_2} \\
&\quad +\frac{c_1p_1(\mu_1-\mu_0)\mu_2+c_2p_2(\mu_2-\mu_0)\mu_1}{\mu_1\mu_2} \\
&\quad +\frac{3c_1c_2(p_1-p_2)}{5\kappa_2\mu_1\mu_2}\left(\left(1-6B_3+6B_3^2\right)p_2\mu_1\kappa_2\right. \\
&\quad \left.-2(1-2B_3)^2(p_1-p_2)\mu_1\mu_2+p_1\kappa_2\left(\mu_1-4B_3\mu_1+6B_3^2\mu_1-2B_3\mu_2\right)\right) \\
&\quad +\frac{c_1c_2(p_1-p_2)^2}{175\kappa_1\kappa_2\mu_1\mu_2}\left[\left(4\eta_1(5-9B_3)^2+5\zeta_1(1-6B_3)^2\right)\kappa_1\kappa_2(\mu_1-\mu_2)\right. \\
&\quad \left.+210\zeta_1(1-2B_3)^2(\kappa_1-\kappa_2)\mu_1\mu_2\right]
\end{aligned}
\tag{4.144}
$$

with

$$B_3 = \frac{3\kappa_0 + 2\mu_0}{3\kappa_0 + 4\mu_0} \tag{4.145}$$

Extremization of (4.144) leads to the lower bound of the effective shear modulus

$$\boxed{\mu_-^{(3)} = (c_1\mu_1 + c_2\mu_2) - \frac{c_1c_2(\mu_1 - \mu_2)^2}{c_1\mu_2 + c_2\mu_1 + \tilde{\mu}_-}} \tag{4.146a}$$

$$\boxed{\tilde{\mu}_- = \frac{\mu_-^*(9\kappa_-^* + 8\mu_-^*)}{6\left(\langle\kappa\rangle_\zeta + 2\mu_-^{**}\right)}} \tag{4.146b}$$

$$\begin{aligned}
\mu_-^* &= \sqrt{\left\langle\mu^{-1}\right\rangle_\zeta^{-1}\left\langle\mu^{-1}\right\rangle_\eta^{-1}} \\
\kappa_-^* &= \left(\frac{5}{21}\sqrt{\frac{\left\langle\mu^{-1}\right\rangle_\zeta^{-1}}{\left\langle\mu^{-1}\right\rangle_\eta^{-1}}} + \frac{16}{21}\sqrt{\frac{\left\langle\mu^{-1}\right\rangle_\eta^{-1}}{\left\langle\mu^{-1}\right\rangle_\zeta^{-1}}}\right)\langle\kappa\rangle_\zeta \\
\mu_-^{**} &= \frac{1}{21}\left\langle\mu^{-1}\right\rangle_\zeta^{-1} + \frac{20}{21}\left\langle\mu^{-1}\right\rangle_\eta^{-1}
\end{aligned} \tag{4.146c, d, e}$$

which is exactly symmetric to the upper bound expression (4.133).

By equating the upper bound (4.133) to the lower bound (4.146), there are four solutions of the geometric parameters obtained as $(\zeta_1, \eta_1) = (0, 0), (0, 1), (1, 0)$, and $(1, 1)$. With any pair of these four, the effective bulk and shear moduli are exact. The first and fourth pairs actually correspond to realization of the Hashin-Shtrikman bounds. As there are no reference moduli physically existent in the composite, the exact result corresponding to either the second or the third pairs is not attainable. Further discussion is given in Section 4.5.

4.3.3 Third-order bounds in 2D elasticity

The third-order bounds in 2D elasticity were derived by Silnutzer (1972) and Milton (1982) in the plane strain case. A typical example of application is fiber reinforced laminates.

Bulk Modulus in the Plane Strain Case

In the plane strain case, by replacing $\boldsymbol{S}_0$ and $\boldsymbol{L}_0$ in Subsection 4.3.1 with the circular Eshelby tensor (4.44) and the elastic moduli (4.45), the first, second, and fourth terms of the upper bound energy functional (3.27) are obtained, respectively, as

$$\frac{1}{2}\left(c_1\kappa_1^{\varepsilon}+c_2\kappa_2^{\varepsilon}\right)e_0^2,$$

$$\frac{1}{2}c_1c_2\frac{(\Delta p)^2}{\kappa_0^{\varepsilon}+\mu_0}, \tag{4.147a, b, c}$$

$$c_1c_2\frac{(\mu_2-\mu_0)(\Delta p)^2}{\left(\kappa_0^{\varepsilon}+\mu_0\right)^2}$$

With respect to the third term, by noting
Property

$$\Gamma_{11nn}^{\infty}+\Gamma_{22nn}^{\infty}=0 \tag{4.148}$$

the plane strain counterparts of (4.93), (4.96), and (4.100) are given, respectively, as

$$\frac{c_1c_2}{2}2(\mu_2-\mu_0)(\Delta p)^2\frac{1}{2\pi}\int_0^{2\pi}\sum_{i=1,j=1}^{2}\left(\hat{\Gamma}_{ijnn}^{\infty}(\theta)\right)^2d\theta=c_1c_2\frac{(\mu_2-\mu_0)}{\left(\kappa_0^{\varepsilon}+\mu_0\right)^2}(\Delta p)^2 \tag{4.149}$$

$$\begin{aligned}\frac{1}{2}\Delta\boldsymbol{p}\left(\boldsymbol{\Gamma}^{\infty}c_{11}(\boldsymbol{L}_1-\boldsymbol{L}_2)\boldsymbol{\Gamma}^{\infty}c_{11}\right)\Delta\boldsymbol{p}&=\frac{1}{2}c_1^2c_2^2\Delta\boldsymbol{p}\left(\boldsymbol{S}_0\boldsymbol{L}_0^{-1}(\boldsymbol{L}_1-\boldsymbol{L}_2)\boldsymbol{S}_0\boldsymbol{L}_0^{-1}\right)\Delta\boldsymbol{p}\\&=\frac{1}{2}c_1^2c_2^2\frac{\kappa_1^{\varepsilon}-\kappa_2^{\varepsilon}}{\left(\kappa_0^{\varepsilon}+\mu_0\right)^2}(\Delta p)^2\end{aligned} \tag{4.150}$$

$$\int_0^{2\pi}\sum_{i=1,j=1}^{2}\Gamma_{ijnn}^{\infty}(r,\theta)\Gamma_{ijnn}^{\infty}(s,\theta+\theta')d\theta=\frac{1}{\pi\left(\kappa_0^{\varepsilon}+\mu_0\right)^2}\frac{\cos 2\theta'}{r^2s^2} \tag{4.151}$$

Note that the modified Green function in the plane strain case is derived from (A2.7). By defining the in-plane bulk parameter of Phase-1 as (Milton, 1982)

Definition

$$\boxed{\zeta_1 = \frac{4}{\pi\, c_1 c_2}\int_0^\infty \frac{dr}{r}\int_0^\infty \frac{ds}{s}\int_0^\pi d\theta \left(c_{111}(r,s,\theta) - \frac{c_{11}(r)c_{11}(s)}{c_1} \right)\cos 2\theta} \tag{4.152}$$

the upper bound energy is obtained as

$$\begin{aligned}\mathcal{U}^+\left(\Delta p, \kappa_0^\varepsilon, \mu_0\right) = {} & \frac{1}{2}\left(c_1\kappa_1^\varepsilon + c_2\kappa_2^\varepsilon\right)e_0^2 + \frac{1}{2}c_1c_2\frac{(\Delta p)^2}{\kappa_0^\varepsilon + \mu_0} + c_1c_2\frac{(\mu_2-\mu_0)(\Delta p)^2}{(\kappa_0^\varepsilon+\mu_0)^2} \\ & + \frac{1}{2}\frac{(\mu_1-\mu_2)(\Delta p)^2}{(\kappa_0^\varepsilon+\mu_0)^2}c_1c_2\zeta_1 + \frac{1}{2}c_1c_2^2\frac{\left(\kappa_1^\varepsilon-\kappa_2^\varepsilon\right)(\Delta p)^2}{\left(\kappa_0^\varepsilon+\mu_0\right)^2} \\ & - c_1c_2\frac{\left(\kappa_1^\varepsilon-\kappa_2^\varepsilon\right)\Delta p}{\kappa_0^\varepsilon+\mu_0}e_0\end{aligned} \tag{4.153}$$

Extremization of (4.153) with respect to Δp leads to the upper bound of the effective bulk modulus

$$\boxed{\kappa_+^{(3)} = \left(c_1\kappa_1^\varepsilon + c_2\kappa_2^\varepsilon\right) - \frac{c_1c_2\left(\kappa_1^\varepsilon-\kappa_2^\varepsilon\right)^2}{c_1\kappa_2^\varepsilon + c_2\kappa_1^\varepsilon + \tilde{\kappa}_+^\varepsilon}} \tag{4.154a}$$

$$\tilde{\kappa}_+^\varepsilon = \left(\mu_1\zeta_1 + \mu_2\zeta_2\right) + \left(\kappa_0^\varepsilon - \kappa_2^\varepsilon\right) + (\mu_2 - \mu_0)$$

where $\zeta_2 = 1 - \zeta_1$. By taking the reference moduli $\kappa_0^\varepsilon = \kappa_2^\varepsilon$ and $\mu_0 = \mu_2$, physically meaning that Phase-2 serves as the matrix, it follows that

$$\boxed{\tilde{\kappa}_+^\varepsilon = \langle \mu \rangle_\zeta} \tag{4.154b}$$

By comparing (4.154) with the HS bounds (4.66–4.67), it is found that $0 \le \zeta_1 \le 1$, $0 \le \zeta_2 \le 1$.

Exchanging of Phase-1 and -2 shows that

$$\boxed{\zeta_2 = -\frac{4}{\pi\, c_1 c_2}\int_0^{\infty}\frac{dr}{r}\int_0^{\infty}\frac{ds}{s}\int_0^{\pi}d\theta\left(c_{222}(r,s,\alpha)-\frac{c_{22}(r)c_{22}(s)}{c_2}\right)\cos 2\theta} \tag{4.155}$$

Similarly, the lower bound energy in the plane strain case is expressed in terms of four arguments as follows:

$$\begin{aligned}
&\mathcal{U}^-\left(p_1,p_2,\kappa_0^\varepsilon,\mu_0\right)\\
&=\frac{1}{2}\left(\kappa_0^\varepsilon-(\kappa_0^\varepsilon)^2\left(\frac{c_1}{\kappa_1^\varepsilon}+\frac{c_2}{\kappa_2^\varepsilon}-\frac{1}{\kappa_0^\varepsilon}\right)\right)e_0^2-\frac{1}{2}\left(\frac{c_1p_1^2}{\kappa_1^\varepsilon}+\frac{c_2p_2^2}{\kappa_2^\varepsilon}\right)\\
&-\kappa_0^\varepsilon\left(c_1p_1\left(\frac{1}{\kappa_1^\varepsilon}-\frac{1}{\kappa_0^\varepsilon}\right)+c_2p_2\left(\frac{1}{\kappa_2^\varepsilon}-\frac{1}{\kappa_0^\varepsilon}\right)\right)e_0\\
&-\frac{1}{2}c_1c_2\frac{(p_1-p_2)^2}{\kappa_0^\varepsilon+\mu_0}+c_1c_2\frac{(1/\kappa_1^\varepsilon-1/\kappa_2^\varepsilon)(\kappa_0^\varepsilon)^2}{\kappa_0^\varepsilon+\mu_0}(p_1-p_2)e_0\\
&+c_1c_2\frac{\left(\kappa_0^\varepsilon\right)^2\left(p_1/\kappa_1^\varepsilon-p_2/\kappa_2^\varepsilon\right)(p_1-p_2)}{\kappa_0^\varepsilon+\mu_0}\\
&-\frac{(1/\mu_1-1/\mu_2)\mu_0^2(p_1-p_2)^2}{2\left(\kappa_0^\varepsilon+\mu_0\right)^2}c_1c_2\zeta_1-\frac{1}{2}c_1c_2^2\frac{\left(1/\kappa_1^\varepsilon-1/\kappa_2^\varepsilon\right)\left(\kappa_0^\varepsilon\right)^2}{\left(\kappa_0^\varepsilon+\mu_0\right)^2}(p_1-p_2)^2
\end{aligned} \tag{4.156}$$

and extremization of which with respect to p_1, p_2, κ_0^ε and μ_0 leads to the lower bound of the effective bulk modulus as

$$\boxed{\kappa_-^{(3)}=\left(c_1\kappa_1^\varepsilon+c_2\kappa_2^\varepsilon\right)-\frac{c_1c_2\left(\kappa_1^\varepsilon-\kappa_2^\varepsilon\right)^2}{c_1\kappa_2^\varepsilon+c_2\kappa_1^\varepsilon+\tilde{\kappa}_-^\varepsilon}}$$

$$\tilde{\kappa}_-^\varepsilon=\left(\frac{\zeta_1}{\mu_1}+\frac{\zeta_2}{\mu_2}+\frac{\mu_2-\mu_0}{\mu_2\mu_0}+\frac{\kappa_0^\varepsilon(\kappa_2^\varepsilon-\kappa_0^\varepsilon)}{\kappa_2^\varepsilon\mu_0^2}\right)^{-1} \tag{4.157a}$$

By taking the reference moduli $\kappa_0^\varepsilon=\kappa_2^\varepsilon$ and $\mu_0=\mu_2$, it follows that

$$\boxed{\tilde{\kappa}_-^\varepsilon=\left\langle\mu^{-1}\right\rangle_\zeta^{-1}} \tag{4.157b}$$

which is exactly symmetric to the upper bound case (4.154).

Shear Modulus in the Plane Strain Case

In the plane strain case, let the reference strain be a pure in-plane shear strain identical to (4.121). The strain energy for the upper and lower bounds are similarly obtained by using the plane strain Green function and the plane strain elastic moduli. With the intermediate steps left to the reader as an exercise, the final bounds are presented here:

$$\boxed{\mu_+^{(3)} = (c_1\mu_1 + c_2\mu_2) - \frac{c_1c_2(\mu_1-\mu_2)^2}{c_1\mu_2 + c_2\mu_1 + \tilde{\mu}_+^{\varepsilon}}} \tag{4.158a}$$

$$\boxed{\tilde{\mu}_+^{\varepsilon} = \frac{1}{2\langle\kappa^{\varepsilon}\rangle_{\zeta}^{-1} + \langle\mu\rangle_{\eta}^{-1}}} \tag{4.158b}$$

$$\boxed{\mu_-^{(3)} = (c_1\mu_1 + c_2\mu_2) - \frac{c_1c_2(\mu_1-\mu_2)^2}{c_1\mu_2 + c_2\mu_1 + \tilde{\mu}_-^{\varepsilon}}} \tag{4.159a}$$

$$\boxed{\tilde{\mu}_-^{\varepsilon} = \frac{1}{2\langle\kappa^{\varepsilon-1}\rangle_{\zeta} + \langle\mu^{-1}\rangle_{\eta}}} \tag{4.159b}$$

where the in-plane shear parameters are defined as (Milton, 1982)

Definition

$$\eta_i = \frac{16}{\pi\, c_1c_2}\int_0^{\infty}\frac{dr}{r}\int_0^{\infty}\frac{ds}{s}\int_0^{\pi}d\theta\left(c_{iii}(r,s,\theta) - \frac{c_{ii}(r)c_{ii}(s)}{c_i}\right)\cos 4\theta \qquad i=1,2 \tag{4.160}$$

with $0 \le \eta_1 \le 1$, $0 \le \eta_2 \le 1$ *and* $\eta_1 + \eta_2 = 1$.

It is noted that, when $\zeta_1 = \eta_1 = 0$ or 1 the bounds (4.158–4.159) reduce to the HS bounds (4.68–4.69) covering the both well-ordered and non-well-ordered cases, which tighten (4.64–4.65) in the two-phase case to (4.68–4.69).

Finally, in the plane stress case, the bounds of the both effective bulk and shear moduli are directly obtained from the above plane strain formulas by replacing the plane strain in-plane bulk modulus with the plane stress in-plane bulk modulus (4.71).

4.3.4 Third-order bounds of the effective conductivity

Let the Poisson's ratio be zero, and it immediately yields $\mu_j = \frac{3}{2}\kappa_j$ in 3D and $\mu_j = \kappa_j^{\varepsilon}$ or κ_j^{σ} in 2D. From (4.107), (4.120), (4.154), and (4.157), the third-order bounds of the effective conductivity reduce to

$$K_+^{(3)} = (c_1K_1 + c_2K_2) - \frac{c_1c_2(K_1 - K_2)^2}{c_1K_2 + c_2K_1 + (d-1)\langle K\rangle_\zeta} \quad (4.161a)$$

$$K_-^{(3)} = (c_1K_1 + c_2K_2) - \frac{c_1c_2(K_1 - K_2)^2}{c_1K_2 + c_2K_1 + (d-1)\langle K^{-1}\rangle_\zeta^{-1}} \quad (4.161b)$$

on two-phase composites, where d = 2 or 3 indicates dimensionality, and the geometric parameters are defined in (4.102) and (4.109) in 3D and (4.152) and (4.155) in 2D.

Note that the lower bound (4.161b) is not optimal in that it is not physically attainable, and a tighter lower bound given next in (4.183) is found to be the best possible one realized by a doubly coated sphere assemblage (Avellaneda et al., 1988), which is particularly employed in evaluation of attainability of the fourth-order bounds in the next section.

4.4 FOURTH-ORDER BOUNDS

4.4.1 Derivation

Suppose a two-phase composite has thermal conductivities K_1 and K_2 for Phase-1 and Phase-2, respectively. Choose the reference conductivity $K0=K2$, and thus the polarization is zero in Phase-2. Based on the infinite series solution of the polarization (3.28), on a conduction problem the second-order trial polarization is given as

$$\boldsymbol{p} = \boldsymbol{p}^{(1)}\chi_1 + K_2\chi_1\boldsymbol{\Gamma}^{\infty}\boldsymbol{p}^{(2)}\chi_1 \quad (4.162)$$

Clearly, application of (4.162) to the even-order HS principle involves the four-point correlation, leading to the fourth-order bounds.

To derive the effective conductivity, apply to an RVE of the composite a unit global temperature gradient $\nabla_1 u_0$ along axis-1. Accordingly, the trial polarization (4.162) is explicitly given in terms of two arguments $p_1^{(1)}$ and $p_1^{(2)}$ as

$$\boldsymbol{p} = \left(p_1^{(1)}\chi_1 + K_2\chi_1\Gamma_{11}^{\infty}p_1^{(2)}\chi_1\right)\mathbf{i} + \left(K_2\chi_1\Gamma_{21}^{\infty}p_1^{(2)}\chi_1\right)\mathbf{j} + \left(K_2\chi_1\Gamma_{31}^{\infty}p_1^{(2)}\chi_1\right)\mathbf{k} \quad (4.163)$$

where **i**, **j**, **k** are the unit vectors in axes-1, -2, and -3, respectively. Substitution of the trial function (4.163) and the modified Green function (A2.12) into the HS functional (3.57) yields the following explicit results, except for a fourth-order term T_{1111}

$$\begin{aligned}
&\frac{1}{2}\left\langle \boldsymbol{p}, \frac{1}{K-K_2}\boldsymbol{p}\right\rangle \\
&= \frac{1}{2}\frac{c_1}{K_1-K_2}\left(p_1^{(1)}\right)^2 + \frac{K_2\Gamma_{11}^{\infty}c_{11}}{K_1-K_2}p_1^{(1)}p_1^{(2)} \\
&\quad + \frac{1}{2}\frac{K_2^2}{K_1-K_2}\sum_{i=1}^{3}\Gamma_{i1}^{\infty}\Gamma_{i1}^{\infty}c_{111}\left(p_1^{(2)}\right)^2 \\
&= \frac{1}{2}\frac{c_1}{K_1-K_2}\left(p_1^{(1)}\right)^2 \\
&\quad + \frac{K_2}{K_1-K_2}\frac{c_1c_2}{3K_2}p_1^{(1)}p_1^{(2)} + \frac{1}{2}\frac{K_2^2}{K_1-K_2}\left(\frac{c_1c_2^2}{9K_2^2} + \frac{2\zeta_1c_1c_2}{9K_2^2}\right)\left(p_1^{(2)}\right)^2
\end{aligned} \tag{4.164}$$

$$\begin{aligned}
&\frac{1}{2}\left\langle \boldsymbol{p}, \boldsymbol{\Gamma}^{\infty}\boldsymbol{p}\right\rangle \\
&= \frac{1}{2}\Gamma_{11}^{\infty}c_{11}\left(p_1^{(1)}\right)^2 + K_2\sum_{i=1}^{3}\Gamma_{i1}^{\infty}\Gamma_{i1}^{\infty}c_{111}p_1^{(1)}p_1^{(2)} \\
&\quad + \frac{1}{2}K_2^2\sum_{i=1}^{3}\sum_{j=1}^{3}\Gamma_{j1}^{\infty}\Gamma_{ji}^{\infty}\Gamma_{i1}^{\infty}c_{1111}\left(p_1^{(2)}\right)^2 \\
&= \frac{1}{2}\frac{c_1c_2}{3K_2}\left(p_1^{(1)}\right)^2 \\
&\quad + K_2\left(\frac{c_1c_2^2}{9K_2^2} + \frac{2\zeta_1c_1c_2}{9K_2^2}\right)p_1^{(1)}p_1^{(2)} + \frac{1}{2}K_2^2\left(\frac{c_1c_2^3}{27K_2^3} + \frac{T_{1111}}{K_2^3}\right)\left(p_1^{(2)}\right)^2
\end{aligned} \tag{4.165}$$

$$\begin{aligned}
\left\langle \boldsymbol{p}, \nabla_1 u_0\right\rangle &= c_1p_1^{(1)} + K_2\Gamma_{11}^{\infty}c_{11}p_1^{(2)} \\
&= c_1p_1^{(1)} + \frac{c_1c_2}{3}p_1^{(2)}
\end{aligned} \tag{4.166}$$

where the triple integral T_{1111} involving the four-point correlation is defined as

$$T_{1111} \equiv K_2^3\sum_{i=1}^{3}\sum_{j=1}^{3}\int_Y \Gamma_{j1}^{\infty}(\boldsymbol{r})\int_Y \Gamma_{ji}^{\infty}(\boldsymbol{s})\int_Y \Gamma_{i1}^{\infty}(\boldsymbol{t})\left(c_{1111}(\boldsymbol{r},\boldsymbol{s},\boldsymbol{t}) - \frac{c_{11}(\boldsymbol{r})c_{11}(\boldsymbol{s})c_{11}(\boldsymbol{t})}{c_1^2}\right)d\boldsymbol{t}d\boldsymbol{s}d\boldsymbol{r}$$

or more concisely

$$T_{1111} \equiv \sum_{i=1}^{3}\sum_{j=1}^{3}\int_{Y}\tilde{\Gamma}_{j1}^{\infty}(\boldsymbol{r})\int_{Y}\tilde{\Gamma}_{ji}^{\infty}(\boldsymbol{s})\int_{Y}\tilde{\Gamma}_{i1}^{\infty}(\boldsymbol{t})\left(c_{1111}(\boldsymbol{r},\boldsymbol{s},\boldsymbol{t})-\frac{c_{11}(\boldsymbol{r})c_{11}(\boldsymbol{s})c_{11}(\boldsymbol{t})}{c_1^2}\right)d\boldsymbol{t}\,d\boldsymbol{s}\,d\boldsymbol{r} \tag{4.167}$$

with $\tilde{\Gamma}_{j1}^{\infty}$ indicating the free space modified Green function characterized with a conductivity of unity.

Similar to those in Section 4.3, the double integral in (4.164) involving the triple correlation is calculated as

$$\begin{aligned}
&\sum_{i=1}^{3}\Gamma_{i1}^{\infty}\Gamma_{i1}^{\infty}c_{111}\\
&=\int_{Y}\int_{Y}\sum_{i=1}^{3}\Gamma_{i1}^{\infty}(\boldsymbol{s})\Gamma_{i1}^{\infty}(\boldsymbol{r})c_{111}(\boldsymbol{r},\boldsymbol{s})\,d\boldsymbol{r}d\boldsymbol{s}\\
&=\int_{Y}\int_{Y}\sum_{i=1}^{3}\Gamma_{i1}^{\infty}(\boldsymbol{s})\Gamma_{i1}^{\infty}(\boldsymbol{r})\left(c_{111}(\boldsymbol{r},\boldsymbol{s})-\frac{c_{11}(\boldsymbol{r})c_{11}(\boldsymbol{s})}{c_1}\right)d\boldsymbol{r}d\boldsymbol{s}\\
&\quad+\frac{1}{c_1}\sum_{i=1}^{3}\left(\int_{Y}\Gamma_{i1}^{\infty}(\boldsymbol{r})c_{11}(\boldsymbol{r})d\boldsymbol{r}\right)^2\\
&=\int_0^{\infty}dr\int_0^{\infty}ds\int_0^{\pi}\sin(\theta)\,d\theta\left(\frac{s^2-3s_1^2}{4\pi s^5}\frac{r^2-3r_1^2}{4\pi r^5}+\frac{3s_1s_2}{4\pi s^5}\frac{3r_1r_2}{4\pi r^5}\right.\\
&\quad\left.+\frac{3s_1s_3}{4\pi s^5}\frac{3r_1r_3}{4\pi r^5}\right)c_{111}(r,s,\theta)/(K_2)^2+\frac{c_1c_2^2}{(9K_2^2)}\\
&=\frac{2\zeta_1c_1c_2}{9K_2^2}+\frac{c_1c_2^2}{9K_2^2}
\end{aligned} \tag{4.168}$$

with $r = |\boldsymbol{r}|$, $s = |\boldsymbol{s}|$, and the geometric parameter ζ_1 defined in (4.102). The integrals involving the autocorrelation have the following simple results:

$$\Gamma_{11}c_{11} = \int_{Y}\Gamma_{11}(\boldsymbol{r})c_{11}(\boldsymbol{r})\,d\boldsymbol{r} = \frac{c_1c_2}{3K_2} \tag{4.169}$$

$$\Gamma_{12}c_{11} = \Gamma_{13}c_{11} = 0 \tag{4.170}$$

With the results (4.164–4.166), the total energy is obtained as

$$\mathcal{H}\left(p_1^{(1)},p_1^{(2)}\right)=\frac{1}{2}K_2+c_1p_1^{(1)}+\frac{c_1c_2}{3}p_1^{(2)}-\frac{1}{2}\left(\frac{c_1}{K_1-K_2}+\frac{c_1c_2}{3K_2}\right)\left(p_1^{(1)}\right)^2$$

$$-c_1c_2\left[\frac{1}{3(K_1-K_2)}+\frac{c_2+2\zeta_1}{9K_2}\right]p_1^{(1)}p_1^{(2)}-\frac{1}{2}\left(\frac{c_1c_2^2+2\zeta_1c_1c_2}{9(K_1-K_2)}+\frac{c_1c_2^3}{27K_2}+\frac{T_{1111}}{K_2}\right)\left(p_1^{(2)}\right)^2 \tag{4.171}$$

Extremization of (4.171) with respect to $p_1^{(1)},p_1^{(2)}$ yields that

$$K^{(4)}=\frac{27T_{1111}\Delta K\left(K_1+2\langle K\rangle\right)-2c_1c_2\zeta_1\left[2c_2(1+2c_1+\zeta_1)K_1^2-\left(1-8c_1^2+16c_1+4c_2\zeta_1\right)K_1K_2+2c_2(-5+2c_1+\zeta_1)K_2^2\right]}{27T_{1111}\Delta K\left(2K_2-\langle\hat{K}\rangle\right)-2c_1c_2\zeta_1\left[2c_2(-c_2-\zeta_1)K_1^2+c_2(1-4c_1+4\zeta_1)K_1K_2+\left(10+c_1-2c_2\zeta_1-2c_1^2\right)K_2^2\right]} \tag{4.172}$$

when $p_1^{(1)},p_1^{(2)}$ take the following optimal values

$$p_1^{(1)}=\frac{3\Delta KK_2\left[27T_{1111}\Delta K+2c_1c_2\zeta_1(3K_2-c_2\Delta K)\right]}{4c_1c_2^3\zeta_1(\Delta K)^2-81T_{1111}K_2\Delta K+2c_1c_2^2\zeta_1\Delta K(2\zeta_1\Delta K+3K_2)-9c_2\left(3T_{1111}(\Delta K)^2+2c_1\zeta_1K_2^2\right)} \tag{4.173}$$

$$p_1^{(2)}=\frac{18c_1c_2\zeta_1(\Delta K)^2K_2}{4c_1c_2^3\zeta_1(\Delta K)^2-81T_{111}K_2\Delta K+2c_1c_2^2\zeta_1\Delta K(2\zeta_1\Delta K+3K_2)-9c_2\left(3T_{1111}(\Delta K)^2+2c_1\zeta_1K_2^2\right)} \tag{4.174}$$

with

$$\langle K\rangle=c_1K_1+c_2K_2$$
$$\langle\hat{K}\rangle=c_1K_2+c_2K_1 \tag{4.175a, b, c}$$
$$\Delta K=K_1-K_2$$

According to the HS principle, when $\Delta K=K_1-K_2<0$ the result (4.172) serves as an upper bound, which should always be bounded from above by the third-order upper bound (4.161a). Conversely, when $K_1>K_2$ the result (4.172) serving as a lower bound should be bounded from below by the third-order lower bound (4.161b). The two bounding conditions yield that the variable T_{1111} must be restricted to a range defined by the following equation

$$T_{1111} = (c_2 + \zeta_1 + \zeta_2 \psi) \frac{4c_1c_2\zeta_1}{27} \tag{4.176}$$

with a fourth-order geometric parameter $0 \le \psi \le 1$. With (4.176), the result (4.172) is rewritten in terms of ψ as

$$K^{(4)} = \langle K \rangle - \frac{c_1c_2(\Delta K)^2}{\langle \hat{K} \rangle + 2K_2^*} \tag{4.177a}$$

$$K_2^* = \langle K \rangle_\zeta - \frac{\zeta_1\zeta_2(\Delta K)^2}{\langle \hat{K} \rangle_\zeta + K_2\left(\dfrac{1}{\psi} - 1\right)} \tag{4.177b}$$

with

$$\langle \hat{K} \rangle_\zeta = \zeta_1 K_2 + \zeta_2 K_1 \tag{4.177c}$$

which serves as the upper or lower bound when $K_2 > K_1$ or $K_2 < K_1$.

By exchanging the subscripts 1 and 2, the other geometric parameter is defined as

$$T_{2222} = (c_1 + \zeta_2 + \zeta_1 \psi') \frac{4c_1c_2\zeta_2}{27} \tag{4.178}$$

with

$$T_{2222} \equiv \sum_{i=1}^{3}\sum_{j=1}^{3} \int_Y \tilde{\Gamma}_{j1}^{\infty}(\boldsymbol{r}) \int_Y \tilde{\Gamma}_{j2}^{\infty}(\boldsymbol{s}) \int_Y \tilde{\Gamma}_{i2}^{\infty}(\boldsymbol{t}) \left(c_{2222}(\boldsymbol{r},\boldsymbol{s},\boldsymbol{t}) - \frac{c_{22}(\boldsymbol{r})c_{22}(\boldsymbol{s})c_{22}(\boldsymbol{t})}{c_2^2} \right) d\boldsymbol{t}\, d\boldsymbol{s}\, d\boldsymbol{r} \tag{4.179}$$

In parallel to (4.177), by using K_1 as the reference conductivity, the fourth-order bound is alternatively expressed as

$$K^{(4)} = \langle K \rangle - \frac{c_1c_2(\Delta K)^2}{\langle \hat{K} \rangle + 2K_1^*} \tag{4.180a}$$

$$K_1^* = \langle K \rangle_\zeta - \frac{\zeta_1\zeta_2(\Delta K)^2}{\langle \hat{K} \rangle_\zeta + K_1\left(\dfrac{1}{\psi'} - 1\right)} \tag{4.180b}$$

which serves as the upper or lower bound when $K_1 > K_2$ or $K_1 < K_2$. By noticing that K_2^* in (4.177) should always be greater or smaller than K_1^* whenever $K_1 - K_2$ is negative or positive, it leads to a necessary equality of the two parameters

$$\psi = \psi' \tag{4.181}$$

In summary, the fourth-order bounds are finally expressed as

$$\boxed{K^{(4)} = \langle K \rangle - \frac{c_1c_2(\Delta K)^2}{\langle \hat{K} \rangle + 2K^*}} \tag{4.182a}$$

with

$$\boxed{K^* = \langle K \rangle_\zeta - \frac{\zeta_1\zeta_2(\Delta K)^2}{\langle \hat{K} \rangle_\zeta + \max\{K_1, K_2\}\left(\dfrac{1}{\psi} - 1\right)}} \tag{4.182b}$$

and

$$\boxed{K^* = \langle K \rangle_\zeta - \frac{\zeta_1\zeta_2(\Delta K)^2}{\langle \hat{K} \rangle_\zeta + \min\{K_1, K_2\}\left(\dfrac{1}{\psi} - 1\right)}} \tag{4.182c}$$

serving as the upper and lower bounds, respectively.

When the geometric parameter $\psi = 0$ or 1, the upper and lower bounds in (4.182) become identical, and the effective conductivity is obtained exactly as the third-order upper or lower bound (4.161). When the geometric parameter of the bulk modulus $\zeta_1 = 0$ or 1, the fourth-order bounds (4.182) directly reduce to the HS bounds (4.77–4.78), independent of ψ.

4.4.2 Attainability of ψ

In the bounds (4.182), the geometric parameter ψ ranges between 0 and 1 corresponding to the range between the upper and lower bounds of the third order. In Avellaneda et al. (1988), the third-order lower bound (4.161b) is shown to be physically unattainable, and the best possible lower bound is found to be

$$K_{-}^{(3)*} = \langle K \rangle - \frac{c_1 c_2 (\Delta K)^2}{\langle \hat{K} \rangle + 2\left(\langle K \rangle_\zeta - \dfrac{\zeta_1 \zeta_2 (\Delta K)^2}{\langle \hat{K} \rangle_\zeta + \dfrac{1}{2}\min\{K_1, K_2\}} \right)} \tag{4.183}$$

which is realized with an assemblage of doubly coated spheres. By comparing (4.183) with (4.182c), the improved lower bound (4.183) corresponds to $\psi = 2/3$, indicating that the range of ψ between $2/3$ and 1 is unattainable. As shown in Figure 4.3, the original lowest bound—i.e. (4.161b) with $\psi = 1$, indicated as Point A′—is improved by (4.183) indicated as Point A. When the geometric parameter $\psi = 2/3$, the effective conductivity can be any value in a range in this example represented as a segment of the dashed line between Points A and B.

To compare the first four orders of bounds on conductivity, the formulae are summarized in Table 4.1. To show the improvement of the fourth-order bounds made over the second- and third-order bounds, the bounds of the three orders are graphically shown in Figure 4.4 for a two-phase composite.

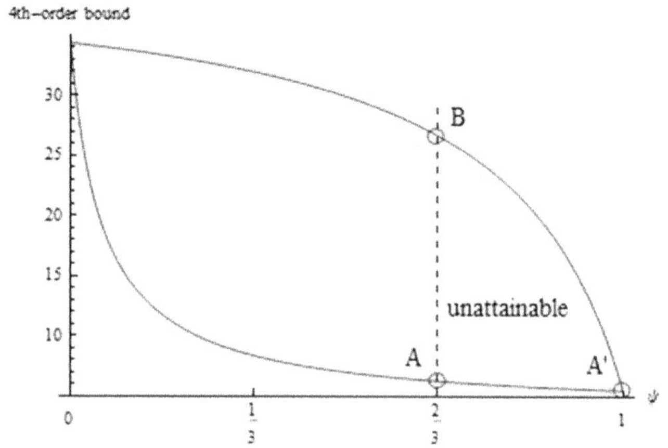

Figure 4.3 Fourth-order bounds (4.182) vs. geometric parameter ψ, with $K_1 = 100$, $K_2 = 1$, $c_1 = 0.5$, and $\zeta_1 = 0.5$.

Table 4.1 Comparison of the first four orders of bounds on conductivity

Order	*Lower bound*	*Upper bound*
1st	$\langle K^{-1} \rangle^{-1}$	$\langle K \rangle$
2nd	$K_-^{(2)} = \langle K \rangle - \dfrac{c_1 c_2 (\Delta K)^2}{\langle \hat{K} \rangle + 2\min\{K_1, K_2\}}$	$K_+^{(2)} = \langle K \rangle - \dfrac{c_1 c_2 (\Delta K)^2}{\langle \hat{K} \rangle + 2\max\{K_1, K_2\}}$
3rd	$K_-^{(3)} = \langle K \rangle - \dfrac{c_1 c_2 (\Delta K)^2}{\langle \hat{K} \rangle + 2\langle K^{-1} \rangle_\zeta^{-1}}$	$K_+^{(3)} = \langle K \rangle - \dfrac{c_1 c_2 (\Delta K)^2}{\langle \hat{K} \rangle + 2\langle K \rangle_\zeta}$
4th	$K_-^{(3)} = \langle K \rangle - \dfrac{c_1 c_2 (\Delta K)^2}{\langle \hat{K} \rangle + 2\left(\langle K \rangle_\zeta - \dfrac{\zeta_1 \zeta_2 (\Delta K)^2}{\langle \hat{K} \rangle_\zeta + \min\{K_1, K_2\}\left(\dfrac{1}{\psi} - 1 \right)} \right)}$	$K_+^{(3)} = \langle K \rangle - \dfrac{c_1 c_2 (\Delta K)^2}{\langle \hat{K} \rangle + 2\left(\langle K \rangle_\zeta - \dfrac{\zeta_1 \zeta_2 (\Delta K)^2}{\langle \hat{K} \rangle_\zeta + \max\{K_1, K_2\}\left(\dfrac{1}{\psi} - 1 \right)} \right)}$

Note: The meanings of the brackets refer to (4.175) and (4.177c).

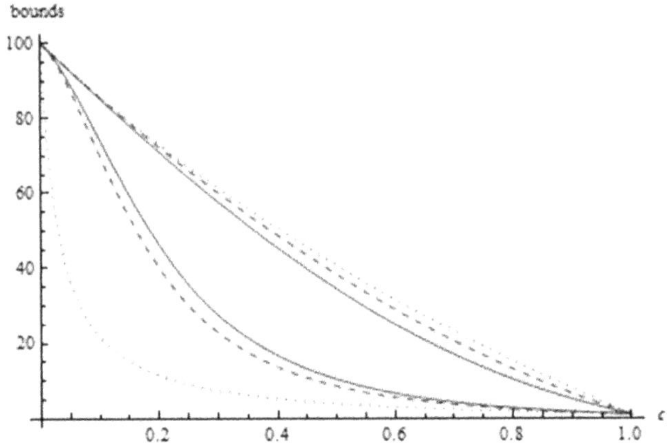

Figure 4.4 Comparison among the second- (dotted), third- (dashed), and fourth-order (solid) bounds, with $K_1 = 1$, $K_2 = 100$, $\zeta_1 = c_1/2$, and $\psi = 2/3$.

4.4.3 Explicit expression of ψ

To find the explicit result of the triple integral (4.167), the rectangular and spherical coordinates are established for the four spatial points of the four-point correlation as follows. In Figure 4.5 the four points are denoted as O, O′, O″, and O‴, with $\overrightarrow{OO'} = \boldsymbol{r}$, $\overrightarrow{OO''} = \boldsymbol{s}$, $\overrightarrow{O'O'''} = \boldsymbol{t}$. With the angles designated in Figure 4.5, the rectangular coordinates of vectors $\boldsymbol{r}$, $\boldsymbol{s}$, and $\boldsymbol{t}$ are expressed in terms of the spherical coordinates as

$$
\begin{aligned}
r_1 &= r\sin(\varphi_1)\cos(\theta_1)\\
r_2 &= r\sin(\varphi_1)\sin(\theta_1)\\
r_3 &= r\cos(\varphi_1)\\
s_1 &= s\big(\cos(\varphi)\sin(\varphi_1)\cos(\theta_1) + \sin(\varphi)\sin(\theta_2)\cos(\varphi_1)\cos(\theta_1) + \sin(\varphi)\cos(\theta_2)\sin(\theta_1)\big)\\
s_2 &= s\big(\cos(\varphi)\sin(\varphi_1)\sin(\theta_1) + \sin(\varphi)\sin(\theta_2)\cos(\varphi_1)\sin(\theta_1) - \sin(\varphi)\cos(\theta_2)\cos(\theta_1)\big)\\
s_3 &= s\big(\cos(\varphi)\cos(\varphi_1) - \sin(\varphi)\sin(\theta_2)\sin(\varphi_1)\big)\\
t_1 &= t\big(\cos(\tilde{\varphi})\sin(\varphi_1)\cos(\theta_1) + \sin(\tilde{\varphi})\sin(\theta_2)\cos(\varphi_1)\cos(\theta_1) + \sin(\tilde{\varphi})\cos(\theta_2+\theta)\sin(\theta_1)\big)\\
t_2 &= t\big(\cos(\tilde{\varphi})\sin(\varphi_1)\sin(\theta_1) + \sin(\tilde{\varphi})\sin(\theta_2+\theta)\cos(\varphi_1)\sin(\theta_1) - \sin(\tilde{\varphi})\cos(\theta_2+\theta)\cos(\theta_1)\big)\\
t_3 &= t\big(\cos(\tilde{\varphi})\cos(\varphi_1) - \sin(\tilde{\varphi})\sin(\theta_2+\theta)\sin(\varphi_1)\big)
\end{aligned}
\tag{4.184}
$$

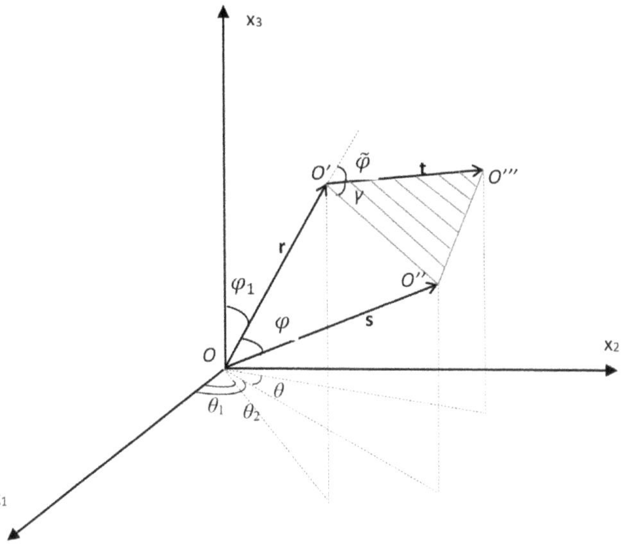

Figure 4.5 Rectangular and spherical coordinates for the four points of the fourth-order correlation.

where $t = |\boldsymbol{t}|$. By using these spherical coordinates, as an example one component of the triple integral (4.167) is expanded here

$$\int_Y \tilde{\Gamma}^{\infty}_{11}(\boldsymbol{r})\int_Y \tilde{\Gamma}^{\infty}_{11}(\boldsymbol{s})\int_Y \tilde{\Gamma}^{\infty}_{11}(\boldsymbol{t})\left(c_{1111}(\boldsymbol{r},\boldsymbol{s},\boldsymbol{t})-\frac{c_{11}(\boldsymbol{r})c_{11}(\boldsymbol{s})c_{11}(\boldsymbol{t})}{c_1^2}\right)d\boldsymbol{t}d\boldsymbol{s}d\boldsymbol{r}$$

$$=\int_Y\int_Y\int_Y \frac{s^2-3s_1^2}{4\pi s^5}\frac{r^2-3r_1^2}{4\pi r^5}\frac{t^2-3t_1^2}{4\pi t^5}\left(c_{1111}(\boldsymbol{r},\boldsymbol{s},\boldsymbol{t})-\frac{c_{11}(\boldsymbol{r})c_{11}(\boldsymbol{s})c_{11}(\boldsymbol{t})}{c_1^2}\right)d\boldsymbol{t}d\boldsymbol{s}d\boldsymbol{r}$$

$$=\int_0^{\pi}\int_0^{2\pi}\int_0^{2\pi}\frac{1-3\left(\cos(\varphi)\sin(\varphi_1)\cos(\theta_1)+\sin(\varphi)\sin(\theta_2)\cos(\varphi_1)\cos(\theta_1)+\sin(\varphi)\cos(\theta_2)\sin(\theta_1)\right)^2}{4\pi s^3}$$

$$\frac{1-3\left(\cos(\tilde{\varphi})\sin(\varphi_1)\cos(\theta_1)+\sin(\tilde{\varphi})\sin(\theta_2)\cos(\varphi_1)\cos(\theta_1)+\sin(\tilde{\varphi})\cos(\theta_2+\theta)\sin(\theta_1)\right)^2}{4\pi t^3}$$

$$\frac{1-3\left(\sin(\varphi_1)\cos(\theta_1)\right)^2}{4\pi r^3}t^2r^2s^2\sin(\varphi_1)d\theta_1 d\theta_2 d\varphi_1$$

$$\int_{0^+}^{\infty}\int_{0^+}^{\infty}\int_{0^+}^{\infty}\int_0^{\pi}\int_0^{\pi}\int_0^{2\pi}\left(c_{1111}(r,s,t,\theta,\tilde{\varphi},\varphi)-\frac{c_{11}(r)c_{11}(s)c_{11}(t)}{c_1^2}\right)\sin(\tilde{\varphi})\sin(\varphi)\,d\theta\,d\tilde{\varphi}\,d\varphi\,dt\,ds\,dr \tag{4.185}$$

In an isotropic random medium, with θ fixed $c_{1111}(\boldsymbol{r}, \boldsymbol{s}, \boldsymbol{t})$ is independent of the orientation of the triangle $OO'O''$, i.e. θ_1, ϕ_2, θ_1, θ_2, φ_1, and thus the first three integrals in (4.185) about θ_1, θ_2, φ_1 can be explicitly obtained. By summing up all the components of the triple integral (4.167), T_{1111} or ψ is explicitly obtained as a six-fold integral

$$T_{1111} = (c_2 + \zeta_1 + \zeta_2\psi)\frac{4c_1c_2\zeta_1}{27}$$
$$= \frac{1}{16\pi}\int_0^\infty \frac{dr}{r}\int_0^\infty \frac{ds}{s}\int_0^\infty \frac{dt}{t}\int_{-1}^1 d\tilde{u}\int_{-1}^1 du\int_0^{2\pi} d\theta\left(c_{1111}(r,s,t,\theta,\tilde{u},u) - \frac{c_{11}(r)c_{11}(s)c_{11}(t)}{c_1^2}\right)$$
$$\left(3\cos(2\theta)(1-u^2)(1-\tilde{u}^2) - 4P_2(u)P_2(\tilde{u}) - 6\cos(\theta)u\tilde{u}\sqrt{(1-u^2)(1-\tilde{u}^2)}\right) \tag{4.186}$$

with the second Legendre polynomials $P_2(u) = \frac{3u^2-1}{2}$, $P_2(\tilde{u}) = \frac{3\tilde{u}^2-1}{2}$ and $u = \cos(\phi)$, $\tilde{u} = \cos(\tilde{\phi})$. Note that the angle θ is a projection of γ the angle between two edges $O'O'''$ and $O'O''$ of the tetrahedron (see Figure 4.5), and the two are related via the following equation:

$$\cos(\gamma) = \frac{(su-r)u_2 + s\cos(\theta)\sqrt{(1-u^2)(1-\tilde{u}^2)}}{\sqrt{r^2+s^2-2rsu}} \tag{4.187}$$

It is finally noted that the derivation, results, and discussion presented in this section have not been previously published.

4.5 FLUID–ANTIFLUID ANNIHILATION

As noted in Section 4.3, there are four pairs of the geometric parameters $(\zeta_1, \eta_1) = (0, 0)$, $(0, 1)$, $(1, 1)$ resulting in the exact effective elastic moduli for a two-phase composite. In this section, we apply these four exact cases to phase transition of matter from a mechanical point of view.

In physics, a *fluid* is defined as a substance that continuously deforms under an applied shear stress, which in the meantime has high resistance to a change of volume. In other words, a fluid has a large bulk modulus and an almost zero shear modulus, i.e. $\kappa >> \mu \sim 0$. A *solid* has high resistance to a change of either volume or shape, i.e. $\kappa \sim \mu >> 0$. Between a fluid and a solid, a plastic or mud-like state can be similarly defined as $\kappa >> \mu > 0$. A void state in terms of rigidities is simply $\kappa = \mu \sim 0$. When a solid is completely mixed with a fluid, for example, a mixture of sand and water with surface force neglected, there are four cases of mixing results that are theoretically exact

Table 4.2 Mixture of a solid and a fluid

Case		*Solid* ($\mu_1 \sim \kappa_1 >> 0$) + *Fluid* ($\kappa_2 >> \mu_2 \sim 0$)	*Phase transition*
$\zeta_1 = 0$	$\eta_1 = 0$	$\kappa_+^{(3)} = \kappa_-^{(3)} = \langle \kappa^{-1} \rangle^{-1}$ $\mu_+^{(3)} = \mu_-^{(3)} = \dfrac{2\mu_1 + 3\langle \mu \rangle}{2\mu_1 \langle \mu^{-1} \rangle + 3}$	Fluid/Mud/Solid
	$\eta_1 = 1$	$\kappa_+^{(3)} = \kappa_-^{(3)} = \langle \kappa^{-1} \rangle^{-1}$ $\mu_+^{(3)} = \mu_-^{(3)} = \dfrac{24 c_1 \mu_1 \kappa_2}{24\kappa_2 + c_2(40\mu_1 + 21\kappa_2)}$	Fluid/Mud/Solid
$\zeta_1 = 1$	$\eta_1 = 0$	$\kappa_+^{(3)} = \kappa_-^{(3)} = \dfrac{3\kappa_1\kappa_2 + 4\mu_1 \langle \kappa \rangle}{3\langle \hat{\kappa} \rangle + 4\mu_1}$ $\mu_+^{(3)} = \mu_-^{(3)} = \dfrac{15 c_1 \mu_1 \kappa_1}{15\kappa_1 + c_2(4\mu_1 + 42\kappa_1)}$	Fluid/Mud/Solid
	$\eta_1 = 1$	$\kappa_+^{(3)} = \kappa_-^{(3)} = \dfrac{3\kappa_1\kappa_2 + 4\mu_1 \langle \kappa \rangle}{3\langle \hat{\kappa} \rangle + 4\mu_1}$ $\mu_+^{(3)} = \mu_-^{(3)} = \dfrac{c_1 \mu_1}{1 + \dfrac{6 c_2(\kappa_1 + 2\mu_1)}{9\kappa_1 + 8\mu_1}}$	Fluid/Mud/Solid

as presented in Table 4.2. With increase of solid content c_1 from 0 to 1, in all the four cases the mixture goes through a mechanical phase transition process from a fluid state to a mud-like state, and finally to a solid state. This phenomenon is well observed in daily experience when mixing powder with water.

Next we consider a peculiar substance, a so-called anti-rubber or dilatational medium with its Poisson's ratio approaching –1, which can be artificially fabricated by assembling components of a certain two-phase composite with hexagonal symmetry (Milton, 1992). Another way is to assemble Hoberman spheres of all sizes self-similarly like the Hashin coated sphere assemblage (Figure 4.1). Given the Poisson's ratio in terms of the bulk and shear moduli

$$\nu = \frac{3\kappa - 2\mu}{2(3\kappa + \mu)}$$

it is clear the Poisson's ratio being –1 corresponds to the elastic moduli $\mu >> \kappa \sim 0$, which is opposite to a fluid in that a dilatational medium has high

Table 4.3 Mixture of a fluid and an anti-fluid

Case		*Fluid ($\kappa_1 >> \mu_1 \sim 0$) + Anti-Fluid ($\mu_2 >> \kappa_2 \sim 0$)*	*Phase transition*
$\zeta_1 = 0$	$\eta_1 = 0$	$\kappa_+^{(3)} = \kappa_-^{(3)} = \dfrac{4c_1\mu_2\kappa_1}{3c_2\kappa_1 + 4\mu_2}$ $\mu_+^{(3)} = \mu_-^{(3)} = \dfrac{3\mu_1 + 2\langle\mu\rangle}{3\mu_1\langle\mu^{-1}\rangle + 2}$	Anti-Fluid/ Anti-Mud/Solid/ Mud/Fluid
	$\eta_1 = 1$	$\kappa_+^{(3)} = \kappa_-^{(3)} = \dfrac{4c_1\mu_2\kappa_1}{3c_2\kappa_1 + 4\mu_2}$ $\mu_+^{(3)} = \mu_-^{(3)} = \dfrac{4\mu_1 + c_2(15\kappa_2 + 56\mu_1)}{4\langle\hat{\mu}\rangle + 15\kappa_2 + 56\mu_1}\mu_2$	Anti-Fluid/ Anti-Mud/Solid/ Mud/Fluid
$\zeta_1 = 1$	$\eta_1 = 0$	$\kappa_+^{(3)} = \kappa_-^{(3)} = \dfrac{3\kappa_2 + 4c_1\mu_1}{3\langle\hat{\kappa}\rangle + 4\mu_1}\kappa_1$ $\mu_+^{(3)} = \mu_-^{(3)} = \dfrac{24\kappa_1\langle\mu\rangle + \mu_1(21\kappa_1 + 40\mu_2)}{24\kappa_1 + c_1(21\kappa_1 + 40\mu_2)}$	Anti-Fluid/ Anti-Mud/**Void**/ Mud/Fluid
	$\eta_1 = 1$	$\kappa_+^{(3)} = \kappa_-^{(3)} = \dfrac{3\kappa_2 + 4c_1\mu_1}{3\langle\hat{\kappa}\rangle + 4\mu_1}\kappa_1$ $\mu_+^{(3)} = \mu_-^{(3)} = \dfrac{14\mu_1 + 21c_2\mu_1}{14\langle\hat{\mu}\rangle + 21\mu_1}\mu_2$	Anti-Fluid/**Void**/ Fluid

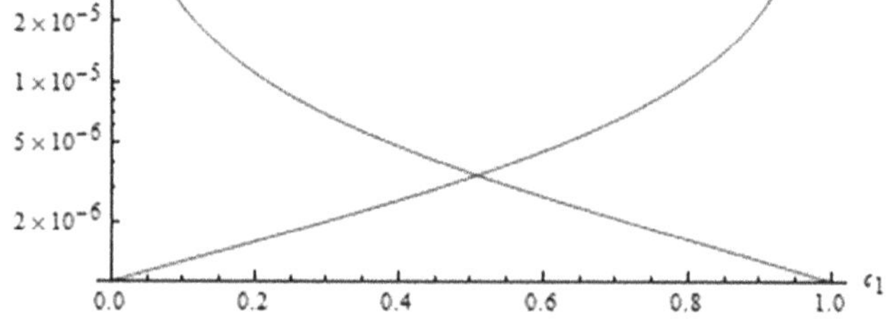

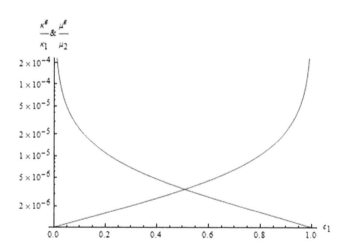

Figure 4.6 Effective bulk and shear moduli vs. fluid content c_1, for a mixture of a fluid and an anti-fluid, with $\kappa_2 = \mu_1 = 10^{-6}\kappa_1 = 10^{-6}\mu_2$, $\zeta_2 = 10^{-6}$, $\eta_2 = 10^{-6}$.

resistance to a change of shape but no resistance to a volumetric change. Such a state of matter with $\mu >> \kappa \sim 0$ is hereby particularly named as an *anti-fluid* state, as we are going to mix a fluid and an anti-fluid to find any peculiarity. Following this definition, an *anti-mud* can be similarly defined as an intermediate between a solid and an anti-fluid as $\mu >> \kappa > 0$.

The results of a fluid and anti-fluid mixture are listed in Table 4.3. In the first and second cases, with increase of fluid content c_1 the mixture goes through a phase transition process from an original anti-fluid to an anti-mud, a solid, a mud, and finally to a fluid when c_1 approaches 1, which is similar to those cases in Table 4.2. A most interesting and peculiar result is obtained in the third and fourth cases, in that the mixture experiences a void state. When mixed in such a way with the geometric parameters $(\zeta_1, \eta_1) = (1, 0)$ or $(1,1)$, in a certain range of fluid content, a fluid and an anti-fluid annihilate each other resulting in a neutral void with approximately zero rigidities in both volumetric and shear deformation. In an example shown in Figure 4.6, within a fluid content range between 0.05 to 0.95, the bulk and shear moduli of the mixture are at the order of 10^{-4} or less of those of the original fluid and anti-fluid, which is practically a void in terms of elastic rigidities. Note that such a mixture has been proved to be physically attainable in the fourth case $(\zeta_1, \eta_1) = (1, 1)$ with finite-rank laminates (Francfort & Murat, 1986) and reiterated cell structures (Lukkassen, 1999). With regard to an experimental setup, one can imagine in a fluid evenly mixed with a certain type of elastic Hoberman spheres of various sizes, the experimental parameters are tuned to see whether a longitudinal wave propagates in this mixture very slowly or nearly impossible.

Chapter 5

Ellipsoidal bound

In Chapter 4, variational bounds are analytically derived based on morphological information expressed in terms of correlation functions. Correlation functions, however, are not the most efficient tool to describe the shape of particles, or more specifically aspect ratios of spheroidal inclusions, which is one of the most essential morphological features in a particulate composite. This limitation can be explained through the relationship between morphological features and correlation functions, as graphically demonstrated in Chapter 2. In Chapter 4, the Hashin-Shtrikman bounds of the effective bulk modulus and conductivity are shown to be morphologically associated with spherical inclusions. In application, many natural and engineering composites contain nonspherical inclusions with the aspect ratios deviating much from the unity value of spheres (e.g. fibers and cracks). In the fast-growing field of nanotechnology, nanofillers are distinctively characterized with extremely large or small aspect ratios. Given the aspect ratios of the inclusions in a composite, it is of both theoretical and practical interest to know the upper and lower bounds of the physical properties.

Following the first variational estimate made on ellipsoidal inclusions (Ponte Castañeda & Willis, 1995), a so-called ellipsoidal bound is generally formulated by introducing two essential morphological patterns, isotropic mixture and anisotropic mixture (Xu, 2012b; Xu & Stefanou, 2012a). More importantly, the formulation of the ellipsoidal bound applies to morphologies close to occurrence of percolation, thereby enabling prediction of continuum percolation threshold, a long intriguing problem in physics and engineering materials. Theoretical prediction of the continuum percolation threshold is separately described in Chapter 6.

5.1 FORMULATION OF THE ELLIPSOIDAL BOUND

5.1.1 Morphological model

An RVE domain is composed of a hosting matrix and a number of randomly distributed nonoverlapping ellipsoidal inclusions. The random field is statistically homogeneous and ergodic. A *component* is hereby specifically defined as a group of ellipsoids in the RVE sharing identical elastic moduli and a common shape, size, and orientation. In an $\tilde{N}$-component composite, denote the $\tilde{N}$th component the hosting matrix and $\chi_n^{(i)}$ the domain for the ith inclusion of Component-n, i.e.

$$\chi_n^{(i)}(\boldsymbol{x},\vartheta)=\begin{cases}1 & \boldsymbol{x}\in D_n^{(i)}\\ 0 & \boldsymbol{x}\notin D_n^{(i)}\end{cases} \tag{5.1}$$

with $n = 1, 2, ..., \tilde{N} - 1$. The random field for the elastic moduli of such an $\tilde{N}$-component composite is therefore written as

$$\boldsymbol{L}(\boldsymbol{x},\vartheta)=\boldsymbol{L}_0+\sum_{n=1}^{\tilde{N}-1}(\boldsymbol{L}_n-\boldsymbol{L}_0)\chi_n(\boldsymbol{x},\vartheta) \tag{5.2a}$$

with

$$\chi_n(\boldsymbol{x},\vartheta)=\sum_{i=1}^{\tilde{N}_n}\chi_n^{(i)}(\boldsymbol{x},\vartheta) \tag{5.2b}$$

where $\boldsymbol{L}_0 = \boldsymbol{L}_{\tilde{N}}$ and $\boldsymbol{L}_n$ denote the elastic moduli of the matrix and Component-n, respectively, and $\tilde{N}_n$ the number of the inclusions in Component-n. Note that, to apply the HS principle as shown next, $\boldsymbol{L}_n - \boldsymbol{L}_0$ should be positive or negative semidefinite for any component n.

The morphological model is hereby constructed by fixing the shape, size, and orientation of all the inclusions, while allowing the center of an inclusion be randomly located and correlated with those of the other inclusions due to the nonoverlapping, or so-called nonpenetrating, condition. One can imagine such a model by randomly throwing inclusions into an RVE domain subject to the nonoverlapping condition.

Denote $\boldsymbol{x}_n^{(i)}$ the center position of the ith inclusion in Component-n. A random point field for the centers of all the inclusions in Component-n is represented as

$$h_n(\boldsymbol{x},\vartheta)=\sum_{i=1}^{\tilde{N}_n}\delta\left(\boldsymbol{x}-\boldsymbol{x}_n^{(i)}(\vartheta)\right)$$

the volume integral of which over the RVE domain yields the number of the inclusions, i.e.

$$\tilde{\mathrm{N}}_n = \langle h_n(\boldsymbol{x}, \vartheta) \rangle \tag{5.3}$$

Define an indicator function $\tilde{\chi}_n(\boldsymbol{x})$ to represent the domain of an ellipsoid in Component-n with the origin of the coordinates shifted to O′, the center of the ellipsoid (Figure 5.1). The domain for all the constituent ellipsoids in Component-n is expressed in terms of the following convolution

$$\chi_n(\boldsymbol{x}, \vartheta) = \int_{\mathrm{Y}} \tilde{\chi}_n(\boldsymbol{x} - \boldsymbol{x}', \vartheta) h_n(\boldsymbol{x}', \vartheta)\, d\boldsymbol{x}' \tag{5.4}$$

Given the volume of an inclusion in Component-n as $\tilde{c}_n^{(i)}$, the volume of Component-n is therefore obtained as

$$\begin{aligned} \langle \chi_n(\boldsymbol{x}, \vartheta) \rangle &= \tilde{c}_n \\ &= \tilde{\mathrm{N}}_n \tilde{c}_n^{(i)} \end{aligned}$$

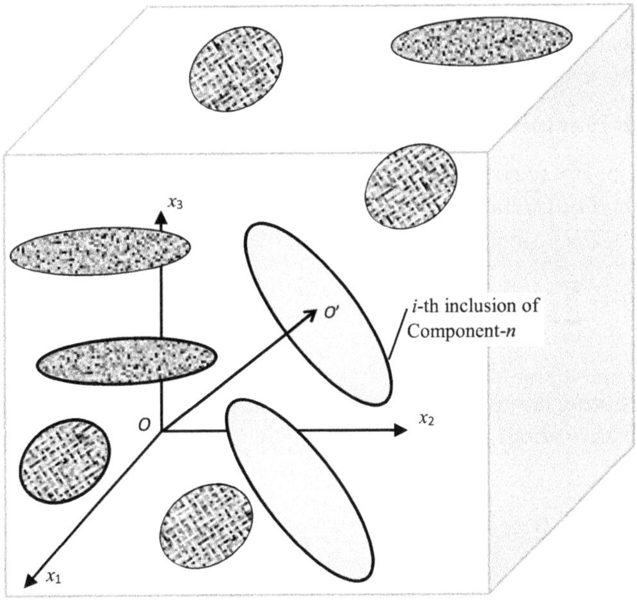

Figure 5.1 Morphological model of randomly distributed inclusions in an RVE.

Similar to (4.9–4.10), the intercomponent autocorrelation function

$$\tilde{c}_{nm}(\boldsymbol{x}-\boldsymbol{x}') = \langle \chi_n(\boldsymbol{x}',\vartheta)\chi_m(\boldsymbol{x}',\vartheta)\rangle \\ = \overline{\chi_n(\boldsymbol{x},\vartheta)\chi_m(\boldsymbol{x}',\vartheta)} \tag{5.5}$$

can be expressed in terms of the correlation coefficient $\tilde{\rho}_{nm}(\boldsymbol{x})$, as

$$\tilde{c}_{nm}(\boldsymbol{x}) = \left(\tilde{c}_n(\delta_{nm}-\tilde{c}_m)\right)\tilde{\rho}_{nm}(\boldsymbol{x})+\tilde{c}_n\tilde{c}_m \tag{5.6}$$

Substitution of (5.4) into the intercomponent autocorrelation (5.5) yields

$$\tilde{c}_{nm}(\boldsymbol{x}-\boldsymbol{x}') = \int_{\mathrm{Y}}\int_{\mathrm{Y}} \tilde{\chi}_n(\boldsymbol{x}-\boldsymbol{y})\tilde{\chi}_m(\boldsymbol{x}'-\boldsymbol{y}')\ \tilde{c}_{h,mn}(\boldsymbol{y}-\boldsymbol{y}')d\boldsymbol{y}\,d\boldsymbol{y}' \tag{5.7}$$

$$\tilde{c}_{h,nm}(\boldsymbol{y}-\boldsymbol{y}') = \langle h_n(\boldsymbol{y},\vartheta)h_m(\boldsymbol{y}',\vartheta)\rangle \\ = \overline{h_n(\boldsymbol{y},\vartheta)h_m(\boldsymbol{y}',\vartheta)} \tag{5.8}$$

where the interpoint autocorrelation $\tilde{c}_{h,mn}$ is assumed to be statistically homogeneous.

5.1.2 Derivation of the ellipsoidal bound

To apply the HS variational principle, similar to (4.11), a component-wise constant trial polarization is constructed as

$$\boldsymbol{p}(\boldsymbol{x},\vartheta) = \sum_{n=1}^{\tilde{N}-1}\chi_n(\boldsymbol{x},\vartheta)\boldsymbol{p}_n$$

By substituting the trial polarization into the HS functional (3.47) and equating its derivative about $\boldsymbol{p}_n$ to zero, it results in the following $\tilde{N}-1$ equations,

$$(\boldsymbol{L}_n-\boldsymbol{L}_0)^{-1}\boldsymbol{p}_n\tilde{c}_n + \sum_{m=1}^{\tilde{N}-1}\int_{\mathbf{Y}}\boldsymbol{\Gamma}^{\infty}(\boldsymbol{x}-\boldsymbol{x}')\tilde{c}_{nm}(\boldsymbol{x}-\boldsymbol{x}')d\boldsymbol{x}'\boldsymbol{p}_m = \boldsymbol{\varepsilon}_0\tilde{c}_n \tag{5.9}$$

with $n = 1,2,\ldots,\tilde{N}-1$. By substituting (5.6–5.7) into (5.9), and taking Property (4.20) into account, (5.9) reduces to

$$(\boldsymbol{L}_n - \boldsymbol{L}_0)^{-1}\boldsymbol{p}_n + \sum_{m=1}^{\tilde{N}-1} \int_Y \boldsymbol{\Gamma}^{\infty}(\boldsymbol{x})\,\tilde{\rho}_{nm}(\boldsymbol{x})\,d\boldsymbol{x}\;(\delta_{mn} - \tilde{c}_m)\boldsymbol{p}_m = \boldsymbol{\varepsilon}_0 \tag{5.10}$$

with $n = 1,2,\ldots,\tilde{N}-1$.

When the correlation coefficient $\tilde{\rho}_{nm}(\boldsymbol{x})$ has its contour lines follow an elliptic shape, the random distribution of Component-m and Component-n in the composite holds a so-called ellipsoidal symmetry (Willis, 1977). In such a case, the integral $\int_Y \boldsymbol{\Gamma}^{\infty}(\boldsymbol{x})\,\tilde{\rho}_{nm}(\boldsymbol{x})\,d\boldsymbol{x}$ has the following analytical result (Willis, 1981).

Property

Given that the reference elastic moduli $\mathbf{L}_0$ *is isotropic and the correlation coefficient* $\rho(x)$ *holds the ellipsoidal symmetry, i.e.* $\rho(\boldsymbol{x}) = \rho_0(|\boldsymbol{Q}\boldsymbol{x}|)$ *with* ρ_0 *being isotropic and* $\boldsymbol{Q}$ *a real nonsingular matrix to transform an ellipsoid into a sphere, it follows that*

$$\int_Y \boldsymbol{\Gamma}^{\infty}(\boldsymbol{x})\,\rho(\boldsymbol{x})\,d\boldsymbol{x} = \boldsymbol{S}\boldsymbol{L}_0^{-1} \tag{5.11}$$

where $\mathbf{S}$ *is the ellipsoidal Eshelby tensor characterized by the shape of* $\rho(\boldsymbol{x})$ *(Appendix III).*

Proof: The Eshelby's solution (1957) is mathematically expressed in terms of the modified Green function as

$$\int_{|\boldsymbol{Q}\boldsymbol{x}|\le r_1} \boldsymbol{\Gamma}^{\infty}(\boldsymbol{x}' - \boldsymbol{x})\,d\boldsymbol{x} = \boldsymbol{S}\boldsymbol{L}_0^{-1}$$

for any point $\boldsymbol{x}'$ within the same ellipsoid defined as $|\boldsymbol{Q}\boldsymbol{x}'| \le r_1$ with r_1 a real positive number, which is further specialized into

$$\int_{|\boldsymbol{Q}\boldsymbol{x}|\le r_1} \boldsymbol{\Gamma}^{\infty}(\boldsymbol{x})\,d\boldsymbol{x} = \boldsymbol{S}\boldsymbol{L}_0^{-1} \tag{5.12}$$

where the ellipsoidal Eshelby tensor $\boldsymbol{S}$ is characterized by the transformation matrix $\boldsymbol{Q}$, i.e. the shape of $\rho(\boldsymbol{x})$. Since the right-hand side of (5.12) is independent of the value r_1, we obviously have

$$\int_{r_1<|\boldsymbol{Q}\boldsymbol{x}|<r_2} \boldsymbol{\Gamma}^{\infty}(\boldsymbol{x})\,d\boldsymbol{x} = 0 \tag{5.13}$$

for any real positive values $0 < r_1 < r_2$. Since the RVE domain is sufficiently large for the value of $\rho(\boldsymbol{x})$ to decay to zero, the shape of the integration domain can be effectively set to conform to the ellipsoidal shape defined by $\rho(\boldsymbol{x})$ without affecting the integral result. Such an ellipsoidal domain is then divided into an infinite number of ellipsoidal layers within each $\rho(\boldsymbol{x})$ has a constant value. Equation (5.13) indicates the integral over each of these layers is zero, except for the integral on the innermost domain, i.e. (5.12) with r_1 approaching zero. Since in this infinitesimal ellipsoid $\rho(0) = 1$, Equation (5.11) is established. □

In a random composite the assumed ellipsoidal symmetry is not necessarily satisfied. In such a case, a concept of the generalized Eshelby tensor $\boldsymbol{S}_{nm}$ is introduced to represent the following integral:

$$\int_Y \boldsymbol{\Gamma}^{\infty}(\boldsymbol{x}-\boldsymbol{x}')\tilde{c}_{nm}(\boldsymbol{x}-\boldsymbol{x}')\,d\boldsymbol{x}' = -\tilde{c}_n\tilde{c}_m\boldsymbol{S}_{nm}\boldsymbol{L}_0^{-1} \qquad n \neq m \tag{5.14}$$

When ellipsoidal inclusions are aligned and distributed with the same ellipsoidal shape as the inclusion shape, the previously generalized Eshelby tensor reduces to $\boldsymbol{S}_{nm} = \boldsymbol{S}_n = \boldsymbol{S}_m$. In this special simple case, the result of Mori-Tanaka estimates (Mori & Tanaka, 1973; Benveniste, 1987) is consistent with the ellipsoidal bound, e.g. the HS bound of the effective bulk modulus with spherical inclusions, and Eq. (5.95) for a solid containing randomly located parallel cracks. In other cases, the Mori-Tanaka estimates use a simple ellipsoidal Eshelby tensor in place of $\boldsymbol{S}_{nm}$, yielding results inconsistent with microstructure (Ponte Castañeda & Willis, 1995).

When $n = m$, the left-hand side of (5.14) is expanded by using (5.4) as

$$\begin{aligned}
&\int_Y \boldsymbol{\Gamma}^{\infty}(\boldsymbol{x}-\boldsymbol{x}')\tilde{c}_{nn}(\boldsymbol{x}-\boldsymbol{x}')\,d\boldsymbol{x}' \\
&= \int_Y \boldsymbol{\Gamma}^{\infty}(\boldsymbol{x}-\boldsymbol{x}')\,d\boldsymbol{x}' \int_Y \tilde{\chi}_n(\boldsymbol{x}-\boldsymbol{x}'')\tilde{\chi}_n(\boldsymbol{x}'-\boldsymbol{x}'') \ \langle h_n(\boldsymbol{x}'',\vartheta)\rangle\, d\boldsymbol{x}'' \\
&= \tilde{\mathrm{N}}_n \int_Y \boldsymbol{\Gamma}^{\infty}(\boldsymbol{x}-\boldsymbol{x}')\,d\boldsymbol{x}' \int_Y \tilde{\chi}_n(\boldsymbol{x}-\boldsymbol{x}'')\tilde{\chi}_n(\boldsymbol{x}'-\boldsymbol{x}'')\ d\boldsymbol{x}''
\end{aligned} \tag{5.15}$$

Let $\boldsymbol{y} = \boldsymbol{x} - \boldsymbol{x}''$, $\boldsymbol{z} = \boldsymbol{x}' - \boldsymbol{x}''$, and (5.15) reduces to

$$\begin{aligned}
&\tilde{\mathrm{N}}_n \int_Y \boldsymbol{\Gamma}^{\infty}(\boldsymbol{y}-\boldsymbol{z})\tilde{\chi}_n(\boldsymbol{z})\,d\boldsymbol{z} \int_Y \tilde{\chi}_n(\boldsymbol{y})\,d\boldsymbol{y} \\
&= \tilde{\mathrm{N}}_n \int_{\tilde{D}_n}\int_{\tilde{D}_n} \boldsymbol{\Gamma}^{\infty}(\boldsymbol{y}-\boldsymbol{z})\,d\boldsymbol{z}\,d\boldsymbol{y} \\
&= \tilde{c}_n\boldsymbol{S}_n\boldsymbol{L}_0^{-1}
\end{aligned} \tag{5.16}$$

where $\tilde{D}_n$ denotes a single ellipsoidal domain of Component-n. The last equality of (5.16) follows exactly the Eshelby's solution (Eshelby, 1957). Note that the Eshelby tensor $\boldsymbol{S}_n$ in (5.16) is associated with not only the aspect ratio but also the particular orientation of Component-n.

By substituting (5.14) and (5.16) into (5.9), the polarization is solved as

$$\boldsymbol{p}_n = (\boldsymbol{L}_n - \boldsymbol{L}_0)\boldsymbol{A}_n \left(\boldsymbol{I} - \sum_{m=1}^{\tilde{N}-1} \tilde{c}_m \boldsymbol{S}_{nm} \left(\boldsymbol{L}_0^{-1} \boldsymbol{L}_m - \boldsymbol{I} \right) \boldsymbol{A}_m \right)^{-1} \boldsymbol{\varepsilon}_0 \tag{5.17}$$

with the concentration tensor

$$\boxed{\boldsymbol{A}_n = \left(\boldsymbol{I} + \boldsymbol{S}_n \left(\boldsymbol{L}_0^{-1} \boldsymbol{L}_n - \boldsymbol{I} \right) \right)^{-1}} \tag{5.18}$$

Define an "average" Eshelby tensor $\tilde{\boldsymbol{S}}_n$ via the following equivalence:

$$\tilde{\boldsymbol{S}}_n \sum_{m=1}^{\tilde{N}-1} \tilde{c}_m \left(\boldsymbol{L}_0^{-1} \boldsymbol{L}_m - \boldsymbol{I} \right) \boldsymbol{A}_m = \sum_{m=1}^{\tilde{N}-1} \tilde{c}_m \boldsymbol{S}_{nm} \left(\boldsymbol{L}_0^{-1} \boldsymbol{L}_m - \boldsymbol{I} \right) \boldsymbol{A}_m \tag{5.19}$$

Subsequently, (5.17) becomes

$$\boldsymbol{p}_n = (\boldsymbol{L}_n - \boldsymbol{L}_0)\boldsymbol{A}_n \left(\boldsymbol{I} - \tilde{\boldsymbol{S}}_n \sum_{m=1}^{\tilde{N}-1} \tilde{c}_m \left(\boldsymbol{L}_0^{-1} \boldsymbol{L}_m - \boldsymbol{I} \right) \boldsymbol{A}_m \right)^{-1} \boldsymbol{\varepsilon}_0 \tag{5.20}$$

By substituting (5.20) into the HS functional (3.47), the ellipsoidal bound is finally derived as

$$\boldsymbol{L}^{EB} = \boldsymbol{L}_0 + \sum_{n=1}^{\tilde{N}-1} \tilde{c}_n \, \boldsymbol{L}_n^A \left(\boldsymbol{I} - \tilde{\boldsymbol{S}}_n \boldsymbol{L}_0^{-1} \sum_{m=1}^{\tilde{N}-1} \tilde{c}_m \, \boldsymbol{L}_m^A \right)^{-1} \tag{5.21}$$

with

$$\boxed{\boldsymbol{L}_m^A = (\boldsymbol{L}_m - \boldsymbol{L}_0)\boldsymbol{A}_m} \tag{5.22}$$

Note that, by grouping all the inclusions of a common orientation into one component, the original, expression of the ellipsoidal bound (see Xu, 2012b) is simplified into (5.21). Compared with (5.17), the expression of the stress polarization, the term (5.22) is thus named as *polarization moduli*.

Note that the polarization moduli can have negative diagonal entries. The ellipsoidal bound (5.21) serves as the upper or lower bound when $\boldsymbol{L}_n - \boldsymbol{L}_0$ is negative or positive semidefinite for any $n = 1,2, \cdots, \tilde{N} - 1$.

With regard to a conduction problem, the ellipsoidal bound is expressed as

$$K^{EB}\boldsymbol{I} = K_0\boldsymbol{I} + \sum_{n=1}^{\tilde{N}-1} \tilde{c}_n \boldsymbol{K}_n^A \left(\boldsymbol{I} - \frac{1}{K_0} \tilde{\boldsymbol{S}}_n \sum_{m=1}^{\tilde{N}-1} \tilde{c}_m \boldsymbol{K}_m^A \right)^{-1} \tag{5.23}$$

with

$$\boldsymbol{K}_m^A = (K_m - K_0)\boldsymbol{A}_m \tag{5.24}$$

$$\boldsymbol{A}_m = \left(\boldsymbol{I} + \boldsymbol{S}_m \left(\frac{K_m}{K_0} - 1 \right) \right)^{-1} \tag{5.25}$$

The average Eshelby tensor $\tilde{\boldsymbol{S}}_n$ in (5.23) is defined via the following equivalence:

$$\tilde{\boldsymbol{S}}_n \sum_{m=1}^{\tilde{N}-1} \tilde{c}_m \left(\frac{K_m}{K_0} - 1 \right) \boldsymbol{A}_m = \sum_{m=1}^{N-1} \tilde{c}_m \boldsymbol{S}_{nm} \left(\frac{K_m}{K_0} - 1 \right) \boldsymbol{A}_m \tag{5.26}$$

5.2 ASYMPTOTIC RESULTS OF THE ELLIPSOIDAL BOUND

Given the probability distribution of the inclusion orientations in an RVE all the terms in the ellipsoidal bound (5.21) become explicit, except for the average Eshelby tensor $\tilde{\boldsymbol{S}}_n$. Hereby we define two morphological patterns that lead to an explicit expression of $\tilde{\boldsymbol{S}}_n$. The first morphological pattern is called *isotropic mixture*, resulting in $\tilde{\boldsymbol{S}}_n = \boldsymbol{S}_0$, the spherical Eshelby tensor. The second pattern is called *anisotropic mixture*, in which case the average Eshelby tensor corresponds to an ellipsoidal Eshelby tensor. Based on these two underlying morphological patterns, the ellipsoidal bound is analytically estimated for the effective elastic moduli. The ellipsoidal bound on conductivity is discussed together with percolation threshold in Chapter 6.

5.2.1 Morphological pattern of isotropic mixture

There are two morphological examples proved to be of isotropic mixture. The first one is Hashin-type self-similar sphere assemblage introduced in Ponte Castañeda and Willis (1995). As shown in Figure 5.2a, all the inclusions sharing a common aspect ratio are randomly oriented and distributed with spherical symmetry similar to the Hashin coated sphere assemblage (recall Figure 4.1). The diameter of a sphere is equal to the longest axis of the inclusion within. Consequently, the random orientation of an inclusion is independent from both the center and the orientation of any other inclusion. According to the spherical symmetry, the interpoint autocorrelation $c_{h,mn}(\boldsymbol{y} - \boldsymbol{y}')$ is isotropic, and there is a spherical exclusion zone specifying the minimum distance between the centers of any two inclusions in which $c_h = 0$. Similar to (5.6), the interpoint autocorrelation is given in terms of the correlation coefficient as

$$\tilde{c}_{h,nm}(\boldsymbol{x}) = -\overline{h_n}\,\overline{h_m}\rho_{h,nm}(\boldsymbol{x}) + \overline{h_n}\,\overline{h_m} \qquad n \neq m \tag{5.27}$$

which, by taking (5.3) into account, reduces to

$$\tilde{c}_{h,mn}(\boldsymbol{x}) = \tilde{\mathrm{N}}_n\tilde{\mathrm{N}}_m\left(1 - \tilde{\rho}_{h,mn}(\boldsymbol{x})\right) \qquad n \neq m \tag{5.28}$$

By applying (5.7) and (5.27–5.28), Equation (5.14) is rewritten as

$$\begin{aligned}
&\int_{\mathrm{Y}} \boldsymbol{\Gamma}^{\infty}(\boldsymbol{x} - \boldsymbol{x}')\tilde{c}_{nm}(\boldsymbol{x} - \boldsymbol{x}')\,d\boldsymbol{x}' \\
&= \int_{\mathrm{Y}} \boldsymbol{\Gamma}^{\infty}(\boldsymbol{x} - \boldsymbol{x}')\,d\boldsymbol{x}' \int_{\mathrm{Y}}\int_{\mathrm{Y}} \tilde{\chi}_n(\boldsymbol{x} - \boldsymbol{y})\tilde{\chi}_m(\boldsymbol{x}' - \boldsymbol{y}')\,\tilde{c}_{h,,nm}(\boldsymbol{x} - \boldsymbol{x}')\,d\boldsymbol{y}\,d\boldsymbol{y}' \\
&= \tilde{\mathrm{N}}_n\tilde{\mathrm{N}}_m \int_{\mathrm{Y}} \boldsymbol{\Gamma}^{\infty}(\boldsymbol{x} - \boldsymbol{x}')\,d\boldsymbol{x}' \int_{\mathrm{Y}}\int_{\mathrm{Y}} \tilde{\chi}_n(\boldsymbol{x} - \boldsymbol{y})\tilde{\chi}_m(\boldsymbol{x}' - \boldsymbol{y}')\left(1 - \tilde{\rho}_{h,,nm}(\boldsymbol{x} - \boldsymbol{x}')\right)d\boldsymbol{y}\,d\boldsymbol{y}'
\end{aligned} \tag{5.29}$$

Let $\boldsymbol{z} = \boldsymbol{x} - \boldsymbol{y}$, $\boldsymbol{z}' = \boldsymbol{x}' - \boldsymbol{y}'$, and $\boldsymbol{z}'' = \boldsymbol{x} - \boldsymbol{x}'$. By employing Property (4.14) or (5.11) and Property (4.20), (5.29) is simplified into

$$\begin{aligned}
&\tilde{\mathrm{N}}_n\tilde{\mathrm{N}}_m \int_{\mathrm{Y}}\int_{\mathrm{Y}} \tilde{\chi}_n(\boldsymbol{z})\tilde{\chi}_m(\boldsymbol{z}')\,d\boldsymbol{z}\,d\boldsymbol{z}' \int_{\mathrm{Y}} \boldsymbol{\Gamma}^{\infty}(\boldsymbol{z}'')\left(1 - \tilde{\rho}_{h,,nm}(\boldsymbol{z}'')\right)d\boldsymbol{z}'' \\
&= -\tilde{c}_n\tilde{c}_m\mathbf{S}_0\boldsymbol{L}_0^{-1}
\end{aligned} \tag{5.30}$$

when $n \neq m$.

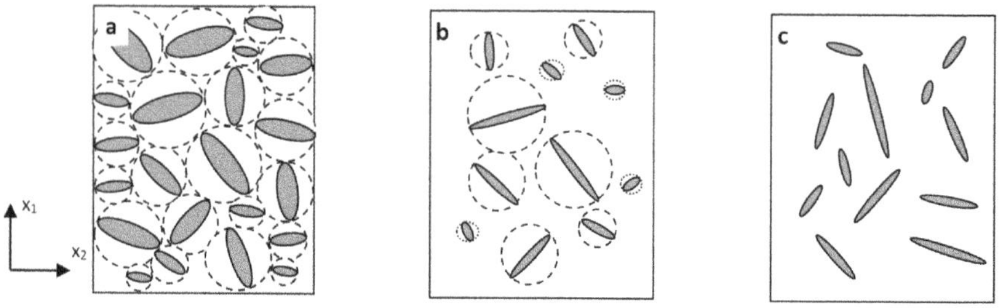

Figure 5.2 Morphological examples of isotropic mixture (a – Hashin-type self-similar sphere assemblage; b – spherical exclusion; c – complete dispersion). (From Xu, X.F. and G. Stefanou: Explicit bounds on elastic moduli of solids containing isotropic mixture of cracks and voids. *Fatigue Fract. Eng Mater. Struct.* 2012a. 35. 708–717. Copyright Wiley-VCH Verlag GmbH & Co. KGaA. Reproduced with permission. Permission conveyed through Copyright Clearance Center, Inc.)

The second morphological example leading to (5.30) is called the spherical exclusion model shown in Figure 5.2b. As a generalization of the Hashin-type self-similar sphere assemblage, the spherical exclusion model similarly specifies a spherical exclusion zone with the diameter equal to the longest axis of the ellipsoid within. All the inclusions are randomly located and randomly oriented, subjected to the condition that no spherical exclusion zones are allowed to overlap with each other. The difference from the Hashin-type self-similar sphere assemblage is that in the spherical exclusion model the RVE domain does not need to be completely filled with a self-similar assemblage. With such an arrangement, the previously demonstrated equations (5.27–5.30) are equally established. It should be noted that, in continuum percolation theory, there is a concept of *excluded volume* defined as "the volume around an object into which the center of another similar object is not allowed to enter if overlapping of the two objects is to be avoided" (Balberg, Binenbaum, & Wagner, 1984), which is different from the spherical exclusion described here.

The condition (5.30), achieved via the special arrangement of an exclusion zone surrounding each individual inclusion, is just one way to satisfy (5.19) and thereby to realize the pattern of isotropic mixture with $\tilde{\boldsymbol{S}}_n = \boldsymbol{S}_0$. Due to the existence of such an exclusion zone, it is clear that both the self-similar sphere assemblage and the spherical exclusion morphology are subjected to an upper limit on the volume fraction of inclusions. A so-called complete dispersion morphology (Figure 5.2c) is further defined that satisfies $\tilde{\boldsymbol{S}}_n = \boldsymbol{S}_0$ but is free of the upper limit on the volume fraction, which is of most interest to percolation (discussed in Chapter 6).

With the result (5.30) obtained, according to (5.19) the average Eshelby tensor $\tilde{\boldsymbol{S}}_n$ is found to be exactly $\boldsymbol{S}_0$, and thereby the ellipsoidal bound (5.21) reduces to

$$\boldsymbol{L}^{EB} = \boldsymbol{L}_0 + \sum_{n=1}^{\tilde{N}-1} \tilde{c}_n \, \boldsymbol{L}_n^A \left(\boldsymbol{I} - \boldsymbol{S}_0 \boldsymbol{L}_0^{-1} \sum_{m=1}^{\tilde{N}-1} \tilde{c}_m \, \boldsymbol{L}_m^A \right)^{-1} \tag{5.31}$$

with the polarization moduli $\boldsymbol{L}_n^A$ given in (5.22) and (5.18).

In fact, the definition of a component given at the beginning of this chapter can be effectively extended to cover all different sizes of ellipsoids sharing identical elastic moduli, a common aspect ratio, and a common orientation, while all the results obtained above remain unchanged. In a multiphase composite, when all the inclusions in a phase not only share identical elastic moduli and a common aspect ratio but also follow an isotropic distribution of random orientations, we call it specifically a *multiphase isotropic composite*. In such an N-phase isotropic composite, the component-based sum $\sum_{m=1}^{\tilde{N}-1} \tilde{c}_m \, \boldsymbol{L}_m^A$ in (5.21) can be regrouped into the following phase-based sum:

$$\sum_{m=1}^{\tilde{N}-1} \tilde{c}_m \, \boldsymbol{L}_m^A = \sum_{n=1}^{N-1} c_n \, \overline{\boldsymbol{L}_n^A} \tag{5.32}$$

where $\overline{\boldsymbol{L}_n^A}$ denotes orientational averaging of the *polarization moduli* in Phase-n, and c_n the volume fraction of Phase-n. Accordingly, (5.31) is simplified into the ellipsoidal bound of an N-phase isotropic composite as

$$\boxed{\boldsymbol{L}^{EB} = \boldsymbol{L}_0 + \sum_{n=1}^{N-1} c_n \, \overline{\boldsymbol{L}_n^A} \left(\boldsymbol{I} - \boldsymbol{S}_0 \boldsymbol{L}_0^{-1} \sum_{m=1}^{N-1} c_m \, \overline{\boldsymbol{L}_m^A} \right)^{-1}} \tag{5.33}$$

with the two-phase version

$$\boxed{\boldsymbol{L}^{EB} = \boldsymbol{L}_0 + \left(\frac{1}{c_1} \, \overline{\boldsymbol{L}_1^A}^{-1} - \boldsymbol{S}_0 \boldsymbol{L}_0^{-1} \right)^{-1}} \tag{5.34}$$

Given the positive or negative semidefiniteness of $\boldsymbol{L} - \boldsymbol{L}_0$, the ellipsoidal bound (5.33) rigorously applies to a morphology of isotropic mixture, which however is generally not attainable except for the spherical HS bounds. When the elastic moduli of all the N phases are isotropic, the ellipsoidal bound (5.33) is explicitly expressed for the effective bulk and shear moduli, respectively, as (Xu, 2012b)

$$\kappa^{EB} = \kappa_0 + \frac{\sum_{n=1}^{N-1} c_n \overline{\kappa_n^A}}{1 - \frac{1}{\frac{4}{3}\mu_0 + \kappa_0} \sum_{n=1}^{N-1} c_n \overline{\kappa_n^A}} \tag{5.35a}$$

$$\mu^{EB} = \mu_0 + \frac{\sum_{n=1}^{N-1} c_n \overline{\mu_n^A}}{1 - \frac{6(\kappa_0 + 2\mu_0)}{5\mu_0(3\kappa_0 + 4\mu_0)} \sum_{n=1}^{N-1} c_n \overline{\mu_n^A}} \tag{5.35b}$$

where the polarization moduli $\overline{\boldsymbol{L}_n^A} \sim \left(3\overline{\kappa_n^A}, 2\overline{\mu_n^A}\right)$. When specialized to a two-phase isotropic composite, (5.35a & b) reduce, respectively, to

$$\kappa^{EB} = \kappa_0 + \left(\frac{1}{c_1\overline{\kappa_1^A}} - \frac{1}{\frac{4}{3}\mu_0 + \kappa_0}\right)^{-1} \tag{5.36a}$$

$$\mu^{EB} = \mu_0 + \left(\frac{1}{c_1\overline{\mu_1^A}} - \frac{6(\kappa_0 + 2\mu_0)}{5\mu_0(3\kappa_0 + 4\mu_0)}\right)^{-1} \tag{5.36b}$$

In case of spheroidal inclusions that are characterized with the isotropic elastic moduli, according to (5.18), the polarization moduli

$$\begin{aligned} \boldsymbol{L}_n^A &= (\boldsymbol{L}_n - \boldsymbol{L}_0)\boldsymbol{A}_n \\ &= \left((\boldsymbol{L}_n - \boldsymbol{L}_0)^{-1} + \boldsymbol{S}_n \boldsymbol{L}_0^{-1}\right)^{-1} \end{aligned}$$

is transversely isotropic with respect to the ellipsoidal axes. By applying the rotation matrix with the three-dimensional Euler angles, it is found that the average polarization moduli are isotropic with the bulk and shear moduli given as follows:

$$\overline{\kappa_n^A} = \frac{1}{9}\boldsymbol{L}_{n,iijj}^A \qquad \overline{\mu_n^A} = \frac{1}{10}\left(\boldsymbol{L}_{n,ijij}^A - \frac{1}{3}\boldsymbol{L}_{n,iijj}^A\right) \tag{5.37}$$

With the ellipsoidal Eshelby tensor S analytically available (see Appendix III), the two moduli in (5.37) are explicit in terms of the aspect ratio of constituent inclusions, though in a long expression. The ellipsoidal bound (5.33) is calculated conveniently by using symbolic software such as *Mathematic, Maple*, etc. In the following three subsections, the ellipsoidal bound (5.33) is applied to three limiting cases, i.e. spheres, rods, and platelets, among which the case of spheres as a verification recovers the multiphase HS bounds derived in Chapter 4.

As a closing remark of this subsection, it is emphasized that the ellipsoidal bound (5.33) remains valid until the condition of isotropic mixture breaks down upon occurrence of percolation.

5.2.2 Randomly dispersed spheres

Suppose the inclusions in a composite are all spherical, i.e. $\boldsymbol{S}_n = \boldsymbol{S}_0$, and by applying (5.37) it follows that

$$\begin{aligned}\overline{\kappa_n^A} &= \kappa_n^A \\ &= \left((\kappa_n - \kappa_0)^{-1} + \left(\frac{4}{3}\mu_0 + \kappa_0 \right)^{-1} \right)^{-1}\end{aligned} \tag{5.38}$$

$$\begin{aligned}\overline{\mu_n^A} &= \mu_n^A \\ &= \left((\mu_n - \mu_0)^{-1} + \frac{6(\kappa_0 + 2\mu_0)}{5\mu_0(3\kappa_0 + 4\mu_0)} \right)^{-1}\end{aligned} \tag{5.39}$$

Substitution of (5.38–5.39) into the bounds (5.35a & b) yields the spherical bounds for a composite containing multiphase spheres,

$$\kappa^{EB} = \kappa_0 + \cfrac{1}{\cfrac{1}{\displaystyle\sum_{n=1}^{N-1} \cfrac{c_n}{\cfrac{1}{\kappa_n - \kappa_0} + \cfrac{1}{\frac{4}{3}\mu_0 + \kappa_0}}} - \cfrac{1}{\frac{4}{3}\mu_0 + \kappa_0}} \tag{5.40}$$

$$\mu^{EB} = \mu_0 + \cfrac{1}{\cfrac{1}{\sum_{n=1}^{N-1} \cfrac{c_n}{\cfrac{1}{\mu_n - \mu_0} + \cfrac{6(\kappa_0 + 2\mu_0)}{5\mu_0(3\kappa_0 + 4\mu_0)}}} - \cfrac{6(\kappa_0 + 2\mu_0)}{5\mu_0(3\kappa_0 + 4\mu_0)}} \tag{5.41}$$

which, by noting $\sum_{n=1}^{N} c_n = 1$, $\kappa_0 = \kappa_N$ and $\mu_0 = \mu_N$, are in fact equivalent to the multiphase HS bounds (4.31–4.32) in terms of the reference moduli.

5.2.3 Randomly dispersed needle-like rods

Denote a_1, a_2, and a_3 the three semi-axes of a spheroid, and define the aspect ratio η via $a_3 = \eta a_1 = \eta a_2$, with prolate and oblate spheroids corresponding to $\eta > 1$ and $\eta < 1$, respectively. A needle-like rod is modelled in the limiting case with the aspect ratio $\eta \to \infty$. By substituting (A3.3) the needle-like Eshelby tensor into (5.37), the average polarization moduli in the ellipsoidal bound (5.36) are asymptotically obtained as (Xu, 2012b)

$$\boxed{\lim_{\eta \to \infty} \overline{\kappa_1^A} = \frac{(\kappa_1 - \kappa_0)(3\kappa_1 + 3\mu_0 + \mu_1)}{(3\kappa_0 + 3\mu_0 + \mu_1)}} \tag{5.42}$$

$$\boxed{\lim_{\eta \to \infty} \overline{\mu_1^A} = \frac{(\mu_1 - \mu_0)\left[3\kappa_0(\mu_0 + \mu_1)\left(3\kappa_1(9\mu_0 + \mu_1) + 4\mu_0(7\mu_0 + 3\mu_1)\right) + \mu_0\left(3\kappa_1(7\mu_0 + \mu_1)(3\mu_0 + 7\mu_1) + 8\mu_0\left(8\mu_0^2 + 23\mu_0\mu_1 + 9\mu_1^2\right)\right)\right]}{5(\mu_0 + \mu_1)\left(3(\kappa_1 + \mu_0) + \mu_1\right)\left(3\kappa_0(\mu_0 + \mu_1) + \mu_0(\mu_0 + 7\mu_1)\right)}} \tag{5.43}$$

Define the density of the needle-like rods as

$$\alpha = \frac{4}{3}\pi\, N_f \overline{a_3^3} \tag{5.44}$$

with N_f the number of the fillers in the RVE, and a_3 a random variable representing the half length of a rod. The volume fraction of the rods is expressed in terms of the density as

$$c_1 = \frac{\alpha}{\eta^2} \tag{5.45}$$

Note that the definition (5.44) is physically meaningful in that the density α represents the volume fraction of a_3-sized spheres distributed as the self-similar sphere assemblage (Figure 5.2a). When either the self-similar sphere assemblage or the spherical exclusion model is applied, the maximum of the density α is 1, and thereby the upper bound of the volume fraction c_1 equals $\frac{1}{\eta^2}$. Such a bound clearly is below the value of percolation threshold, visually self-explanatory in Figures 5.2a and 5.2b. With the extension made by the complete dispersion (Figure 5.2c), the ellipsoidal bounds (5.36) with (5.42–5.43) can be applied to a situation when the volume fraction is beyond $\frac{1}{\eta^2}$ prior to occurrence of percolation.

The bounds (5.36) with (5.42–5.43) are depicted in Figures 5.3 and 5.4, respectively, to illustrate the effect of the Poisson's ratio on the effective

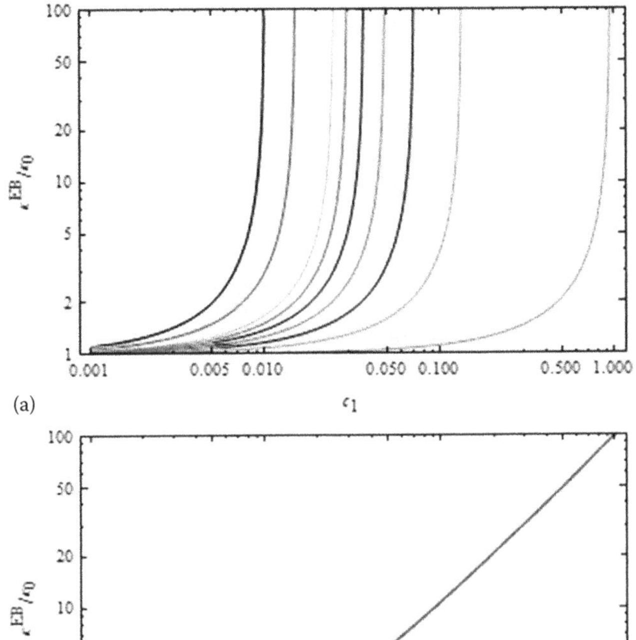

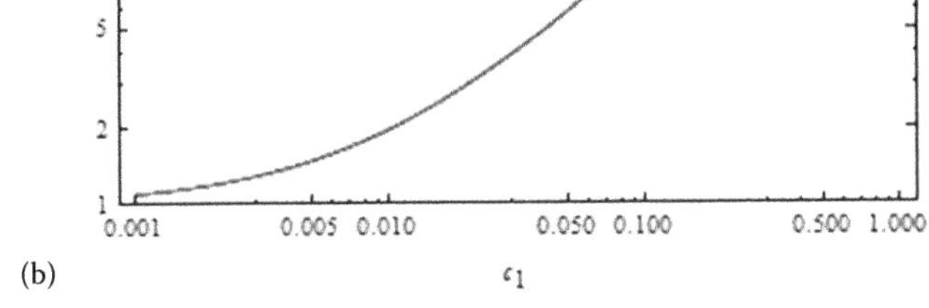

Figure 5.3 Lower bound of bulk modulus vs. volume fraction of needle-like fillers with $\kappa_1 = 100\kappa_0$; (a) $\nu_0 = 0.499$; (b) $\nu_0 = -0.999$; with $\nu_1 = -0.999, -0.5, -0.1, 0, 0.1, 0.2, 0.3, 0.4, 0.499$ for the curves from left to right; *(Continued)*

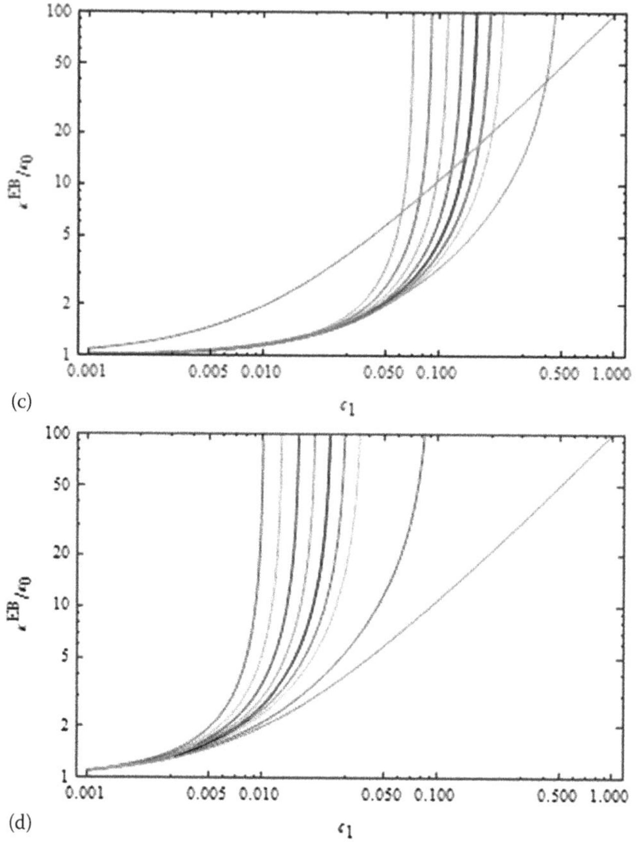

Figure 5.3 (Continued) Lower bound of the effective bulk modulus vs. volume fraction of needle-like fillers with $\kappa_I = 100\kappa_0$; (c) $\nu_I = 0.3$; (d) $\nu_I = -0.999$; with $\nu_0 = -0.999$, -0.5, -0.1, 0, 0.1, 0.2, 0.3, 0.4, 0.499 for the curves from right to left.

bulk and shear moduli of a needle-reinforced composite. It should be noted that in the figures the right ends of the curves are invalid beyond the occurrence of percolation (e.g. those parts of the curves clearly violate the classical bounds such as the HS bounds).

With the ratio fixed between the two bulk moduli of two phases, given the Poisson's ratio of the matrix is rubber-like close to 0.5, Figure 5.3a shows the effective bulk modulus increases with decrease of the Poisson's ratio of the needles, while in Figure 5.3b this effect becomes negligible when the matrix is extreme auxetic with the Poisson's ratio approaching –1. With the Poisson's ratio of the needles fixed, Figures 5.3c and 5.3d indicate that the effective bulk modulus increases with rise of the Poisson's ratio of the matrix. It is concluded that, with respect to the reinforcing effect on

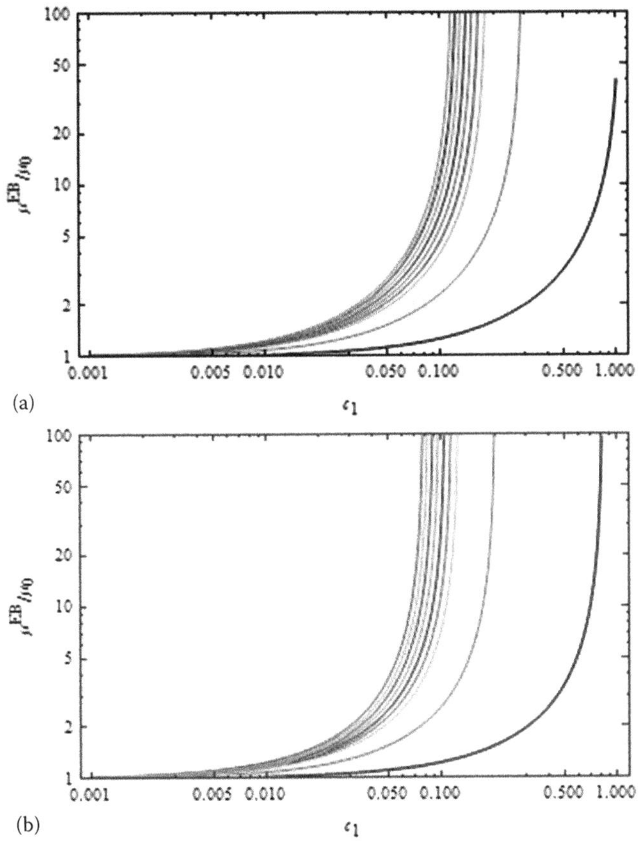

Figure 5.4 Lower bound of the effective shear modulus vs. volume fraction of needle-like fillers with $\mu_1 = 100\mu_0$; (a) $\nu_0 = 0.499$; (b) $\nu_0 = -0.999$; with $\nu_1 = -0.999, -0.5, -0.1, 0, 0.1, 0.2, 0.3, 0.4, 0.499$ for the curves from right to left; *(Continued)*

the bulk modulus, optimum design of a needle-reinforced composite is to have a rubber-type matrix reinforced with extreme auxetic needles, while the opposite is to have an extreme auxetic matrix reinforced with rubber-like needles.

The effect of the Poisson's ratio is shown in Figure 5.4 on the shear modulus, which is opposite to Figure 5.3 on the bulk modulus. With respect to the reinforcing effect on the effective shear modulus, optimum design is to have an extreme auxetic matrix reinforced with rubber-like needles, and the worst is a rubber-like matrix reinforced with extreme auxetic needles.

To verify the asymptotic results (5.42–5.43), the bounds are applied to a two-phase nanocomposite containing carbon nanotubes (CNTs). Assume the elastic moduli of the matrix and CNTs are both isotropic, with the

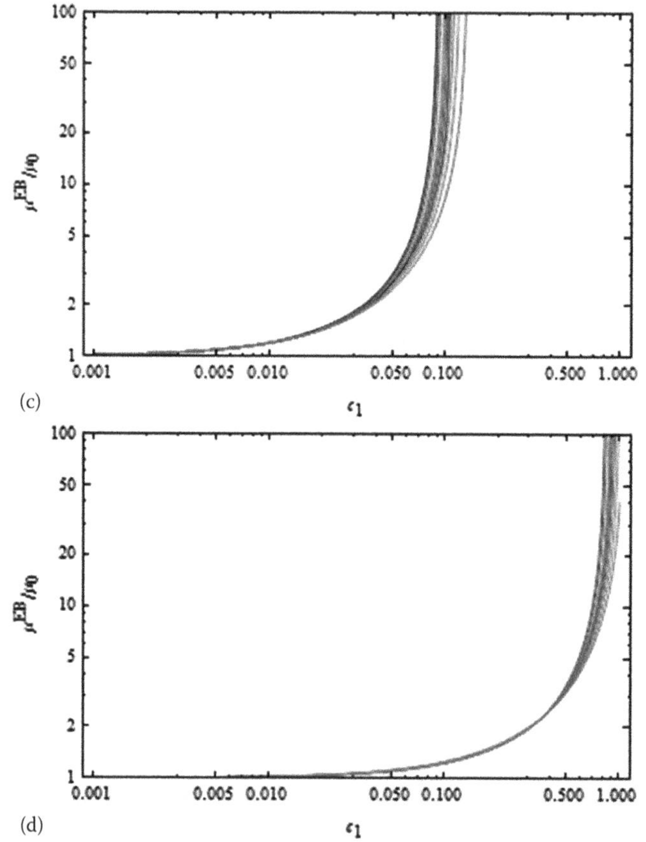

Figure 5.4 (Continued) Lower bound of the effective shear modulus vs. volume fraction of needle-like fillers with $\mu_1 = 100\mu_0$; (c) $\nu_1 = 0.3$; (d) $\nu_1 = -0.999$ and $\nu_0 = -0.999$, -0.5, -0.1, 0, 0.1, 0.2, 0.3, 0.4, 0.499 for the curves from left to right.

Poisson's ratios $\upsilon_0 = \upsilon_1 = 0.3$ and the Young's moduli $E_1 = 200\ E_0$. The aspect ratio of CNTs is given as $\eta = 100$, which is exactly used in the numerical result of (5.37), while in the bounds (5.42–5.43) the aspect ratio is asymptotically taken to be infinitely large. Figure 5.5 shows that the effective Young's modulus predicted by the asymptotic bounds is very close to that of the numerical result. The HS lower bound is also shown in the figure to indicate one order of magnitude improved by the ellipsoidal bound with respect to the percolation threshold, i.e. the volume fraction of CNTs when the effective Young's modulus fast approaches the Young's modulus of CNTs (Xu, 2012b).

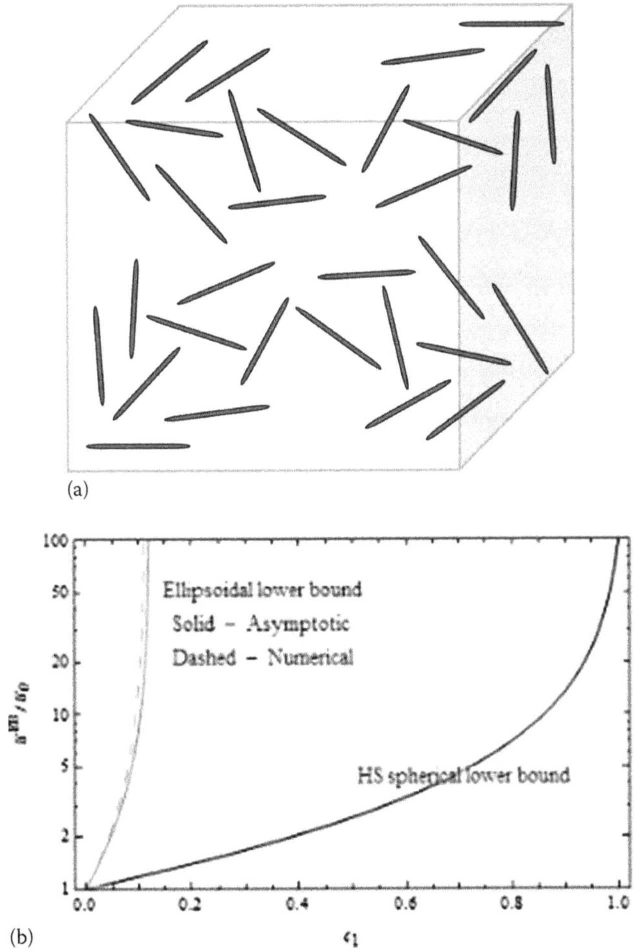

Figure 5.5 (a) schematic of an RVE for a carbon nanotube reinforced composite; (b) Comparison of the effective Young's moduli predicted by the ellipsoidal lower bound, the HS spherical lower bound, and numerical analysis.

5.2.4 Randomly dispersed disk-like platelets

A disk-like platelet is modelled in the limiting case with the aspect ratio $\eta \to 0$. By taking the limit $\eta \to 0$ in the Eshelby tensor (A3.4) and substituting the tensor into (5.37), it results in the following asymptotic results for the average polarization moduli in the bounds (5.36) (Xu, 2012b):

$$\boxed{\overline{\kappa_1^A} = \frac{(\kappa_1 - \kappa_0)(3\kappa_0 + 4\mu_1)}{3\kappa_1 + 4\mu_1}} \tag{5.46}$$

$$\boxed{\overline{\mu_1^A} = \frac{(\mu_1 - \mu_0)\left(6\kappa_1\mu_0 + 9\kappa_1\mu_1 + 12\mu_0\mu_1 + 8\mu_1^2\right)}{5\mu_1(3\kappa_1 + 4\mu_1)}} \tag{5.47}$$

Similar to (5.44), define the density of disk-like platelets as

$$\alpha = \frac{4}{3}\pi\, N_f \overline{a_1^3} \tag{5.48}$$

which gives the volume fraction of the platelets in terms of the density as

$$c_1 = \eta\alpha \tag{5.49}$$

When the volume fraction of the disk-like platelets approaches zero, the ellipsoidal bounds (5.36) with (5.46–5.47) become trivial, degenerating to the elastic moduli of the matrix phase. There are two non-trivial cases when the contrast ratio between the moduli of the inclusions and those of the matrix tends to infinite and zero, respectively. Denote that

$$\gamma_{\max} = \max\left(\frac{\mu_1}{\mu_0}, \frac{\kappa_1}{\kappa_0}\right), \qquad \gamma_{\min} = \min\left(\frac{\mu_1}{\mu_0}, \frac{\kappa_1}{\kappa_0}\right) \tag{5.50}$$

When the contrast ratio $\gamma_{\min}$ is sufficiently large beyond the order $O(1/\eta)$, from (5.37) we obtain the following double limits:

$$\lim_{\eta\to 0} \lim_{\substack{\gamma_{\min}\to\eta^q \\ q<-1}} \eta\overline{\kappa_1^A} = \frac{16\kappa_0(1-2\nu_0)(1-\nu_0)}{3\pi(1+\nu_0)(3-4\nu_0)} \tag{5.51}$$

$$\lim_{\eta\to 0} \lim_{\substack{\gamma_{\min}\to\eta^q \\ q<-1}} \eta\overline{\mu_1^A} = \frac{8\mu_0(1-\nu_0)(43-56\nu_0)}{15\pi(3-4\nu_0)(7-8\nu_0)} \tag{5.52}$$

Substitution of (5.51–5.52) and (5.49) into (5.36) yields the following asymptotic bounds for a high-contrast composite containing a small volume fraction of reinforcing platelets (Xu, 2012b):

$$\kappa^{EB}/\kappa_0 = 1 + \frac{48(1-\nu_0)(1-2\nu_0)\alpha}{(1+\nu_0)\left(9\pi(3-4\nu_0)-16\alpha(1-2\nu_0)\right)}, \quad \alpha < \frac{9\pi(3-4\nu_0)}{16(1-2\nu_0)} \tag{5.53}$$

$$\mu^{EB}/\mu_0 = 1 + \frac{120(1-\nu_0)(43-56\nu_0)\alpha}{225\pi(3-4\nu_0)(7-8\nu_0)-16\alpha(4-5\nu_0)(43-56\nu_0)}, \quad \alpha < \frac{225\pi(3-4\nu_0)(7-8\nu_0)}{16(4-5\nu_0)(43-56\nu_0)} \tag{5.54}$$

where the density α should be below a certain value to ensure the results in (5.53–5.54) is consistently greater than unity. Take a graphene nanoplatelet-reinforced composite as an example, and suppose the aspect ratio η = 0.001, the Poisson's ratios $\nu_0 = \nu_1 = 0.3$, the volume fraction $c_1 = 0.005$. As shown in Figure 5.6, the ellipsoidal bound (5.36) with (5.46) indicated as a thick line becomes invalid in this high-contrast and small-aspect ratio case, which is corrected by the bound (5.53) indicated as a thin line, especially when the contrast ratio is one order or more beyond $O(1/\eta)$.

In the second non-trivial case, let the volume fraction of the disk-like platelets and the contrast ratio both approach zero, and the medium corresponds to a solid containing randomly scattered penny-shaped cracks, which is a major topic of damage and fracture mechanics. Section 5.3 is devoted to the discussion of this topic.

The ellipsoidal bounds of the effective bulk and shear moduli with (5.46–5.47) are depicted in Figures 5.7 and 5.8, respectively, to illustrate the effect of the Poisson's ratio on the effective moduli of a disk-reinforced composite. When the Poisson's ratio of the matrix and the ratio between the two bulk moduli are fixed, Figures 5.7a and 5.7b show that the effective bulk modulus increases with decrease of the Poisson's ratio of the disks. When the Poisson's ratio of the disks is fixed, the effective bulk modulus increases with rise of the Poisson's ratio of the disks, as shown in Figure 5.7c and 5.7d. Regarding the reinforcing effect on the bulk modulus, optimum design is to have a rubber-like matrix reinforced with extreme auxetic disks, with the worst being the opposite, which is identical to the needles case. In Figure 5.8 it is shown that the disk-like platelets are very effective in reinforcing the effective shear modulus, and the effect is not sensitive to any change of the Poisson's ratio from the either phase.

In conclusion, to reinforce the effective bulk modulus of a composite, optimal fillers are either needle-like rods or disk-like platelets, with the Poisson's ratio as auxetic as possible. With respect to the shear reinforcing effect, optimal fillers are disk-like platelets only, with the Poisson's ratio as large or as rubber-like as possible.

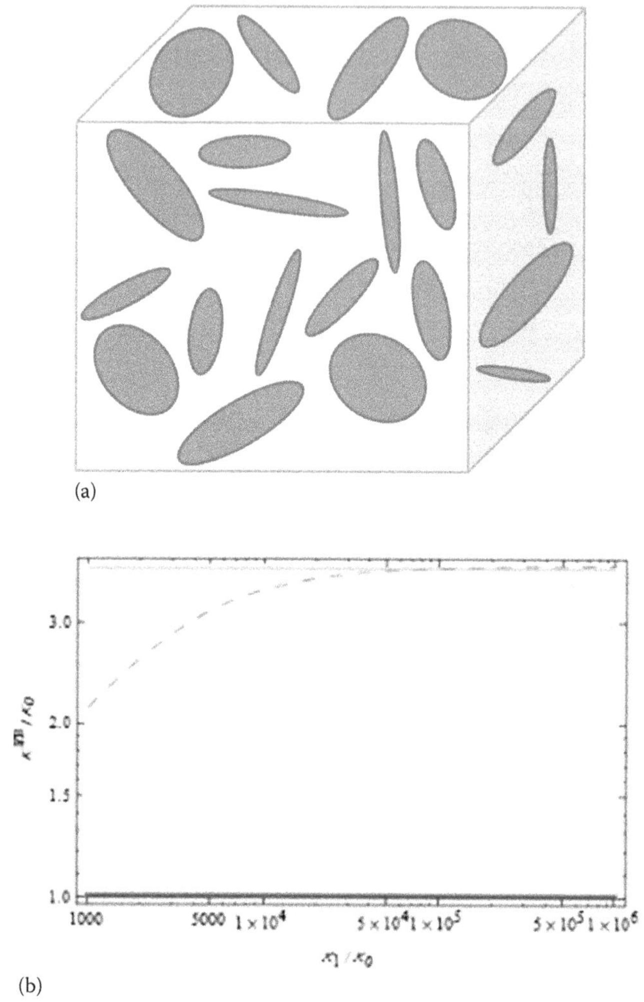

Figure 5.6 (a) Schematic of the RVE for a nanoplatelet-reinforced composite; (b) Comparison of the effective bulk moduli predicted by the numerical result (dashed line) and two asymptotic bounds (thin line – (5.53), thick line – (5.46)), with $\nu_0 = \nu_1 = 0.3$, $\eta = 0.001$, and $c_1 = 0.005$.

5.2.5 Ellipsoidal bound of anisotropic mixture

The pattern of isotropic mixture previously described can be extended to anisotropic mixture, in which case the average Eshelby tensor in (5.19) corresponds to the ellipsoidal Eshelby tensor $\boldsymbol{S}$, with the spherical Eshelby tensor $\mathbf{S}_0$ of isotropic mixture being a special case. Accordingly, the ellipsoidal bound (5.31) is generalized as (Xu, 2012b)

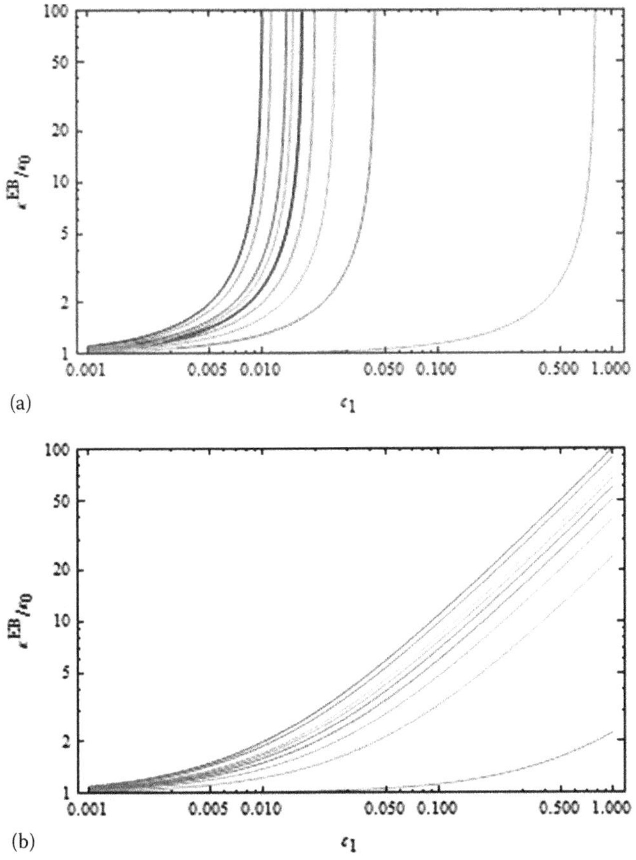

Figure 5.7 Lower bound of the effective bulk modulus vs. volume fraction of disk-like fillers with $\kappa_1 = 100\kappa_0$; (a) $\nu_0 = 0.499$; (b) $\nu_0 = -0.999$; with $\nu_1 = -0.999, -0.5, -0.1, 0, 0.1, 0.2, 0.3, 0.4, 0.499$ for the curves from left to right; *(Continued)*

$$\boxed{\boldsymbol{L}^{EB} = \boldsymbol{L}_0 + \sum_{n=1}^{N-1} c_n \overline{\boldsymbol{L}_n^A}\left(\boldsymbol{I} - \boldsymbol{S}\boldsymbol{L}_0^{-1}\sum_{m=1}^{N-1} c_m \overline{\boldsymbol{L}_m^A}\right)^{-1}} \tag{5.55}$$

The two morphological examples of isotropic mixture—the self-similar sphere assemblage and the spherical exclusion—are extended to anisotropic mixture by changing an outer spherical coating into an outer ellipsoidal coating, as shown in Figures 5.9a and 5.9b. Note that, given the orientation of an inclusion, the size of the inclusion and its outer coating are made compatible with each other, e.g. by letting the longest axis of an inclusion be less than the shortest axis of its outer coating. Similar to those of

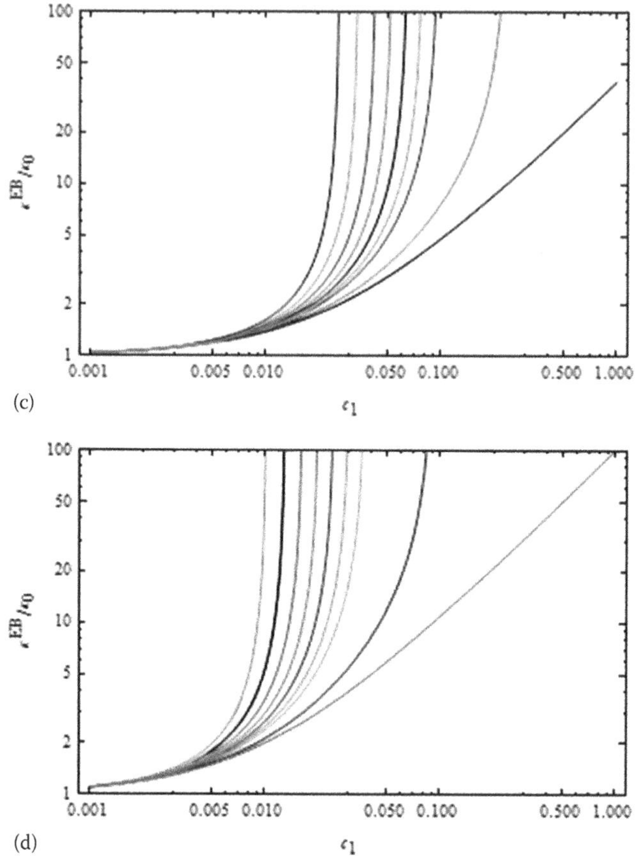

Figure 5.7 (Continued) Lower bound of the effective bulk modulus vs. volume fraction of disk-like fillers with $\kappa_1 = 100\kappa_0$; (c) $\nu_1 = 0.3$; (d) $\nu_1 = -0.999$; with $\nu_0 = -0.999$, -0.5, -0.1, 0, 0.1, 0.2, 0.3, 0.4, 0.499 for the curves from right to left.

the isotropic mixture pattern, the self-similar ellipsoid assemblage and the ellipsoidal exclusion model are subjected to an upper limit on the volume fraction of inclusions. Clustering morphology of anisotropic mixture, as the counterpart of the complete dispersion in isotropic mixture, is free from the restriction of the volume fraction, as shown in Figure 5.9c. The clustering morphology plays an important role in percolation, which is further discussed in Chapter 6.

With the Eshelby tensor $\boldsymbol{S}$ analytically available in Appendix III, and the average polarization moduli $\boldsymbol{L}_n^A$ given in (5.37), the ellipsoidal bound (5.55) of anisotropic mixture is calculated conveniently by using symbolic software such as *Mathmatica*. Examples are provided next on the effective elastic moduli of cracked media, and in Chapter 6 on percolation.

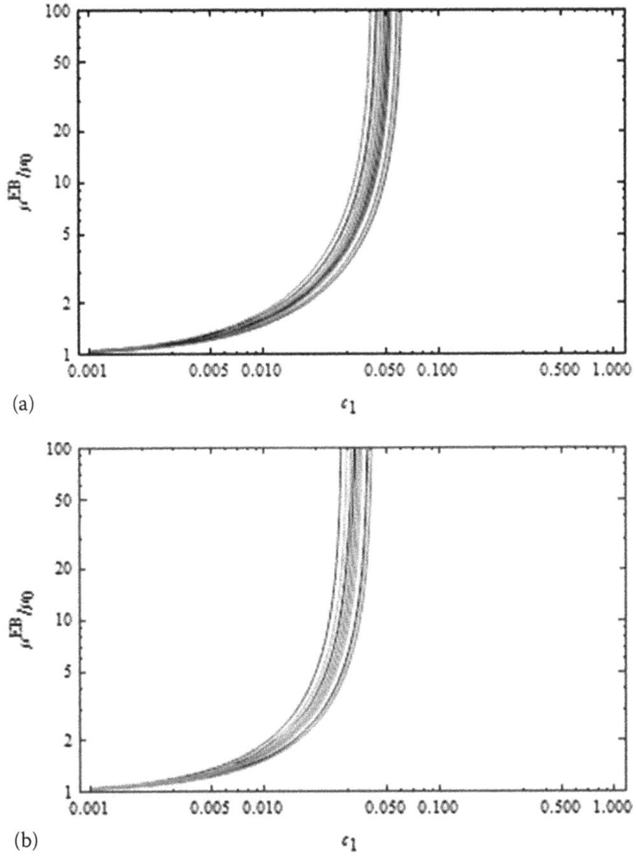

Figure 5.8 Lower bound of the effective shear modulus vs. volume fraction of disk-like fillers with $\mu_1 = 100\mu_0$; (a) $\nu_0 = 0.499$; (b) $\nu_0 = -0.999$; with $\nu_1 = -0.999, -0.5, -0.1, 0, 0.1, 0.2, 0.3, 0.4, 0.499$ for the curves from right to left; *(Continued)*

5.3 EFFECTIVE ELASTIC MODULI OF CRACKED MEDIA

In damage and fracture mechanics, a number of approximation schemes have been proposed to evaluate the effective elastic moduli of a randomly cracked body, e.g. noninteractive scheme (Bristow, 1960), self-consistent scheme (Budiansky & O'Connell, 1976; Horii & Nemat-Nasser, 1985), differential scheme (Zimmerman, 1985, 1991; Hashin, 1988), Mori-Tanaka estimates (Benveniste, 1987; Pan & Weng, 1995), and direct interaction method (Ju & Chen, 1994). Comparatively, only few attempts (Ponte Castañeda & Willis, 1995; Gibiansky & Torquato, 1996; Markov, 1998 Xu & Stefanou, 2012a) have been made on development of variational bounds on randomly cracked solids.

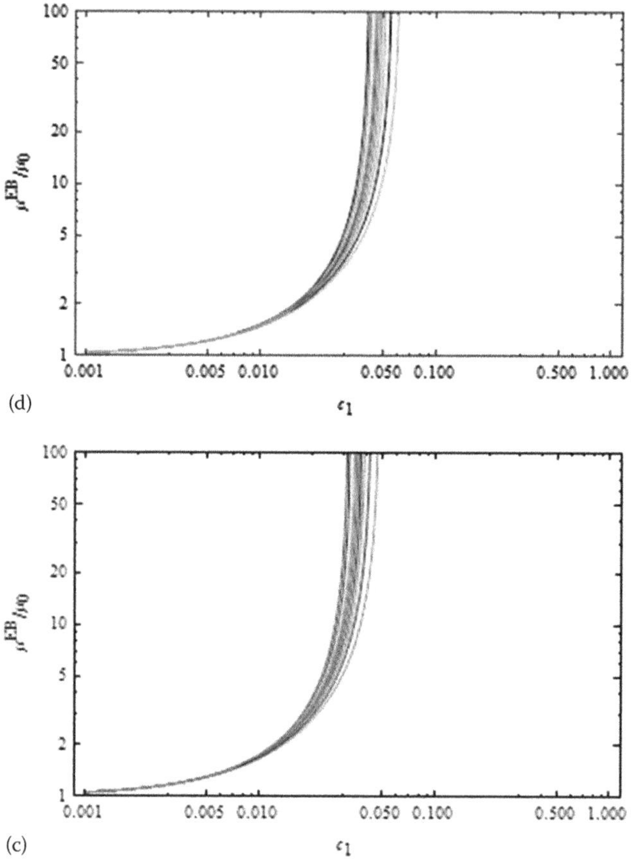

Figure 5.8 (Continued) Lower bound of the effective shear modulus vs. volume fraction of disk-like fillers with $\mu_1 = 100\mu_0$; (c) $\nu_1 = 0.3$; (d) $\nu_1 = -0.999$ and $\nu_0 = -0.999$, -0.5, -0.1, 0, 0.1, 0.2, 0.3, 0.4, 0.499 for the curves from left to right.

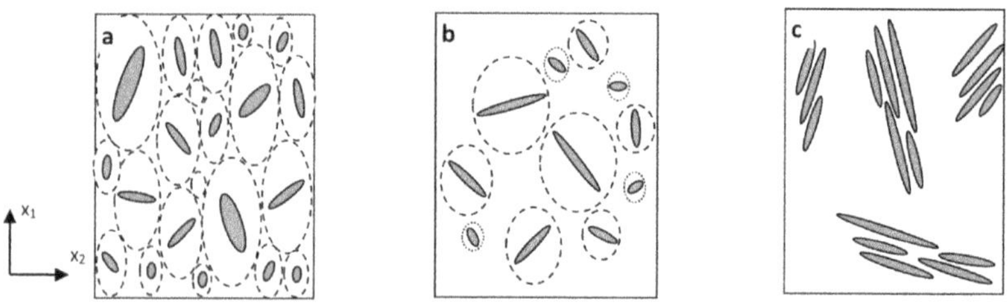

Figure 5.9 Morphological examples of anisotropic mixture (a – self-similar ellipsoid assemblage; b – ellipsoidal exclusion; c – clustering).

As described in Subsection 5.3.3, when the volume fraction of reinforcing platelets approaches zero, the bounds (5.36) with (5.46–5.47) degenerate to the elastic moduli of a hosting matrix, and thus become invalid. This was once a major issue in fracture mechanics when directly applying classical micromechanics theory to cracked media, since the cracks in a solid mechanics model have a zero volume fraction (Kachanov, 1993). As shown in the derivation of the bounds (5.53–5.54), the use of a double limit successfully resolves the singularity issue. In this section, we apply this technique and the Eshelby tensor to derive the effective elastic moduli of cracked media for a variety of crack distributions.

In Subsection 5.3.1, as a verification the Green function approach is first applied to a single crack problem to recover the classical Griffith's energy criterion. In 5.3.2, by assuming no interaction among cracks, the single crack solution is extended to a dilutely cracked solid, yielding an upper bound linear upon the crack density. In Subsections 5.3.3 and 5.3.4, the ellipsoidal bound is applied to a solid containing non-dilute cracks and voids, respectively, yielding an upper bound nonlinear upon the density of cracks or voids.

5.3.1 Recovery of Griffith criterion

A unit volume solid contains a single crack subject to a global strain $\boldsymbol{\varepsilon}_0$. Suppose the crack is sufficiently far away from all the boundaries, and thereby the free space Green function applies. Following exactly the procedure described in Chapter 3, the single crack boundary value problem is decomposed into a reference problem and a fluctuation problem. With the reference moduli identical to $\boldsymbol{L}_0$ the elastic moduli of the solid, by applying the Hashin-Shtrikman functional (3.47) the potential energy of the solid is obtained as

$$\mathcal{H}(\boldsymbol{p}) = \frac{1}{2}\langle \boldsymbol{\varepsilon}_0, \boldsymbol{L}_0\boldsymbol{\varepsilon}_0 \rangle - \frac{1}{2}\langle \boldsymbol{p}, \boldsymbol{\Gamma}^{\infty}\boldsymbol{p} \rangle + \frac{1}{2}\langle \boldsymbol{p}, \boldsymbol{L}_0^{-1}\boldsymbol{p} \rangle + \langle \boldsymbol{\varepsilon}_0, \boldsymbol{p} \rangle \tag{5.56}$$

Construct a constant field for the trial polarization as

$$\boldsymbol{p}(\boldsymbol{x}) = \chi_c(\boldsymbol{x})\boldsymbol{p}_c \tag{5.57}$$

where the indicator function $\chi_c(\boldsymbol{x})$ corresponds to D_c the domain of the crack. By substituting (5.57) into (5.56), the potential energy becomes

$$\mathcal{H}(\boldsymbol{p}_c) = \frac{1}{2}\boldsymbol{\varepsilon}_0\boldsymbol{L}_0\boldsymbol{\varepsilon}_0 - V_c\left(\frac{1}{2}\boldsymbol{p}_c\boldsymbol{S}_c\boldsymbol{L}_0^{-1}\boldsymbol{p}_c - \frac{1}{2}\boldsymbol{p}_c\boldsymbol{L}_0^{-1}\boldsymbol{p}_c - \boldsymbol{\varepsilon}_0\boldsymbol{p}_c\right) \tag{5.58}$$

where V_c denotes the volume of the crack and the Eshelby's solution (5.11) is employed, i.e.

$$\int_{D_c} \boldsymbol{\Gamma}^{\infty}(\boldsymbol{x}' - \boldsymbol{x})\, d\boldsymbol{x} = \boldsymbol{S}_c \boldsymbol{L}_0^{-1} \qquad \boldsymbol{x}' \in D_c \tag{5.59}$$

with $\boldsymbol{S}_c$ denoting the disk-like or slit-like Eshelby tensor (A3.4). To find $\boldsymbol{p}_c$, let the derivative of (5.58) about $\boldsymbol{p}_c$ be zero, i.e.

$$\frac{\partial \mathcal{H}(\boldsymbol{p}_c)}{\partial \boldsymbol{p}_c} = 0 \tag{5.60}$$

and it yields

$$\boldsymbol{p}_c = \boldsymbol{L}^A \boldsymbol{\varepsilon}_0 \tag{5.61}$$

where the polarization moduli

$$\boldsymbol{L}^A = -\boldsymbol{L}_0 (\boldsymbol{I} - \boldsymbol{S}_c)^{-1} \tag{5.62}$$

which is a special case of (5.22) with $N = 2$ and the subscript 1 omitted. Note the Eshelby's solution is exact for the single crack problem, and therefore the HS functional (5.56) is exact for the potential energy, which is finally obtained as

$$\mathcal{H} = \frac{1}{2} \boldsymbol{\varepsilon}_0 \boldsymbol{L}_0 \boldsymbol{\varepsilon}_0 + \frac{V_c}{2} \boldsymbol{\varepsilon}_0 \boldsymbol{L}^A \boldsymbol{\varepsilon}_0 \tag{5.63a}$$

or alternatively

$$\mathcal{H} = \frac{1}{2} \boldsymbol{\sigma}_0 \boldsymbol{L}_0^{-1} \boldsymbol{\sigma}_0 + \frac{V_c}{2} \left(\boldsymbol{L}_0^{-1} \boldsymbol{\sigma}_0\right) \boldsymbol{L}^A \left(\boldsymbol{L}_0^{-1} \boldsymbol{\sigma}_0\right) \tag{5.63b}$$

The second term on the right-hand side of (5.63a) is understood as the strain energy released from the formation of the crack surface. In the following, we derive the Griffith criterion in both the single penny-shaped crack case and the single slit-like crack case.

A Penny-Shaped Crack

A penny-shaped crack with its principal half axes $a_2 = a_3 = a$ has the aspect ratio $\eta = \frac{a_1}{a} \to 0$ and the crack volume $V_c = \frac{4\pi}{3}\eta a^3$. Given the corresponding penny-shaped Eshelby tensor (A3.4), the components of the polarization moduli (5.62) are explicitly obtained as

$$\begin{aligned}
L_{1111}^{A} &= -\frac{8\mu_0(1-\nu_0)^3}{\pi\eta(1-2\nu_0)^2} \\
L_{1122}^{A} = L_{1133}^{A} = L_{2211}^{A} = L_{3311}^{A} &= -\frac{8\mu_0(1-\nu_0)^2\nu_0}{\pi\eta(1-2\nu_0)^2} \\
L_{2222}^{A} = L_{3333}^{A} = L_{3322}^{A} = L_{2233}^{A} &= -\frac{8\mu_0(1-\nu_0)\nu_0^2}{\pi\eta(1-2\nu_0)^2} \\
L_{1212}^{A} = L_{1313}^{A} &= -\frac{4\mu_0(1-\nu_0)}{\pi\eta(2-\nu_0)} \\
L_{2323}^{A} &= -\mu_0
\end{aligned} \tag{5.64a–e}$$

where ν_0 is the Poisson's ratio of the solid.

By substituting (5.64) into (5.63), the potential energy functions $\mathcal{H}_t$ and $\mathcal{H}_s$ are obtained for the cracked body subject to an external tensile stress $\boldsymbol{\sigma}_{0,11}$ and an external shear stress $\boldsymbol{\sigma}_{0,12}$, respectively, as

$$\mathcal{H}_t = \frac{\sigma_{0,11}^2}{4(1+\nu_0)\mu_0} - \frac{4(1-\nu_0)}{3\mu_0}a^3\sigma_{0,11} \tag{5.65}$$

$$\mathcal{H}_s = \frac{\sigma_{0,12}^2}{2\mu_0} - \frac{8(1-\nu_0)}{3(2-\nu_0)\mu_0}a^3\sigma_{0,12}^2 \tag{5.66}$$

Denote γ_s the crack surface energy density, and we have the energy rate balance equation as

$$\frac{\partial}{\partial a}\left(\mathcal{H} + 2\pi a^2\gamma_S\right) = 0 \tag{5.67}$$

substitution of (5.65–5.66) into which recovers the classical Griffith's solution on the critical tensile stress and the critical shear stress, respectively, as

$$\sigma_{0,11}^{c} = \sqrt{\frac{\pi\mu_0\gamma_S}{(1-\nu_0)a}} \tag{5.68}$$

$$\sigma_{0,12}^{c} = \sqrt{\frac{\pi\mu_0(2-\nu_0)\gamma_S}{2(1-\nu_0)a}} \tag{5.69}$$

for a penny-shaped crack with its size a.

A Single Slit-Like Crack

A plane strain slit-like crack with its principal half axes $a_1 \to 0$, $a_2 = a$, $a_3 \to \infty$ has the aspect ratio $\eta = \frac{a_1}{a} \to 0$ and the volume $V_c = \pi\eta a_3 a^2$. Given the corresponding elliptic cylinder Eshelby tensor (A3.3), the components of the polarization moduli (5.62) are explicitly obtained as

$$\begin{aligned}
L_{1111}^{A} &= -\frac{4\mu_0(1-\nu_0)^3}{\eta(1-2\nu_0)^2}\\
L_{1122}^{A} &= L_{1133}^{A} = L_{2211}^{A} = L_{3311}^{A} = -\frac{4\mu_0(1-\nu_0)^2\nu_0}{\eta(1-2\nu_0)^2}\\
L_{2222}^{A} &= L_{3333}^{A} = L_{3322}^{A} = L_{2233}^{A} = -\frac{4\mu_0(1-\nu_0)\nu_0^2}{\eta(1-2\nu_0)^2}\\
L_{1212}^{A} &= -\frac{\mu_0(1-\nu_0)}{\eta}\\
L_{1313}^{A} &= -\left(1+\frac{1}{\eta}\right)\mu_0\\
L_{2323}^{A} &= -\mu_0
\end{aligned} \tag{5.70a–f}$$

By substituting (5.70) into (5.63), the two potential energy functions are obtained, respectively, as

$$\mathcal{H}_t = \frac{\sigma_{0,11}^2}{4(1+\nu_0)\mu_0} - \frac{1-\nu_0}{2\mu_0}\pi a^2 a_3 \sigma_{0,11}^2 \tag{5.71}$$

$$\mathcal{H}_s = \frac{\sigma_{0,12}^2}{2\mu_0} - \frac{1-\nu_0}{2\mu_0}\pi a^2 a_3 \sigma_{0,12}^2 \tag{5.72}$$

for the cracked body subject to an external tensile stress $\boldsymbol{\sigma}_{0,11}$ and an external shear stress $\boldsymbol{\sigma}_{0,12}$. By using the energy rate balance equation (5.67), the classical solution is recovered as

$$\sigma_{0,11}^{c} = \sigma_{0,12}^{c} = \sqrt{\frac{4\mu_0\gamma_S}{\pi(1-\nu_0)a}} \tag{5.73}$$

on the critical tensile stress and the critical shear stress of a planar slit-like crack with its size a. In the plane stress case, by simply replacing ν_0 with $\nu_0/(1+\nu_0)$, the critical stresses are obtained as

$$\sigma_{0,11}^{c} = \sigma_{0,12}^{c} = \sqrt{\frac{2E_0\gamma_S}{\pi a}} \tag{5.74}$$

where the Young's modulus $E_0 = 2\,\mu_0(1+\nu_0)$.

5.3.2 Effective elastic moduli of a dilutely cracked solid

In a dilutely cracked solid, the interaction among cracks is assumed negligible, and the potential energy derived in the preceding subsection is directly applicable. Given N_c the number of cracks in the unit volume RVE, the potential energy (5.63) is rewritten as

$$\mathcal{H} = \frac{1}{2}\boldsymbol{\varepsilon}_0 \boldsymbol{L}_0 \boldsymbol{\varepsilon}_0 + \frac{N_c\overline{V_c}}{2}\boldsymbol{\varepsilon}_0 \boldsymbol{L}^A \boldsymbol{\varepsilon}_0 \tag{5.75}$$

where the overbar indicates ensemble averaging of the cracks of various sizes.

Similar to the density of disk-like platelets (5.48), the density of cracks is defined as

$$\alpha = \begin{cases} N_c \dfrac{4\pi}{3}\overline{a^3} & \text{3D penny-shaped cracks} \\ N_c \pi \overline{a^2} & \text{2D slit-like cracks} \end{cases} \tag{5.76}$$

where a denotes the major radius of an ellipsoid, which varies as a random variable. Compared to the traditional definition of the crack density (Bristow, 1960; Budianski & O'Connell, 1976), the definition (5.76) adds a factor $4\pi/3$ to the penny-shaped case and a factor π to the slit-like case, which is physically meaningful in that the density α represents the volume fraction of a-sized spheres in the Hashin-type self-similar sphere assemblage (Figure 5.2a).

With the crack density given in terms of the crack volume as

$$\alpha = N_c \frac{\overline{V_c}}{\eta} \tag{5.77}$$

(5.75) becomes

$$\mathcal{H} = \frac{1}{2}\boldsymbol{\varepsilon}_0 \boldsymbol{L}_0 \boldsymbol{\varepsilon}_0 + \frac{\eta\alpha}{2}\boldsymbol{\varepsilon}_0 \boldsymbol{L}^A \boldsymbol{\varepsilon}_0 \tag{5.78}$$

The effective elastic moduli tensor for a dilutely cracked solid immediately follows as

$$\boldsymbol{L}^e = \boldsymbol{L}_0 + \eta\alpha \boldsymbol{L}^A \tag{5.79}$$

i.e.

$$\boxed{\boldsymbol{L}^e = \boldsymbol{L}_0 - \eta\alpha \boldsymbol{L}_0 (\boldsymbol{I} - \boldsymbol{S}_c)^{-1}} \tag{5.80}$$

which is linear upon the crack density. By further taking random orientation into account, (5.79) is generalized as

$$\boxed{\boldsymbol{L}^e = \boldsymbol{L}_0 + \eta\alpha \overline{\boldsymbol{L}^A}} \tag{5.81}$$

Exactly like (5.37), the average polarization moduli are isotropic with the bulk and shear moduli given as

$$\overline{\kappa^A} = \frac{1}{9} L^A_{iijj} \qquad \overline{\mu^A} = \frac{1}{10}\left(L^A_{ijij} - \frac{1}{3} L^A_{iijj} \right) \tag{5.82a, b}$$

With the Eshelby tensor $\boldsymbol{S}_c$ given in (A3.4), (5.80–5.82) are explicitly expressed for the following crack distribution patterns.

Dilutely Distributed Parallel Penny-Shaped Cracks $\left(\alpha = N_c \frac{4\pi}{3}\overline{a^3}\right)$

$$
\begin{aligned}
&L^e_{1111} = \mu_0\left[\frac{2(1-\nu_0)}{1-2\nu_0} - \frac{8(1-\nu_0)^3}{\pi(1-2\nu_0)^2}\alpha\right]\\
&L^e_{1122} = L^e_{1133} = L^e_{2211} = L^e_{3311} = \mu_0\left[\frac{2\nu_0}{1-2\nu_0} - \frac{8(1-\nu_0)^2\nu_0}{\pi(1-2\nu_0)^2}\alpha\right]\\
&L^e_{2222} = L^e_{3333} = \mu_0\left[\frac{2(1-\nu_0)}{1-2\nu_0} - \frac{8(1-\nu_0)\nu_0^2}{\pi(1-2\nu_0)^2}\alpha\right]\\
&L^e_{3322} = L^e_{2233} = \mu_0\left[\frac{2-\nu_0}{1-2\nu_0} - \frac{8(1-\nu_0)\nu_0^2}{\pi(1-2\nu_0)^2}\alpha\right]\\
&L^e_{1212} = L^e_{1313} = \mu_0\left[1 - \frac{4(1-\nu_0)}{\pi(2-\nu_0)}\alpha\right]\\
&L^e_{2323} = \mu_0
\end{aligned}
\tag{5.83a–f}
$$

Dilutely Distributed Parallel Slit-Like Cracks (Plane Strain) $\left(\alpha = N_c\pi\overline{a^2}\right)$

$$
\begin{aligned}
&L^e_{1111} = \mu_0\left[\frac{2(1-\nu_0)}{1-2\nu_0} - \frac{4(1-\nu_0)^3}{(1-2\nu_0)^2}\alpha\right]\\
&L^e_{1122} = L^e_{1133} = L^e_{2211} = L^e_{3311} = \mu_0\left[\frac{2\nu_0}{1-2\nu_0} - \frac{4(1-\nu_0)^2\nu_0}{(1-2\nu_0)^2}\alpha\right]\\
&L^e_{2222} = L^e_{3333} = \mu_0\left[\frac{2(1-\nu_0)}{1-2\nu_0} - \frac{8(1-\nu_0)\nu_0^2}{(1-2\nu_0)^2}\alpha\right]\\
&L^e_{3322} = L^e_{2233} = \mu_0\left[\frac{2\nu_0}{1-2\nu_0} - \frac{4(1-\nu_0)\nu_0^2}{(1-2\nu_0)^2}\alpha\right]\\
&L^e_{1212} = \mu_0\left[1-(1-\nu_0)\alpha\right]\\
&L^e_{1313} = \mu_0(1-\alpha)\\
&L^e_{2323} = \mu_0
\end{aligned}
\tag{5.84a–g}
$$

Dilutely Distributed Parallel Slit-Like Cracks (Plane Stress) $\left(\alpha = N\pi\overline{a^2}\right)$

$$\boxed{\begin{aligned}
L^e_{1111} &= \mu_0\left[\frac{2}{1-\nu_0} - \frac{4}{(1-\nu_0)^2(1+\nu_0)}\alpha\right] \\
L^e_{1122} = L^e_{2211} = L^e_{3311} = L^e_{1133} &= \mu_0\left[\frac{2\nu_0}{1-\nu_0} - \frac{4\nu_0}{(1-\nu_0)^2(1+\nu_0)}\alpha\right] \\
L^e_{2222} &= \mu_0\left[\frac{2}{1-\nu_0} - \frac{4\nu_0^2}{(1-\nu_0)^2(1+\nu_0)}\alpha\right] \\
L^e_{2233} = L^e_{3322} &= \mu_0\left[\frac{2\nu_0}{1-\nu_0} - \frac{4\nu_0^2}{(1-\nu_0)^2(1+\nu_0)}\alpha\right] \\
L^e_{1212} &= \mu_0\left[1 - \frac{1}{(1+\nu_0)}\alpha\right]
\end{aligned}} \tag{5.85a–e}$$

Dilutely Distributed and Randomly Oriented Penny-Shaped Cracks $\left(\alpha = N_c\dfrac{4\pi}{3}\overline{a^3}\right)$

The bulk modulus and the shear modulus of the average polarization moduli are obtained from (5.82), respectively, as

$$\overline{\kappa^A} = -\frac{8\mu_0(1-\nu_0)(1+\nu_0)^2}{9\pi\eta(1-2\nu_0)^2} \tag{5.86}$$

$$\overline{\mu^A} = -\frac{8\mu_0(1-\nu_0)(5-\nu_0)}{15\pi\eta(2-\nu_0)} \tag{5.87}$$

substitution of which into (5.81) yields the effective bulk and shear moduli as

$$\boxed{\kappa^e = \mu_0\left(\frac{2(1+\nu_0)}{3(1-2\nu_0)} - \frac{8(1-\nu_0)(1+\nu_0)^2}{9\pi(1-2\nu_0)^2}\alpha\right)} \tag{5.88}$$

$$\boxed{\mu^e = \mu_0\left(1 - \frac{8(1-\nu_0)(5-\nu_0)}{15\pi(2-\nu_0)}\alpha\right)} \tag{5.89}$$

Dilutely Distributed and Randomly Oriented Slit-Like Cracks (Plane Strain) $\left(\alpha = N_c \pi \overline{a^2}\right)$

From (5.82) and (5.81), the average polarization moduli and the effective moduli are explicitly obtained, respectively, as

$$\begin{aligned}
\overline{L}^A_{1111} &= \overline{L}^A_{2222} = -\frac{2\mu_0(1-\nu_0)\left(1-2\nu_0+2\nu_0^2\right)}{\eta(1-2\nu_0)^2} \\
\overline{L}^A_{1122} &= \overline{L}^A_{2211} = -\frac{4\mu_0(1-\nu_0)^2\nu_0}{\eta(1-2\nu_0)^2} \\
\overline{L}^A_{3333} &= -\frac{4\mu_0(1-\nu_0)\nu_0^2}{\eta(1-2\nu_0)^2} \\
\overline{L}^A_{1133} &= \overline{L}^A_{3311} = \overline{L}^A_{2233} = \overline{L}^A_{3322} = -\frac{2\mu_0(1-\nu_0)\nu_0}{\eta(1-2\nu_0)^2} \\
\overline{L}^A_{1212} &= -\frac{\mu_0(1-\nu_0)}{\eta} \\
\overline{L}^A_{1313} &= -\left(1+\frac{1}{\eta}\right)\mu_0 \\
\overline{L}^A_{2323} &= -\mu_0
\end{aligned} \tag{5.90a–g}$$

$$\boxed{\begin{aligned}
L^e_{1111} &= L^e_{2222} = \mu_0\left(\frac{2(1-\nu_0)}{1-2\nu_0} - \frac{2(1-\nu_0)\left(1-2\nu_0+2\nu_0^2\right)}{(1-2\nu_0)^2}\alpha\right) \\
L^e_{1122} &= L^e_{2211} = \mu_0\left(\frac{2\nu_0}{1-2\nu_0} - \frac{4(1-\nu_0)^2\nu_0}{(1-2\nu_0)^2}\alpha\right) \\
L^e_{3333} &= \mu_0\left(\frac{2(1-\nu_0)}{1-2\nu_0} - \frac{4(1-\nu_0)\nu_0^2}{(1-2\nu_0)^2}\alpha\right) \\
L^e_{1133} &= L^e_{3311} = L^e_{2233} = L^e_{3322} = \mu_0\left(\frac{2\nu_0}{1-2\nu_0} - \frac{2(1-\nu_0)\nu_0}{(1-2\nu_0)^2}\alpha\right) \\
L^e_{1212} &= \mu_0\left(1-(1-\nu_0)\alpha\right) \\
L^e_{1313} &= \mu_0(1-\alpha) \\
L^e_{2323} &= \mu_0
\end{aligned}} \tag{5.91a–g}$$

Dilutely Distributed and Randomly Oriented Slit-Like Cracks (Plane Stress) $\left(\alpha = N_c \pi \overline{a^2}\right)$

By replacing ν_0 with $\nu_0/(1 + \nu_0)$ in (5.91), the components of the effective moduli are obtained as

$$
\begin{aligned}
L^e_{1111} &= L^e_{2222} = \mu_0\left(\frac{2}{1-\nu_0} - \frac{2(1+\nu_0^2)}{(1+\nu_0)(1-\nu_0)^2}\alpha\right) \\
L^e_{1122} &= L^e_{2211} = \mu_0\left(\frac{2\nu_0}{1-\nu_0} - \frac{4\nu_0}{(1+\nu_0)(1-\nu_0)^2}\alpha\right) \\
L^e_{1133} &= L^e_{2233} = L^e_{3311} = L^e_{3322} = \mu_0\left(\frac{2\nu_0}{1-\nu_0} - \frac{2\nu_0}{(1-\nu_0)^2}\alpha\right) \\
L^e_{1212} &= \mu_0\left(1 - \frac{1}{1+\nu_0}\alpha\right)
\end{aligned}
\qquad (5.92a\text{–}d)
$$

5.3.3 Effective elastic moduli of a non-dilutely cracked solid

Substituting $N = 2$, $\boldsymbol{L}_1 = 0$, and $c_1 = \eta\alpha$ into (5.55), the ellipsoidal bound on a solid containing isotropic distribution of cracks is obtained as

$$
\boldsymbol{L}^{EB} = \boldsymbol{L}_0 + \eta\alpha\overline{\boldsymbol{L}}^A\left(\mathrm{I} - \eta\alpha\boldsymbol{S}\boldsymbol{L}_0^{-1}\overline{\boldsymbol{L}}^A\right)^{-1} \qquad (5.93)
$$

where the subscript 1 is omitted. By comparing (5.81) with (5.93), it is noticed that the dilute case solution corresponds to the first-order truncation on the expansion of the non-dilute-case solution (5.93) about $\eta\alpha$. Below some explicit results of (5.93) are presented for a variety of crack distribution patterns, which serve as the rigorous upper bounds to these patterns with a specified crack density. In literature there are a few approximations made to estimate the effective elastic moduli of a cracked solid, while the underlying crack patterns of these approximations are normally unknown. By comparing these approximations against the upper bounds obtained below, certain insights are gained about applicability of these approximations. For example, if an approximation for the effective bulk modulus is found to be greater than the upper bound (5.102) in the isotropic mixture case below, then it is inferred that the underlying crack pattern for the approximation corresponds to incomplete mixture (i.e. certain clustering).

Randomly Distributed Parallel Penny-Shaped Cracks $\left(\alpha = N_c \frac{4\pi}{3}\overline{a^3}\right)$

With regard to randomly located parallel cracks, the components of the average polarization moduli $\overline{\boldsymbol{L}}^A = \boldsymbol{L}^A$ are given in (5.64). In the isotropic

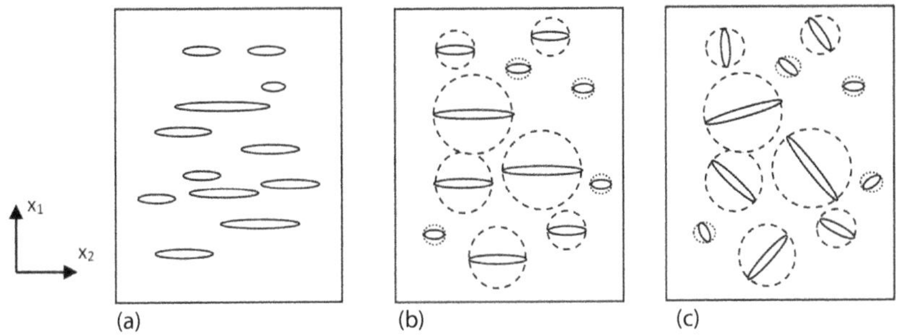

Figure 5.10 (a) Anisotropic mixture of randomly distributed parallel cracks; (b) spherical exclusion model as isotropic mixture of randomly distributed parallel cracks; (c) spherical exclusion model as isotropic mixture of randomly distributed and randomly oriented cracks. (From Xu, X.F. and G. Stefanou, *Int. J. Multiscale Comput. Eng.*, 9: 347–363, 2011. With permission from Begell House Inc. Publishers.)

mixture case, the spatial distribution of the centers of parallel cracks is statistically isotropic, with the corresponding spherical exclusion model shown in Figure 5.10b. Given the average Eshelby tensor in (5.93) $\boldsymbol{S} = \boldsymbol{S}_0$, the components of the upper bound (5.93) are obtained as

$$
\begin{aligned}
L_{1111}^{EB} &= \mu_0\left(\frac{2(1-\nu_0)}{1-2\nu_0} - \frac{120(1-\nu_0)^3}{(1-2\nu_0)\left(15\pi(1-2\nu_0)+4\alpha\left(7-14\nu_0+15\nu_0^2\right)\right)}\alpha\right)\\
L_{1122}^{EB} = L_{1133}^{EB} = L_{2211}^{EB} = L_{3311}^{EB} &= \mu_0\left(\frac{2\nu_0}{1-2\nu_0} - \frac{120(1-\nu_0)^2\nu_0}{(1-2\nu_0)\left(15\pi(1-2\nu_0)+4\alpha\left(7-14\nu_0+15\nu_0^2\right)\right)}\alpha\right)\\
L_{2222}^{EB} = L_{3333}^{EB} &= \mu_0\left(\frac{2(1-\nu_0)}{1-2\nu_0} - \frac{120(1-\nu_0)\nu_0^2}{(1-2\nu_0)\left(15\pi(1-2\nu_0)+4\alpha\left(7-14\nu_0+15\nu_0^2\right)\right)}\alpha\right)\\
L_{3322}^{EB} = L_{2233}^{EB} &= \mu_0\left(\frac{2\nu_0}{1-2\nu_0} - \frac{120(1-\nu_0)\nu_0^2}{(1-2\nu_0)\left(15\pi(1-2\nu_0)+4\alpha\left(7-14\nu_0+15\nu_0^2\right)\right)}\alpha\right)\\
L_{1212}^{EB} = L_{1313}^{EB} &= \mu_0\left(1-\frac{60(1-\nu_0)}{15\pi(2-\nu_0)+8\alpha(4-5\nu_0)}\alpha\right)\\
L_{2323}^{EB} &= \mu_0
\end{aligned}
\tag{5.94a–f}
$$

In the anisotropic mixture case, when the average Eshelby tensor $\boldsymbol{S} = \boldsymbol{S}_c$, the effective moduli are analytically available. The corresponding ellipsoidal exclusion model is shown in Figure 5.10a, where the parallel cracks can be very close to each other along the axis-1. The components of the upper bound are obtained as

$$
\begin{aligned}
&L^{EB}_{1111} = \mu_0 \left(\frac{2(1-\nu_0)}{1-2\nu_0} - \frac{4(1-\nu_0)^3}{(1-2\nu_0)\left((1-2\nu_0)+2\alpha(1-\nu_0)^2\right)}\alpha \right) \\
&L^{EB}_{1122} = L^{EB}_{1133} = L^{EB}_{2211} = L^{EB}_{3311} = \mu_0 \left(\frac{2\nu_0}{1-2\nu_0} - \frac{4(1-\nu_0)^2\nu_0}{(1-2\nu_0)\left((1-2\nu_0)+2\alpha(1-\nu_0)^2\right)}\alpha \right) \\
&L^{EB}_{2222} = L^{EB}_{3333} = \mu_0 \left[\frac{2(1-\nu_0)}{1-2\nu_0} - \frac{4(1-\nu_0)\nu_0^2}{(1-2\nu_0)\left[(1-2\nu_0)+2\alpha(1-\nu_0)^2\right]}\alpha \right] \\
&L^{EB}_{3322} = L^{EB}_{2233} = \mu_0 \left(\frac{2(1-\nu_0)}{1-2\nu_0} - \frac{4(1-\nu_0)\nu_0^2}{(1-2\nu_0)\left((1-2\nu_0)+2\alpha(1-\nu_0)^2\right)}\alpha \right) \\
&L^{EB}_{1212} = \mu_0 \left(1 - \frac{1-\nu_0}{1+\alpha(1-\nu_0)}\alpha \right) \\
&L^{EB}_{1313} = \mu_0 \frac{1}{1+\alpha} \\
&L^{EB}_{2323} = \mu_0
\end{aligned}
$$

(5.95a–g)

Randomly Distributed Parallel Slit-Like Cracks (Plane Strain) $\left(\alpha = N_c \pi \overline{a^2}\right)$

The components of the average polarization moduli $\overline{\boldsymbol{L}^A} = \boldsymbol{L}^A$ are given in (5.70). In the isotropic mixture case (Figure 5.10b), the average Eshelby tensor $\boldsymbol{S} = \boldsymbol{S}_0$, and the components of the upper bound (5.93) follow as

$$
\begin{aligned}
&L^{EB}_{1111} = \mu_0 \left(\frac{2(1-\nu_0)}{1-2\nu_0} - \frac{8(1-\nu_0)^3}{(1-2\nu_0)\left(\pi(1-2\nu_0)+4\alpha(1-\nu_0)^2\right)}\alpha \right) \\
&L^{EB}_{1122} = L^{EB}_{1133} = L^{EB}_{2211} = L^{EB}_{3311} = \mu_0 \left(\frac{2\nu_0}{1-2\nu_0} - \frac{8(1-\nu_0)^2\nu_0}{(1-2\nu_0)\left(\pi(1-2\nu_0)+4\alpha(1-\nu_0)^2\right)}\alpha \right) \\
&L^{EB}_{2222} = L^{EB}_{3333} = \mu_0 \left(\frac{2(1-\nu_0)}{1-2\nu_0} - \frac{8(1-\nu_0)\nu_0^2}{(1-2\nu_0)\left(\pi(1-2\nu_0)+4\alpha(1-\nu_0)^2\right)}\alpha \right) \\
&L^{EB}_{3322} = L^{EB}_{2233} = \mu_0 \left(\frac{2\nu_0}{1-2\nu_0} - \frac{8(1-\nu_0)\nu_0^2}{(1-2\nu_0)\left(\pi(1-2\nu_0)+4\alpha(1-\nu_0)^2\right)}\alpha \right) \\
&L^{EB}_{1212} = L^{EB}_{1313} = \mu_0 \left(1 - \frac{4(1-\nu_0)}{\pi(2-\nu_0)+4\alpha(1-\nu_0)}\alpha \right) \\
&L^{EB}_{2323} = \mu_0
\end{aligned}
$$

(5.96a–f)

In the anisotropic mixture case (Figure 5.10a), when the average Eshelby tensor $\boldsymbol{S} = \boldsymbol{S}_c$ the effective moduli are obtained as

$$
\begin{aligned}
L_{1111}^{EB} &= \mu_0\left(\frac{2(1-\nu_0)}{1-2\nu_0} - \frac{60(1-\nu_0)^3}{(1-2\nu_0)\left(15(1-2\nu_0)+2\alpha(7-14\nu_0+15\nu_0^2)\right)}\alpha\right)\\
L_{1122}^{EB} = L_{1133}^{EB} = L_{2211}^{EB} = L_{3311}^{EB} &= \mu_0\left(\frac{2\nu_0}{1-2\nu_0} - \frac{60(1-\nu_0)^2\nu_0}{(1-2\nu_0)\left(15(1-2\nu_0)+2\alpha(7-14\nu_0+15\nu_0^2)\right)}\alpha\right)\\
L_{2222}^{EB} = L_{3333}^{EB} &= \mu_0\left(\frac{2(1-\nu_0)}{1-2\nu_0} - \frac{8(1-\nu_0)\nu_0^2}{(1-2\nu_0)\left(15(1-2\nu_0)+2\alpha(7-14\nu_0+15\nu_0^2)\right)}\alpha\right)\\
L_{3322}^{EB} = L_{2233}^{EB} &= \mu_0\left(\frac{2\nu_0}{1-2\nu_0} - \frac{8(1-\nu_0)\nu_0^2}{(1-2\nu_0)\left(15(1-2\nu_0)+2\alpha(7-14\nu_0+15\nu_0^2)\right)}\alpha\right)\\
L_{1212}^{EB} &= \mu_0\left(1-\frac{15(1-\nu_0)}{15+2\alpha(4-5\nu_0)}\alpha\right)\\
L_{1313}^{EB} &= \mu_0\left(1-\frac{15(1-\nu_0)}{15(1-\nu_0)+2\alpha(4-5\nu_0)}\alpha\right)\\
L_{2323}^{EB} &= \mu_0
\end{aligned}
$$

(5.97a–g)

Randomly Distributed Parallel Slit-Like Cracks (Plane Stress) $\left(\alpha = N_c\pi\overline{a^2}\right)$

By replacing ν_0 with $\nu_0/(1+\nu_0)$ into (5.96) and (5.97), respectively, the effective moduli are obtained in the isotropic mixture case (Figure 5.10b) as

$$
\begin{aligned}
L_{1111}^{EB} &= \mu_0\left(\frac{2}{1-\nu_0} - \frac{60}{(1-\nu_0)\left(15\left(1-\nu_0^2\right)+2\alpha\left(7+8\nu_0^2\right)\right)}\alpha\right)\\
L_{1122}^{EB} = L_{1133}^{EB} = L_{2211}^{EB} = L_{3311}^{EB} &= \mu_0\left(\frac{2\nu_0}{1-\nu_0} - \frac{60\nu_0}{(1-\nu_0)\left(15\left(1-\nu_0^2\right)+2\alpha\left(7+8\nu_0^2\right)\right)}\alpha\right)\\
L_{2222}^{EB} &= \mu_0\left(\frac{2}{1-\nu_0} - \frac{60\nu_0^2}{(1-\nu_0)\left(15\left(1-\nu_0^2\right)+2\alpha\left(7+8\nu_0^2\right)\right)}\alpha\right)\\
L_{2233}^{EB} = L_{3322}^{EB} &= \mu_0\left(\frac{2\nu_0}{1-\nu_0} - \frac{60\nu_0^2}{(1-\nu_0)\left(15\left(1-\nu_0^2\right)+2\alpha\left(7+8\nu_0^2\right)\right)}\alpha\right)\\
L_{1212}^{EB} &= \mu_0\left(1-\frac{15}{15(1+\nu_0)+2\alpha(4-\nu_0)}\alpha\right)
\end{aligned}
$$

(5.98a–e)

and in the anisotropic mixture case (Figure 5.10a) when $\boldsymbol{S} = \boldsymbol{S}_c$, as

$$
\begin{aligned}
L_{1111}^{EB} &= \mu_0\left(\frac{2}{1-\nu_0} - \frac{4}{(1-\nu_0)\left((1-\nu_0)^2+2\alpha\right)}\alpha\right)\\
L_{1122}^{EB} = L_{1133}^{EB} = L_{2211}^{EB} = L_{3311}^{EB} &= \mu_0\left(\frac{2\nu_0}{1-\nu_0} - \frac{4\nu_0}{(1-\nu_0)\left((1-\nu_0)^2+2\alpha\right)}\alpha\right)\\
L_{2222}^{EB} &= \mu_0\left(\frac{2}{1-\nu_0} - \frac{4\nu_0^2}{(1-\nu_0)\left((1-\nu_0)^2+2\alpha\right)}\alpha\right)\\
L_{2233}^{EB} = L_{3322}^{EB} &= \mu_0\left(\frac{2\nu_0}{1-\nu_0} - \frac{4\nu_0^2}{(1-\nu_0)\left((1-\nu_0)^2+2\alpha\right)}\alpha\right)\\
L_{1212}^{EB} &= \mu_0\left(1 - \frac{1}{1+\nu_0+\alpha}\alpha\right)
\end{aligned}
\tag{5.99a–e}
$$

Randomly Distributed and Randomly Oriented Penny-Shaped Cracks $\left(\alpha = N_c \frac{4\pi}{3}\overline{a^3}\right)$

In the isotropic mixture case (Figure 5.10c), the average Eshelby tensor $\boldsymbol{S} = \boldsymbol{S}_0$. The bulk modulus and the shear modulus of the average polarization moduli are obtained from (5.82), respectively, as

$$
\overline{\kappa^A} = -\frac{8\mu_0(1-\nu_0)(1+\nu_0)^2}{\eta(1-2\nu_0)\left(9\pi(1-2\nu_0)+4\alpha(1+\nu_0)^2\right)}
\tag{5.100}
$$

$$
\overline{\mu^A} = -\frac{120\mu_0(1-\nu_0)(5-\nu_0)}{\eta\left(225\pi(2-\nu_0)+16\alpha(5-\nu_0)(4-5\nu_0)\right)}
\tag{5.101}
$$

substitution of which into (5.81) yields the upper bounds for the effective bulk and shear moduli as

$$
\boxed{\kappa^{EB} = \kappa_0\left(1 - \frac{12(1-\nu_0)(1+\nu_0)}{9\pi(1-2\nu_0)+4\alpha(1+\nu_0)^2}\alpha\right)}
\tag{5.102}
$$

$$\boxed{\mu^{EB} = \mu_0\left(1 - \frac{120(1-\nu_0)(5-\nu_0)}{225\pi(2-\nu_0)+16\alpha(5-\nu_0)(4-5\nu_0)}\alpha\right)} \tag{5.103}$$

The formulas (5.102–5.103) were first obtained in Ponte Castañeda and Willis (1995) by using the Hashin-type self-similar sphere assemblage. As compared in Subsection 5.2.1 (Figure 5.2), the morphology presented in the self-similar assemblage model is a special case of the spherical exclusion model (Figure 5.10c).

Randomly Distributed and Randomly Oriented Slit-Like Cracks (Plane Strain) $\left(\alpha = N_c \pi a^2\right)$

In the isotropic mixture case (Figure 5.10c), the average Eshelby tensor $\boldsymbol{S} = \boldsymbol{S}_0$. From (5.82) and (5.81) the average polarization moduli and the effective moduli are explicitly obtained, respectively, as

$$\begin{aligned}
&\bar{L}^A_{1111} = \bar{L}^A_{2222} = -\frac{2\mu_0(1-\nu_0)(1-2\nu_0+2\nu_0^2)}{\eta(1-2\nu_0)^2}\\
&\bar{L}^A_{1122} = \bar{L}^A_{2211} = -\frac{4\mu_0(1-\nu_0)^2\nu_0}{\eta(1-2\nu_0)^2}\\
&\bar{L}^A_{3333} = -\frac{4\mu_0(1-\nu_0)\nu_0^2}{\eta(1-2\nu_0)^2}\\
&\bar{L}^A_{1133} = \bar{L}^A_{3311} = \bar{L}^A_{2233} = \bar{L}^A_{3322} = -\frac{2\mu_0(1-\nu_0)\nu_0}{\eta(1-2\nu_0)^2}\\
&\bar{L}^A_{1212} = -\frac{\mu_0(1-\nu_0)}{\eta}\\
&\bar{L}^A_{1313} = -\left(1+\frac{1}{\eta}\right)\mu_0\\
&\bar{L}^A_{2323} = -\mu_0
\end{aligned} \tag{5.104a–g}$$

$$
\begin{aligned}
L^{EB}_{1111} = L^{EB}_{2222} &= \mu_0\left(\frac{2(1-\nu_0)}{1-2\nu_0} - \frac{30(1-\nu_0)\left(15\left(1-2\nu_0+2\nu_0^2\right)+\alpha\left(7-12\nu_0+7\nu_0^2-10\nu_0^3\right)\right)}{(1-2\nu_0)\left(225(1-2\nu_0)+30\alpha(7-14\nu_0+15\nu_0^2)+4\alpha^2(12-19\nu_0+25\nu_0^2-25\nu_0^3)\right)}\alpha\right)\\
L^{EB}_{1122} = L^{EB}_{2211} &= \mu_0\left(\frac{2\nu_0}{1-2\nu_0} - \frac{30(1-\nu_0)\left(30(1-\nu_0)\nu_0+\alpha\left(1+2\nu_0-7\nu_0^2+10\nu_0^3\right)\right)}{(1-2\nu_0)\left(225(1-2\nu_0)+30\alpha\left(7-14\nu_0+15\nu_0^2\right)+4\alpha^2\left(12-19\nu_0+25\nu_0^2-25\nu_0^3\right)\right)}\alpha\right)\\
L^{EB}_{3333} &= \mu_0\left(\frac{2(1-\nu_0)}{1-2\nu_0} - \frac{60(1-\nu_0)\nu_0^2}{(1-2\nu_0)\left(15(1-2\nu_0)+2\alpha(3-\nu_0+5\nu_0^2)\right)}\alpha\right)\\
L^{EB}_{1133} = L^{EB}_{3311} = L^{EB}_{2233} = L^{EB}_{3322} &= \mu_0\left(\frac{2\nu_0}{1-2\nu_0} - \frac{30(1-\nu_0)\nu_0}{(1-2\nu_0)\left(15(1-2\nu_0)+2\alpha\left(3-\nu_0+5\nu_0^2\right)\right)}\alpha\right)\\
L^{EB}_{1212} &= \mu_0\left(1 - \frac{15(1-\nu_0)}{15+2\alpha(4-5\nu_0)}\alpha\right)\\
L^{EB}_{1313} &= \mu_0\left(1 - \frac{15(1-\nu_0)}{15(1-\nu_0)+2\alpha(4-5\nu_0)}\alpha\right)\\
L^{EB}_{2323} &= \mu_0
\end{aligned}
\tag{5.105a–g}
$$

Randomly Distributed and Randomly Oriented Slit-Like Cracks (Plane Stress) $\left(\alpha = N_c\pi a^2\right)$

By replacing ν_0 with $\nu_0/(1+\nu_0)$ in (5.105), the components of the effective moduli are obtained in the isotropic mixture case (Figure 5.10c) as

$$
\begin{aligned}
L^{EB}_{1111} = L^{EB}_{2222} &= \mu_0\left(\frac{2}{1-\nu_0} - \frac{30\left(15\left(1+\nu_0+\nu_0^2+\nu_0^3\right)+\alpha\left(7+9\nu_0+4\nu_0^2-8\nu_0^3\right)\right)}{225\left(1-\nu_0^2\right)^2+30\alpha\left(7+\nu_0^2-8\nu_0^4\right)+4\alpha^2\left(12+5\nu_0+6\nu_0^2-30\nu_0^3+7\nu_0^4\right)}\alpha\right)\\
L^{EB}_{1122} = L^{EB}_{2211} &= \mu_0\left(\frac{2\nu_0}{1-\nu_0} - \frac{30\left(30(1+\nu_0)\nu_0+\alpha\left(1+5\nu_0+6\nu_0^3\right)\right)}{225\left(1-\nu_0^2\right)^2+30\alpha\left(7+\nu_0^2-8\nu_0^4\right)+4\alpha^2\left(12+5\nu_0+6\nu_0^2-30\nu_0^3+7\nu_0^4\right)}\alpha\right)\\
L^{EB}_{1133} = L^{EB}_{2233} = L^{EB}_{3311} = L^{EB}_{3322} &= \mu_0\left(\frac{2\nu_0}{1-\nu_0} - \frac{30(1+\nu_0)\nu_0}{(1-\nu_0)\left(15\left(1-\nu_0^2\right)+2\alpha\left(3+5\nu_0+7\nu_0^2\right)\right)}\alpha\right)\\
L^{EB}_{1212} &= \mu_0\left(1 - \frac{15}{15(1+\nu_0)+2\alpha(4-\nu_0)}\alpha\right)
\end{aligned}
\tag{5.106a–d}
$$

5.3.4 Effective elastic moduli of a solid containing voids

In this book, a *crack* is defined as a spheroidal cavity with the aspect ratio η approaching zero, while a *void* refers to a spheroidal cavity with the aspect ratio η of a finite value. In damage and fracture mechanics, a spheroidal cavity is oblate with the aspect ratio $\eta < 1$. Given the Eshelby tensor of an oblate spheroid (A3.5), in the isotropic mixture case the upper bounds for the effective bulk and shear moduli are obtained from (5.36–5.37) explicitly in terms of the aspect ratio as

$$
\begin{aligned}
&\kappa^{EB}/\kappa_0 = 1 - \frac{3}{\dfrac{1+\nu_0}{1-\nu_0}\alpha + \dfrac{18(1-2\nu_0)(1+\nu_0)}{\sqrt{1-\eta^2}\,(1-\nu_0^2)} \dfrac{\sum_{i=0}^{2} B_i(\cos^{-1}\eta)^i \sum_{i=0}^{3} C_i(\cos^{-1}\eta)^i}{\sum_{i=0}^{4} A_i(\cos^{-1}\eta)^i}}\alpha \\
&A_0 = \eta(1-\eta^2)^2(1+2\eta^2)\big[32(1-\nu_0^2)+2\eta^8(5-9\nu_0+4\nu_0^2)+4\eta^2(11-25\nu_0+8\nu_0^2+8\nu_0^3) \\
&\qquad +\eta^6(-17+95\nu_0-108\nu_0^2+32\nu_0^3)-\eta^4(15+4\nu_0-100\nu_0^2+64\nu_0^3)\big] \\
&A_1 = (1-\eta^2)^{3/2}\big[32(1-\nu_0^2)-8\eta^2(5+9\nu_0-24\nu_0^2+8\nu_0^3)+4\eta^8(61-219\nu_0+183\nu_0^2-32\nu_0^3)-4\eta^6 \\
&\qquad (35-165\nu_0+244\nu_0^2-96\nu_0^3)-\eta^4(453-620\nu_0+36\nu_0^2+128\nu_0^3)-8\eta^{10}(6+\nu_0-15\nu_0^2+8\nu_0^3)\big] \\
&A_2 = \eta(1-\eta^2)\big[4(-21-25\nu_0+40\nu_0^2+8\nu_0^3)+\eta^2(-197+492\nu_0-612\nu_0^2+256\nu_0^3)-6\eta^4\ (-136+ \\
&\qquad 293\nu_0-200\nu_0^2+64\nu_0^3)-4\eta^6(48-97\nu_0+31\nu_0^2+32\nu_0^3)+4\eta^8(-25+123\nu_0-156\nu_0^2+56\nu_0^3)\big] \\
&A_3 = \eta^2\sqrt{1-\eta^2}\,\big[73+180\nu_0-156\nu_0^2-128\nu_0^3+36\eta^2(10-15\nu_0+8\nu_0^2) \\
&\qquad +12\eta^4(-46+105\nu_0-69\nu_0^2+32\nu_0^3)+8\eta^6(25-72\nu_0+87\nu_0^2-32\nu_0^3)\big] \\
&A_4 = 3\eta^3\big[-7-27\nu_0+12\nu_0^2+32\nu_0^3+\eta^2(-52+48\nu_0+36\nu_0^2-64\nu_0^3)+16\eta^4(2-3\nu_0-3\nu_0^2+2\nu_0^3)\big] \\
&B_0 = \eta-\eta^5(1-\nu_0)-\eta^3\nu_0 \\
&B_1 = \sqrt{1-\eta^2}\,\big[1-2\eta^2(1-\nu_0)\big] \\
&B_2 = -\eta(1-\nu_0) \\
&C_0 = -\eta(1-\eta^2)^{3/2}\big[8(-1+\nu_0)+\eta^2(-1+8\nu_0-8\nu_0^2)+\eta^4(5-15\nu_0+8\nu_0^2)+2\eta^6(-1+\nu_0)\big] \\
&C_1 = (1-\eta^2)\big[8(1-\nu_0)-2\eta^2(15-24\nu_0+8\nu_0^2)+\eta^4(13-23\nu_0+8\nu_0^2)+8\eta^6(-1+\nu_0)\nu_0\big] \\
&C_2 = -\eta\sqrt{1-\eta^2}\,\big[(15-8\nu_0-8\nu_0^2)+\eta^2(-25+39\nu_0-8\nu_0^2)+2\eta^4(5-11\nu_0+8\nu_0^2)\big] \\
&C_3 = \eta^2(1+\nu_0)\big[7-8\nu_0+\eta^2(-4+8\nu_0)\big]
\end{aligned}
\tag{5.107}
$$

$$
\begin{aligned}
&\mu^{EB}/\mu_0 = 1 - \frac{15(1-\nu_0)}{2(4-5\nu_0)\alpha + \dfrac{225}{\eta\sqrt{1-\eta^2}\left(\displaystyle\sum_{i=1}^{4} D_i + \frac{\sum_{i=0}^{2} F_i (\cos^{-1}\eta)^i}{\sum_{i=0}^{3} E_i (\cos^{-1}\eta)^i}\right)}}\,\alpha \\
&D_1 = -\frac{24(1-\eta^2)^2}{\sqrt{1-\eta^2}\left(8+2\eta^4-8\nu_0+\eta^2(-7+25\nu_0)\right)+\eta\left(-7+8\nu_0+\eta^2(4-8\nu_0)\right)\cos^{-1}(\eta)} \\
&D_2 = \frac{12(1-\eta^2)^2}{\eta^2\sqrt{1-\eta^2}\left(4-\nu_0+\eta^2(-1+\nu_0)\right)-\eta\left(2-\nu_0+\eta^2(1+\nu_0)\right)\cos^{-1}(\eta)} \\
&D_3 = \frac{2\sqrt{1-\eta^2}\left(4(1-\nu_0)\nu_0+\eta^2(3-5\nu_0+8\nu_0^2)+4\eta^4(1-\nu_0)\nu_0\right)-2\eta(1+2\eta^2)\cos^{-1}(\eta)}{\eta(1-2\nu_0)^2\left[\eta-\eta^5(1-\nu_0)-\eta^3\nu_0+\sqrt{1-\eta^2}\left(1-2\eta^2(1-\nu_0)\right)\cos^{-1}(\eta)-\eta(1+\nu_0)\left(\cos^{-1}(\eta)\right)^2\right]} \\
&D_4 = \frac{\sqrt{1-\eta^2}\left(-4(1-\nu_0)^2+\eta^2(3-17\nu_0+12\nu_0^2)-\eta^4(2-6\nu_0+8\nu_0^2)\right)+\eta(1-\nu_0)\left(3-4(1-\eta^2)\nu_0\right)\cos^{-1}(\eta)}{\eta(1-2\nu_0)^2\left[\eta-\eta^5(1-\nu_0)-\eta^3\nu_0+\sqrt{1-\eta^2}\left(1-2\eta^2(1-\nu_0)\right)\cos^{-1}(\eta)-\eta(1+\nu_0)\left(\cos^{-1}(\eta)\right)^2\right]} \\
&E_0 = \eta^2(1-\eta^2)^{3/2}(1-2\nu_0)^2\left[-8(1-\nu_0)-\eta^2(1-8\nu_0+8\nu_0^2)+\eta^4(5-15\nu_0+8\nu_0^2)-2\eta^6(1-\nu_0)\right] \\
&E_1 = -\eta(1-\eta^2)(1-2\nu_0)^2\left[8(1-\nu_0)-2\eta^2(15-24\nu_0+8\nu_0^2)+\eta^4(13-23\nu_0+8\nu_0^2)-8\eta^6(1-\nu_0)\nu_0\right] \\
&E_2 = \eta^2\sqrt{1-\eta^2}(1-2\nu_0)^2\left[15-8\nu_0-8\nu_0^2+\eta^2(-25+39\nu_0-8\nu_0^2)+2\eta^4(5-11\nu_0+8\nu_0^2)\right] \\
&E_3 = \eta^3(1-2\nu_0)^2(1+\nu_0)\left[7-8\nu_0+\eta^2(-4+8\nu_0)\right] \\
&F_0 = (1-\eta^2)\left[32(1-\nu_0)\nu_0^2+4\eta^2(8-16\nu_0-\nu_0^2+16\nu_0^3)-\eta^4(15-13\nu_0-92\nu_0^2+128\nu_0^3)-\right. \\
&\qquad \left.4\eta^6(9-46\nu_0+78\nu_0^2-48\nu_0^3)+4\eta^8(7-31\nu_0+48\nu_0^2-24\nu_0^3)\right] \\
&F_1 = -2\eta\sqrt{1-\eta^2}\left[16-64\nu_0+30\nu_0^2+32\nu_0^3+\eta^2(-71+233\nu_0-228\nu_0^2+32\nu_0^3)+\right. \\
&\qquad \left.\eta^4(70-286\nu_0+378\nu_0^2-160\nu_0^3)+12\eta^6(-2+9\nu_0-15\nu_0^2+8\nu_0^3)\right] \\
&F_2 = -\eta^2\left[31-101\nu_0-4\nu_0^2+128\nu_0^3-4\eta^2(20-49\nu_0-5\nu_0^2+64\nu_0^3)+8\eta^4(5-13\nu_0-2\nu_0^2+16\nu_0^3)\right]
\end{aligned}
\tag{5.108}
$$

These equations can be conveniently calculated by using symbolic software to show the relationship between the aspect ratio and the elastic moduli, and thereby provide insight into the effect of cavity shape on ductility or brittleness of deformation modes. When the aspect ratio η is taken the limit to 0 and 1, respectively, these equations reduce exactly to (5.102–5.103) in the case of cracks, and to the Hashin-Shtrikman bounds in the case of spherical voids as

$$\kappa^{EB}/\kappa_0 = 1 - \frac{3(1-\nu_0)}{2(1-2\nu_0)+\alpha(1+\nu_0)}\,\alpha \tag{5.109}$$

$$\mu^{EB}\Big/\mu_0 = 1 - \frac{30(1-\nu_0)}{2(7-5\nu_0)+4\alpha(4-5\nu_0)}\alpha \tag{5.110}$$

Note that in (5.109–5.110) when the density $\alpha = 1$, i.e. the voids fully occupy the space with no solid left in the Hashin coated sphere assemblage, the effective moduli consistently become zero.

Chapter 6

Prediction of percolation threshold

Percolation refers to formation of a long-range global connectivity in a random system when probability of local connectivity reaches beyond a certain value (i.e. percolation threshold). The concept of bond percolation was first introduced in Broadbent and Hammersley (1957), and since then discrete percolation theory has been developed based on lattice models with a wide range of application in statistical physics, material science, geology, biology, complex networks, and epidemiology, etc. In the early 1960s, the discrete percolation based on lattice points was generalized to continuum percolation in modeling of wireless networks (Gilbert, 1961). Compared to its discrete counterpart, the continuum percolation theory presents certain unique advantages in exploring the relationship between geometry of microstructure and percolation properties of a random medium.

In engineering application of composites, the most essential percolation property is percolation threshold, the prediction of which poses a major challenge to existing models. To predict the percolation threshold of a composite, most of the studies resort to Monte Carlo simulation. Among few existing analytical methods aiming at more generic results there are mainly two approaches. One is based on geometric percolation, such as the excluded volume approach (Balberg, Binenbaum, & Wagner, 1984), and the other is based on physical percolation, such as conducting or insulating behavior (Helsing & Helte, 1991). While most of geometric percolation results are obtained from penetrating or overlapping ellipsoids (Garboczi et al., 1995), the approach based on the physical percolation offers the advantage in tackling nonoverlapping inclusions that are geometrically close to most of real composites.

From an application perspective, since the emergence of nanotechnology in the 1990s the topic of percolation on electrical and thermal conductivity of nanocomposites has attracted tremendous amount of research interest. The majority of these research efforts are focused on physical experiments, and there is a lack of theories and models on the continuum percolation to predict the percolation threshold, especially about nanofillers with extremely large or small aspect ratios. Along the direction of the physical percolation approach (Xu, 2012a; Xu, 2012c; Xu & Jie, 2014b), the ellipsoidal bound is applied to estimate percolation thresholds of composites containing various ellipsoidal inclusions. A comparison of the estimates with published experimental results (Pan et al., 2011) shows a great improvement achieved by the analytical prediction. In this chapter, the physical percolation approach is introduced by first presenting the ellipsoidal bound of the effective conductivity, then formulating percolation thresholds of complete dispersion and incomplete dispersion in Sections 6.1 and 6.2, respectively.

6.1 PERCOLATION THRESHOLD OF COMPLETE DISPERSION

6.1.1 Ellipsoidal bound

In the case of isotropic mixture described in Subsection 5.2.1, the counterpart of (5.34) on conductivity is directly written as (Xu, 2012a)

$$\boxed{\frac{K^{EB}}{K_0}\boldsymbol{I} = \boldsymbol{I} + c_1\left(\frac{1}{n-1}\overline{\boldsymbol{A}_1}^{-1} - \frac{c_1}{d}\boldsymbol{I}\right)^{-1}} \tag{6.1}$$

with the concentration tensor

$$\boldsymbol{A}_1 = (\boldsymbol{I} + \boldsymbol{S}_1(n-1))^{-1} \tag{6.2}$$

the contrast ratio

$$n = K_1/K_0 \tag{6.3}$$

and the dimensionality $d = 2$ or 3.

6.1.2 Percolation threshold of a 3D composite

When the spatial distribution of the inclusions is statistically isotropic, we simply have the average concentration tensor

$$\overline{\boldsymbol{A}_1} = \frac{1}{3}(A_{1,11} + A_{1,22} + A_{1,33})\boldsymbol{I} \tag{6.4}$$

By substituting (6.2), (6.4), and the rank-2 Eshelby tensor of spheroids (A4.5) into (6.1), it yields the ellipsoidal bound explicitly in terms of the aspect ratio as (Xu, 2012a)

$$\boxed{K^{EB}\big/K_0 = 1 + \frac{c_1}{\dfrac{1}{(n-1)\overline{A_1}} - \dfrac{c_1}{3}}} \tag{6.5}$$

$$\boxed{\overline{A_1} = \frac{1}{3(n-1)}\left(\frac{2}{S_{11}(\eta)+\dfrac{1}{n-1}} - \frac{1}{2S_{11}(\eta)-\dfrac{n}{n-1}}\right)} \tag{6.6}$$

where S_{11} is the 11-component of the Eshelby tensor on spheroids (A4.5) i.e.

$$S_{11}(\eta) = \frac{\eta^2\sqrt{\eta^2-1}+\eta\cdot gic(\eta)}{2(\eta^2-1)^{3/2}} \tag{6.7}$$

with a generalized inverse cosine function defined as

$$gic(\eta) = \begin{cases} -\cosh^{-1}(\eta), & prolate \quad \eta > 1 \\ \sqrt{-1}\cos^{-1}(\eta), & oblate \quad \eta < 1 \end{cases} \tag{6.8}$$

When the aspect ratio η approaches 1, (6.5) reduces exactly to the HS bounds (4.76–4.77) with d = 3. More specifically, the contrast ratio n greater or less than 1 corresponds to the lower bound (4.77) or the upper bound (4.76), respectively.

In an extreme case when the contrast ration $n \to \infty$, by taking the limit of (6.5) it yields the lower bound for a composite containing extremely conductive inclusions

$$\lim_{n\to\infty} K_-^{EB}/K_0 = 1 + \frac{c_1}{\dfrac{3}{\dfrac{2}{S_{11}(\eta)} - \dfrac{1}{2S_{11}(\eta)-1}} - \dfrac{c_1}{3}} \tag{6.9}$$

Similarly, when n = 0, it yields the upper bound for a composite containing extremely insulative inclusions

$$K_+^{EB}/K_0 = 1 + \frac{c_1}{\dfrac{3}{\dfrac{2}{S_{11}(\eta)-1} - \dfrac{1}{2S_{11}(\eta)}} - \dfrac{c_1}{3}} \tag{6.10}$$

With increase of the volume fraction of the inclusions, at a critical volume fraction c^*, the value of the ellipsoidal bound (6.5) rises rapidly toward the conductivity of the inclusion phase, and the percolation occurs. The critical volume fraction c^* is called *percolation threshold*. Since the bound (6.5) increases extremely fast upon the reach of the threshold, there is practically no difference of threshold prediction whether to have the conductivity ratio (6.5) equal to n or any upper bound such as the Voigt upper bound or the Hashin-Shtrikman upper bound.

By simply equating n to the conductivity ratio (6.5), the percolation threshold is analytically solved as

$$c^* = \frac{9\left[(\eta^2-1)(\eta^2-n)\left(\eta^2(n+1)-2\right)-(n-1)\eta\sqrt{\eta^2-1}(n+n\eta^2-2)gic(\eta)-(n-1)^2\eta^2 gic^2(\eta)\right]}{(n+2)(\eta^2-1)\left[(\eta^2-1)\left(\eta^2(5+n)-4n-2\right)-3(n-1)\eta\sqrt{\eta^2-1}gic(\eta)\right]} \tag{6.11}$$

which reduces to

$$c^* = \frac{9\eta\left(\eta-\eta^3-\sqrt{\eta^2-1}(\eta^2+1)gic(\eta)-\eta\cdot gic^2(\eta)\right)}{(\eta^2-1)\left(4-5\eta^2+\eta^4-3\eta\sqrt{\eta^2-1}gic(\eta)\right)} \tag{6.12}$$

when $n \to \infty$ and

$$c^* = \frac{9\eta\left(2\eta-3\eta^3+\eta^5-2\sqrt{\eta^2-1}gic(\eta)-\eta\cdot gic^2(\eta)\right)}{2(\eta^2-1)\left(2-7\eta^2+5\eta^4+3\eta\sqrt{\eta^2-1}gic(\eta)\right)} \tag{6.13}$$

when $n = 0$.

In many engineering applications especially of nanotechnology, the formula (6.11) can be approximated as (Xu, 2012c)

$$\boxed{c^* = 16\eta^{-1.81}} \tag{6.14}$$

for needle-like rods with $\eta \in [60, 200]$ and $n \geq 10^7$, and

$$\boxed{c^* = 3.5\eta} \tag{6.15}$$

for disk-like platelets with $\eta \in [1/2000, 1/60]$ and $n \geq 10^7$. In the specified range of the aspect ratio, the maximum deviation of the approximations (6.14–15) from the original (6.11) is less than 3%.

The percolation threshold formula (6.11) is graphed for a 3D composite in Figures 6.1 and 6.2, versus the contrast ratio and the aspect ratio, respectively. The figures show that, except for the insulating effect of prolate spheroids, the percolation threshold decreases exponentially with departure

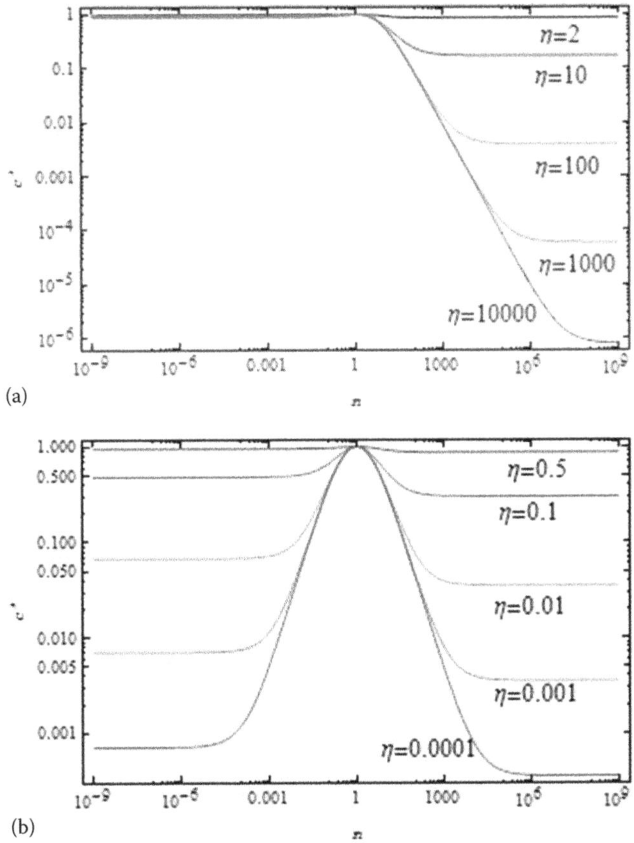

Figure 6.1 Percolation threshold vs. contrast ratio for 3D composites with complete dispersion of prolate (a) and oblate (b) spheroids (Eq. 6.11). (With kind permission from Springer Science+Business Media: *Acta Mech.*, Ellipsoidal bounds and percolation thresholds of transport properties of composites, 223, 2012a, 765–774, Xu, X.F. Permission conveyed through Copyright Clearance Center, Inc.)

of the either ratio from unity until the threshold converges to a certain constant. In Figure 6.1a the poor insulating effect of prolate spheroids is illustrated with the percolation threshold being close to 1 when $n < 1$. Figure 6.1b shows that oblate spheroids are effective in both conducting and insulating, with some more efficiency in conducting. The same behavior observed in Figure 6.2 especially indicates that, within the range of the aspect ratio 10^{-4}–10^{4}, the predicted percolation threshold converges when the contrast ratio reaches beyond 10^{6} or below 10^{-5}.

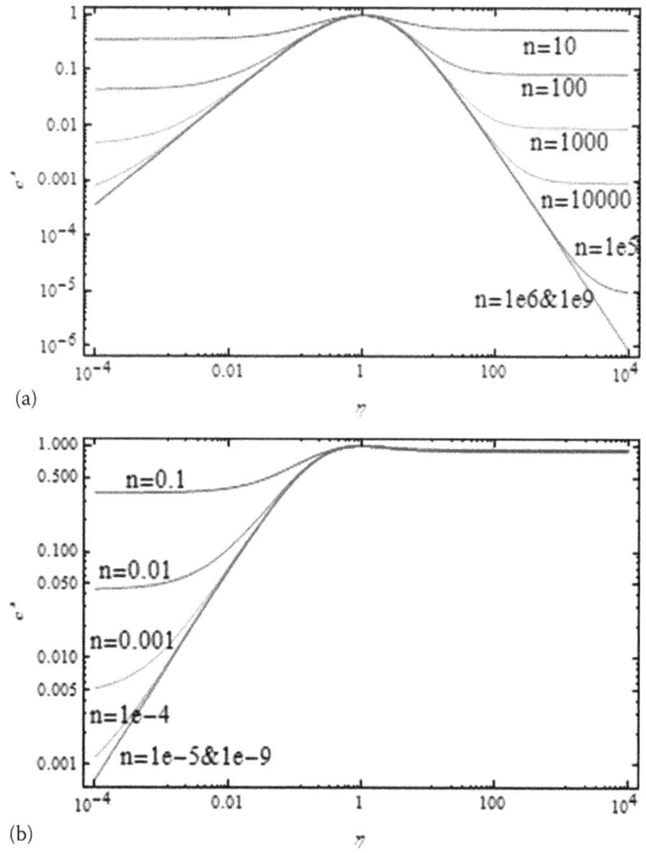

Figure 6.2 Percolation threshold vs. aspect ratio for 3D composites with complete dispersion of conductive (a) and insulative (b) spheroids (Eq. 6.11). (With kind permission from Springer Science+Business Media: *Acta Mech.*, Ellipsoidal bounds and percolation thresholds of transport properties of composites, 223, 2012a, 765–774, Xu, X.F. Permission conveyed through Copyright Clearance Center, Inc.)

When the aspect ratio η approaches infinity and zero, corresponding to needles and disks, respectively, (6.5) reduces to

$$K^{EB}/K_0 = \begin{cases} \dfrac{9(n+1)+2c_1(n-1)(n+5)}{9(n+1)-c_1(n-1)(n+5)} & when\ \eta \to \infty \\ \\ \dfrac{9n-2c_1(n-1)(2n+1)}{9n+c_1(n-1)(2n+1)} & when\ \eta \to 0 \end{cases} \quad (6.16\text{–}17)$$

from which the percolation threshold is solved as

$$c^{\bullet} = \begin{cases} \dfrac{9(1+n)}{(n+2)(n+5)} when\ \eta \to \infty \\ \dfrac{9n}{(n+2)(2n+1)} when\ \eta \to 0 \end{cases} \tag{6.18–19}$$

Equations (6.18–19) asymptotically provide the lower bounds of the percolation threshold, respectively, for any prolate and oblate spheroids. The two formulas indicate that, in a composite with a high contrast ratio, the percolation threshold of needles is about 2 times of that of disks, which in Figure 6.3 is indicated as the gap between two curves when $n >> 1$. Figure 6.3 also shows clearly the asymmetry of needles and the symmetry of disks with respect to conducting and insulating effects.

It is remarked that, when the aspect ratio η approaches infinity or zero, the quantification of the percolation threshold in terms of the volume fraction becomes trivial. According to the formulas $c_1 = \dfrac{\alpha}{\eta^2}$ (5.45) and $c_1 = \eta\alpha$ (5.49), the density of inclusions becomes infinitely large for a finite value volume fraction. In these two limiting cases of the aspect ratio, the percolation threshold should be more appropriately quantified in terms of the density of inclusions, rather than the volume fraction of inclusions. An example below is provided to illustrate the concept.

When the contrast ratio n in (6.5) is of the order $O(1/\eta^q)$with the exponent $q > 1$ and $\eta \to 0$, by taking the limit in (6.5) it yields a nontrivial lower bound

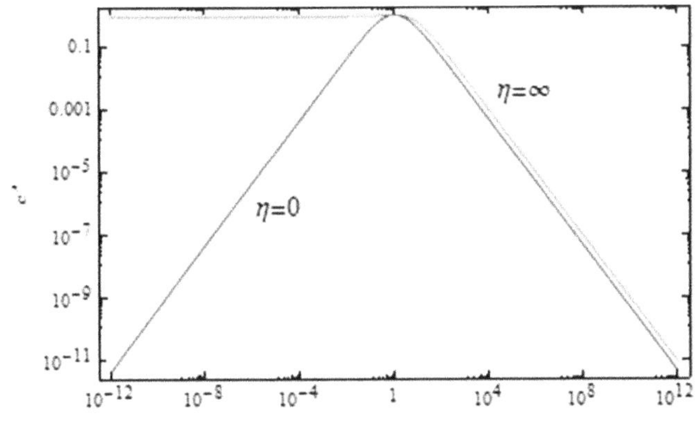

Figure 6.3 Asymptotic limit of percolation threshold for 3D composites containing needles ($\eta \to \infty$) and disks ($\eta \to 0$). (With kind permission from Springer Science+Business Media: *Acta Mech.*, Ellipsoidal bounds and percolation thresholds of transport properties of composites, 223, 2012a, 765–774, Xu, X.F. Permission conveyed through Copyright Clearance Center, Inc.)

$$K_-^{EB}\Big/K_0 = \frac{9\pi + 16\alpha}{9\pi - 8\alpha} \tag{6.20}$$

and the corresponding percolation threshold in terms of the critical density

$$\alpha^* = \frac{9}{8}\pi \tag{6.21}$$

A typical application of (6.20–21) is determination of rock permeability with respect to the crack density (e.g. Gueguen, Chelidze, & Ravalec, 1997). The critical crack density of (6.21) physically means approximately three cracks in each sphere of the Hashin-type self-similar sphere assemblage (Figure 5.2a). At such a critical crack density, a rock is predicted to be completely permeable.

6.1.3 Percolation threshold of a 2D composite

Laminates and thin films are widely used in micro- and nanocomposites. In such a 2D composite, parallel fibrous fillers are modeled as cylinders possessing elliptic cross-sections with the Eshelby tensor given in (A4.2) and the average concentration tensor

$$\bar{\boldsymbol{A}} = \frac{1}{2}(A_{11} + A_{22})\boldsymbol{I} \tag{6.22}$$

In this 2D case, we denote

$$\eta = a_2/a_1 \tag{6.23}$$

the cross-sectional aspect ratio of the fillers. Substitution of (6.22), (6.2), and (A4.2) into (6.1) with $d = 2$ yields the following ellipsoidal bound on the in-plane (axis-1 and axis-2) conductivity of a 2D composite (Xu, 2012c)

$$K^{EB}\Big/K_0 = 1 + \frac{c_1}{\dfrac{2(n+\eta)(1+n\eta)}{(n^2-1)(1+\eta)^2} - \dfrac{c_1}{2}} \tag{6.24}$$

When the cross-sectional aspect ratio $\eta = 1$, i.e. circular cylinders, (6.24) reduces exactly to the HS bound (4.76) or (4.77) with $d = 2$. The percolation threshold is directly solved from (6.24) as

$$c^* = \frac{4(n\eta+1)(n+\eta)}{(n+1)^2(\eta+1)^2} \tag{6.25}$$

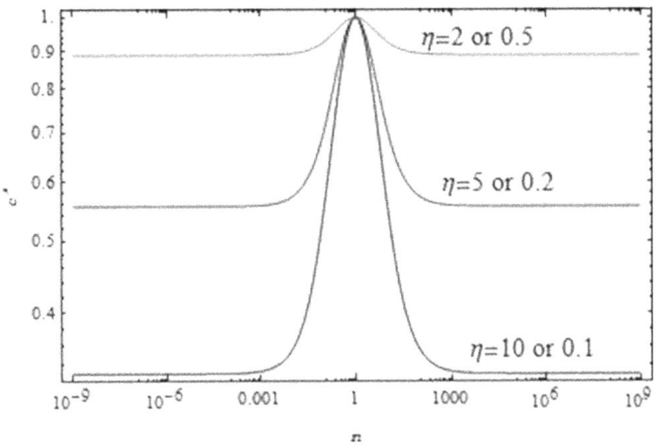

Figure 6.4 Percolation threshold of a 2D composite containing parallel fibrous fillers with an elliptic cross section.

which is completely symmetric about either n or η, as shown in Figure 6.4. When the contrast ratio $n \to \infty$ or 0, (6.25) identically reduces to

$$\boxed{c^* = \frac{4\eta}{(\eta+1)^2}} \tag{6.26}$$

which clearly shows the effect of the cross-sectional aspect ratio on percolation, e.g. the maximum is 1 when $\eta = 1$, corresponding to the morphology of the Hashin coated sphere assemblage.

To take into account the finite thickness of laminates and thin films (i.e. the thickness effect), the dimensional effect has been assessed with the formulas provided in Xu (2012c).

6.1.4 Percolation threshold of fillers with various aspect ratios

In the case of complete dispersion of an N-phase composite, the conductivity counterpart of (5.34) is directly written as

$$\frac{K^{EB}}{K_0}\boldsymbol{I} = \boldsymbol{I} + \left(\overline{\overline{A}}^{-1} - \frac{1}{d}\boldsymbol{I}\right)^{-1} \tag{6.27}$$

or simply

$$\frac{K^{EB}}{K_0} = 1 + \frac{d \cdot \overline{\overline{A}}}{d - \overline{\overline{A}}} \tag{6.28}$$

with the dimensionality $d = 2$ or 3, and

$$\overline{\overline{A}} = \sum_{k=1}^{N-1} c_k (n_k - 1)\overline{A_k} \tag{6.29}$$

or in the continuous version using double integrals

$$\overline{\overline{A}} = (1 - c_0)\iint (n-1)\overline{A_k}(\eta, n) f_{HN}(\eta, n)\, d\eta\, dn \tag{6.30}$$

given $f_{\mathrm{HN}}(\eta, n)$ the bivariate probability density function (PDF), and c_0 the volume fraction of the matrix phase. Similar to (6.6), the average concentration factor $\overline{A}_k(\eta, n)$ in (6.29–30) is obtained from (6.2) and (6.4) or (6.22) as

$$\boxed{\overline{A}_k(\eta, n) = \begin{cases} \dfrac{1}{3(n_k - 1)}\left(\dfrac{2}{S_{11}(\eta_k) + \dfrac{1}{n_k - 1}} - \dfrac{1}{2S_{11}(\eta_k) - \dfrac{n_k}{n_k - 1}}\right) & 3D \\[2ex] \dfrac{(n_k + 1)(1 + \eta_k)^2}{2(n_k + \eta_k)(1 + n_k\eta_k)} & 2D \end{cases}}$$

(6.31–32)

In application, the fillers in a composite characterized with different aspect ratios often share an identical contrast ratio. For example, in a CNT composite the nanotubes with an identical cross section usually have various lengths. In such a case, all the fillers in a composite are considered to be of Phase-1, and the bound (6.27) is rewritten in the form of (6.1) as

$$\frac{K^{EB}}{K_0}\boldsymbol{I} = \boldsymbol{I} + c_1\left(\frac{1}{n-1}\overline{\overline{\boldsymbol{A}_1}}^{-1} - \frac{c_1}{d}\boldsymbol{I}\right)^{-1} \tag{6.33}$$

with

$$\boxed{\overline{\overline{A_1}} = \sum_{k=1}^{N-1} f_k \overline{A_k}} \tag{6.34}$$

or

$$\boxed{\overline{\overline{A_1}} = \int \overline{\overline{A_k}}(\eta) f_H(\eta)\, d\eta} \tag{6.35}$$

where $f_k = c_{1,k}/c_1$ denotes the fraction of Type-k fillers among all the fillers, and $f_H(\eta)$ the probability density function. Let the right side of (6.33) equal to n, and the percolation is solved as

$$c^* = \frac{d}{(n+d-1)\overline{\overline{A_1}}} \tag{6.36}$$

Example

Carbon nanotubes characterized with two aspect ratios, η = 1000 and 100, are randomly dispersed in a 3D composite. Given the contrast ratio of conductivity $n = 10^8$, the percolation threshold calculated from Eq. (6.36) is plotted in Figure 6.5 versus f_1, the percentage of the nanotubes with the aspect ratio 1000 among all the nanotubes in the composite. For comparison, the percolation threshold calculated using the mean aspect ratio is shown as the dashed line, which is always greater than the correct one using Eq. (6.36), e.g. with f_1 = 50% the two thresholds are 0.0117 vol% and 0.0179 vol%, respectively, and the overestimate is more than 50%.

Example

To investigate the electrical conductivity of a polyvinyl-alcohol-based composite, multiwalled nanotubes are dispersed into thin films with a thickness 65μm. The diameters and lengths of the nanotubes are measured as 14 ± 6 nm, and 1.1 ± 0.4 μm, respectively (Hernandez et al., 2008).

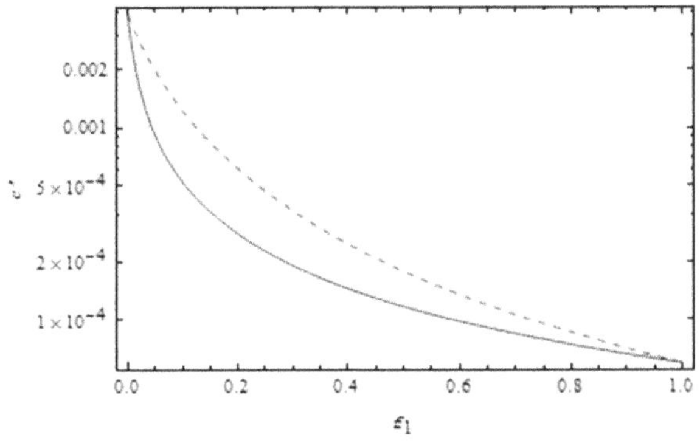

Figure 6.5 Percolation threshold vs. percentage of nanotubes with η_1 = 1000, in a 3D composite containing nantotubes with aspect ratios 1000 and 100, and a contrast ratio 10^8 (continuous – Eq. (6.36), dash – Eq. (6.11) using the mean aspect ratio). (Republished with permission of American Society of Mechanical Engineers (ASME), from Xu, X.F., *J. Eng. Mater. Technol.*, 134: 031008, 2012c. Permission conveyed through Copyright Clearance Center, Inc.)

As the thickness of thin films is much greater than the lengths of the nanotubes, the problem is 3D with the dimensionality $d = 3$. Given the bivariate PDF $f_{XY}(x, y)$ for the diameter and the length, $\overline{\overline{A}}$ in Eq. (6.30) is fast computed using Monte Carlo sampling. For the sake of simplicity assume the diameter and the length of a filler are independent from each other, i.e. $f_{XY}(x, y) = f_X(x)f_Y(y)$, and the two following PDFs are considered.

i. Given $f_X(x)$ a uniform distribution of the diameter in a range $\left[14-6\sqrt{3}, 14+6\sqrt{3}\right]$ and $f_Y(y)$ a uniform distribution of the length in a range $\left[1100-400\sqrt{3}, 1100+400\sqrt{3}\right]$, with $n \geq 10^6$ the optimal percolation threshold is calculated from (6.36) as 0.28vol%; and
ii. Given two truncated normal distributions, i.e. negative samples are discarded and the sampled aspect ratios are further restricted to an interval [10, 500] similar to that of i), with $n \geq 10^6$ the optimal percolation threshold is calculated from (6.36) as 0.32vol%.

The histograms created using 1 million samples are shown in Figure 6.6. Both results compare well with the experimental measurement 0.4vol% (Hernandez et al., 2008), which indicates that the dispersion made in this particular experiment is not far from the optimal, i.e. complete dispersion.

6.1.5 Concluding remark

As a concluding remark of this section, it should be emphasized that the previously discussed percolation threshold is predicted based on two assumptions with opposite effects. The first assumption has been explicitly emphasized as complete dispersion of inclusions, which obviously keeps a predicted threshold minimum. The second assumption is implicitly hidden in the employed ellipsoidal bound, i.e. a continuous spectrum of size distribution for the inclusions from the correlation length ℓ_c down to zero. Take conductive fillers as an example. The derived ellipsoidal bound as the lower bound corresponds to a spatial arrangement of inclusions with a continuous spectrum of size distribution, subjected to the constraints of the inclusion shape and the volume fraction. The second assumption always increases a predicted threshold.

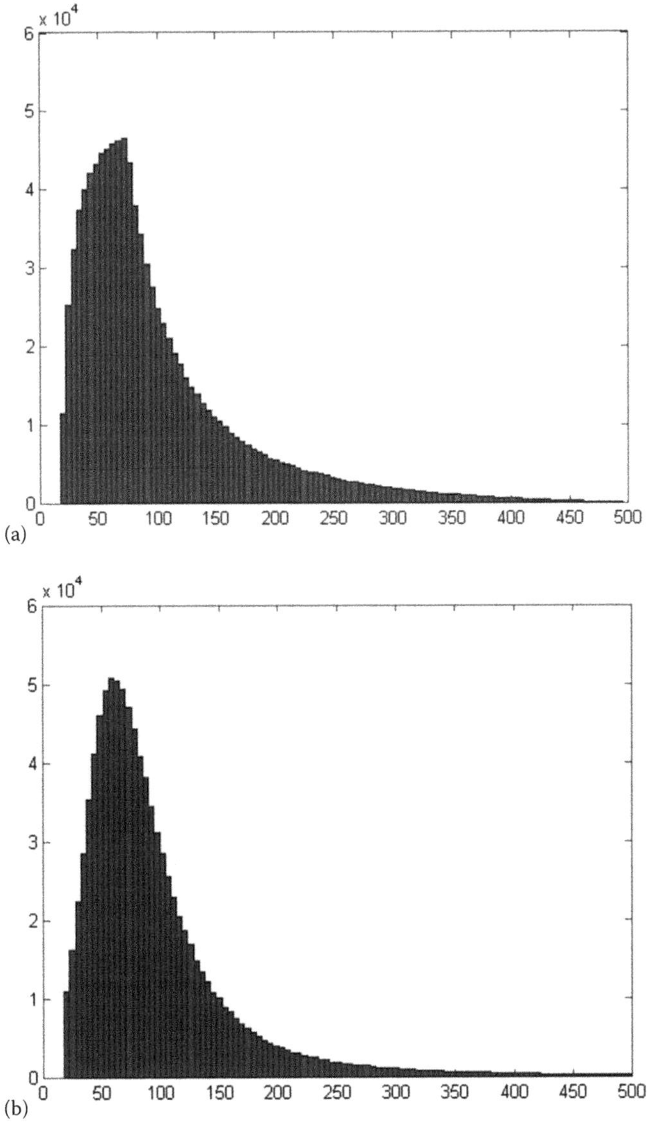

Figure 6.6 Histograms of the aspect ratio, based on 1 million samples with the lengths and the diameters following the uniform distributions (a), and the truncated normal distributions (b). (Republished with permission of American Society of Mechanical Engineers (ASME), from Xu, X.F., *J. Eng. Mater. Technol.*, 134: 031008, 2012c. Permissionconveyed through Copyright Clearance Center, Inc.)

The combined effect of the two assumptions depends on the aspect ratio of fillers. When the aspect ratio is close to unity (i.e. spheres), the effect of complete dispersion is minimum in spheres and the like, while the effect of continuous size distribution becomes overwhelming. For example, given all the spheres with an identical size, the percolation threshold will be over-predicted as 1, if the previously discussed formulae are directly applied without taking into account the additional morphological information about the size being all identical. With the aspect ratio deviated from unity, the effect of complete dispersion quickly surpasses the effect of the continuous size distribution. To most of fibers and platelets with the aspect ratios strongly deviated from 1, the combined effect of the two assumptions always results in an optimal prediction of the percolation threshold. In other words, the formulas derived in this section can be well used as a lower bound benchmark to evaluate the degree of dispersion for a composite of fibers or platelets.

6.2 PERCOLATION THRESHOLD OF INCOMPLETE DISPERSION

6.2.1 Ellipsoidal bound accounting for clustering effect

In the preceding section, the percolation threshold is predicted by assuming complete dispersion, i.e. isotropy of the average Eshelby tensor. In a nanocomposite, nanofillers tend to cluster together due to van der Waals forces. Consequently, dispersion of nanofillers is usually incomplete (see Figure 5.9c). Clustering of nanofillers is characterized as alignment of a group of fillers sharing spatial proximity, i.e. a cluster. In the case of complete dispersion, all fillers are randomly distributed and randomly oriented following the pattern of isotropic mixture (Figure 5.2c), and thereby in (6.1) the average Eshelby tensor is considered isotropic. In the case of clustering, an agglomerate of fillers spatially close to filler i have their orientations aligned with the latter. By denoting such an agglomerate of fillers as a component, say Component-n, the corresponding Eshelby tensor $\tilde{\boldsymbol{S}}_n$ in (5.23) is certainly dependent on the geometry of the agglomerate. When all the agglomerates in a composite are spherical, as shown in Figure 6.7a, the morphology is actually equivalent to the example of isotropic mixture shown in Figure 5.2b by regrouping all the fillers of a component spatially into an agglomerate. Similarly, when the agglomerates have an ellipsoidal shape, as shown in Figure 6.7b, the morphology corresponds to the anisotropic mixture pattern shown in Figure 5.9c. Clearly, the average Eshelby tensor $\boldsymbol{S}$ in (5.55) is determined by the aspect ratio of the agglomerates denoted as $\tilde{\eta}$, which is also called a clustering parameter (Xu & Jie, 2014b). When the clustering parameter

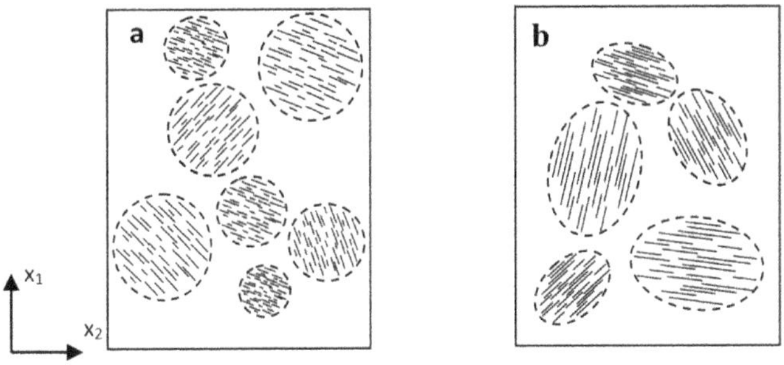

Figure 6.7 Morphological models for agglomerates of (a) isotropic mixture and (b) anisotropic mixture.

$\tilde{\eta}=1$, $\boldsymbol{S}$ becomes the spherical Eshelby tensor indicating complete dispersion. Since the assumption of complete dispersion corresponds to $\tilde{\eta}$ being 1, in case of nanotubes a value of $\tilde{\eta}$ less than 1 has no physical meaning, and the greater it is, the higher degree of clustering occurs. In case of nanoplatelets, conversely, a value of $\tilde{\eta}$ is never greater than 1, and the less it is, the higher degree of clustering occurs.

Based on the above description, the component-based expression (5.23) is rewritten into a two-phase bound with arguments of the two aspect ratios and Euler angles $\boldsymbol{\theta}$ explicitly included as

$$\frac{K^{EB}}{K_0}\boldsymbol{I}=\boldsymbol{I}+c_1\mathbf{A}_1(\eta,\boldsymbol{\theta})\overline{\left(\frac{1}{n-1}\boldsymbol{I}-c_1\mathbf{S}_1(\tilde{\eta},\boldsymbol{\theta})\ \overline{\mathbf{A}_1}(\eta)\right)}^{-1} \tag{6.37}$$

where $\mathbf{A}_1$ and $\mathbf{S}_1$ are exactly aligned with each other. By substituting (6.2), (6.4), and the Eshelby tensor of spheroids (A4.5) into (6.37), the ellipsoidal bound is explicitly expressed in terms of the clustering parameter $\tilde{\eta}$ as (Xu & Jie, 2014b)

$$\boxed{K^{EB}\Big/K_0=1+\frac{c_1}{3}\left[\frac{2}{\dfrac{1}{n-1}+S_{11}(\eta)-\dfrac{c_1}{6}S_{11}(\tilde{\eta})\left(\dfrac{n+2}{n-2(n-1)S_{11}(\eta)}+3\right)}+\frac{1}{\dfrac{n}{n-1}-2S_{11}(\eta)+\dfrac{c_1}{3}(2S_{11}(\tilde{\eta})-1)\left(\dfrac{2(n+2)}{1+(n-1)S_{11}(\eta)}-3\right)}\right]} \tag{6.38}$$

with S_{11} given in (A4.5).

When the clustering parameter $\tilde{\eta}$ approaches unity, (6.38) reduces exactly to (6.5–6.6) of the case of complete dispersion. By letting the bound (6.38) equal to the contrast ratio n, the percolation threshold is solved with the expression omitted due to its excessive length. The calculation can be conveniently handled by symbolic mathematics software. In Figure 6.8, the percolation threshold is graphed versus the clustering parameter with the contrast ratio being 10^{10}. In case of nanotubes (Figure 6.8a), the percolation threshold rises with increase of the clustering parameter $\tilde{\eta}$ from 1, and decrease of the aspect ratio η of individual nanotubes. In case of nanoplatelets (Figure 6.8b), conversely, the percolation threshold rises with decrease

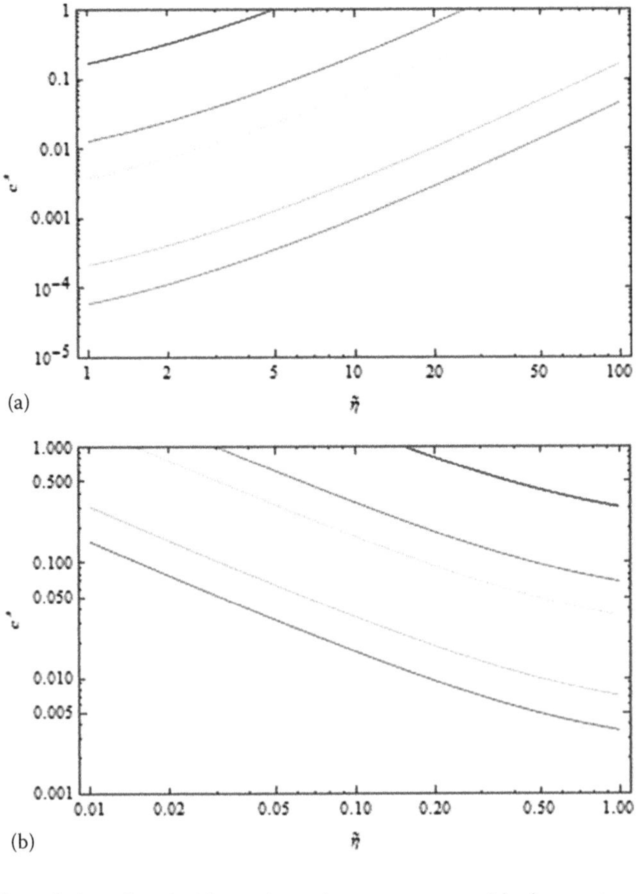

Figure 6.8 Percolation threshold vs. clustering parameter with the contrast ratio $n = 10^{10}$. The curves from top to bottom correspond to the aspect ratio η = 10, 50, 100, 500, 1000 (a) and 1/10, 1/50, 1/100, 1/500, 1/1000 (b), respectively.

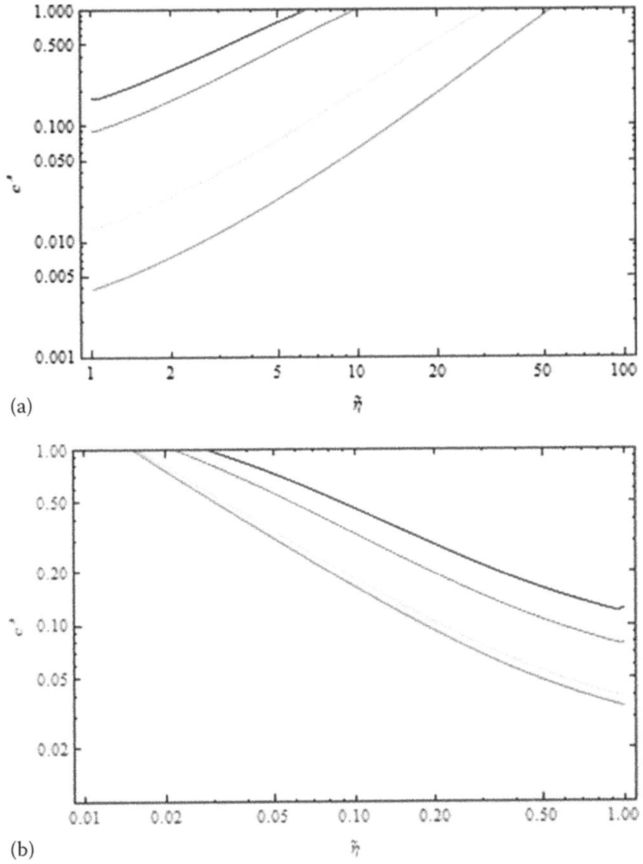

Figure 6.9 Percolation threshold vs. clustering parameter with the aspect ratio η = 100 (a) and 1/100 (b). The curves from top to bottom correspond to the contrast ratio n = 50, 100, 1000, 10^6 and 10^9, respectively, where the bottom two curves overlap with each other.

of the clustering parameter $\tilde{\eta}$ from 1, and increase of the aspect ratio η of individual nanoplatelets. In Figure 6.9, the percolation threshold is graphed versus the clustering parameter with the aspect ratio being 100 (a) and 1/100 (b). The relations are similarly shown between the percolation threshold and the two aspect ratios (i.e. $\tilde{\eta}$ and η).

The straight parallel lines in Figure 6.10 indicate that in particular ranges of the aspect ratio and the contrast ratio, the threshold is well approximated as a power law of the aspect ratio. In the complete dispersion case $\tilde{\eta}$ = 1, the power law approximations are provided as Eq. (6.14)

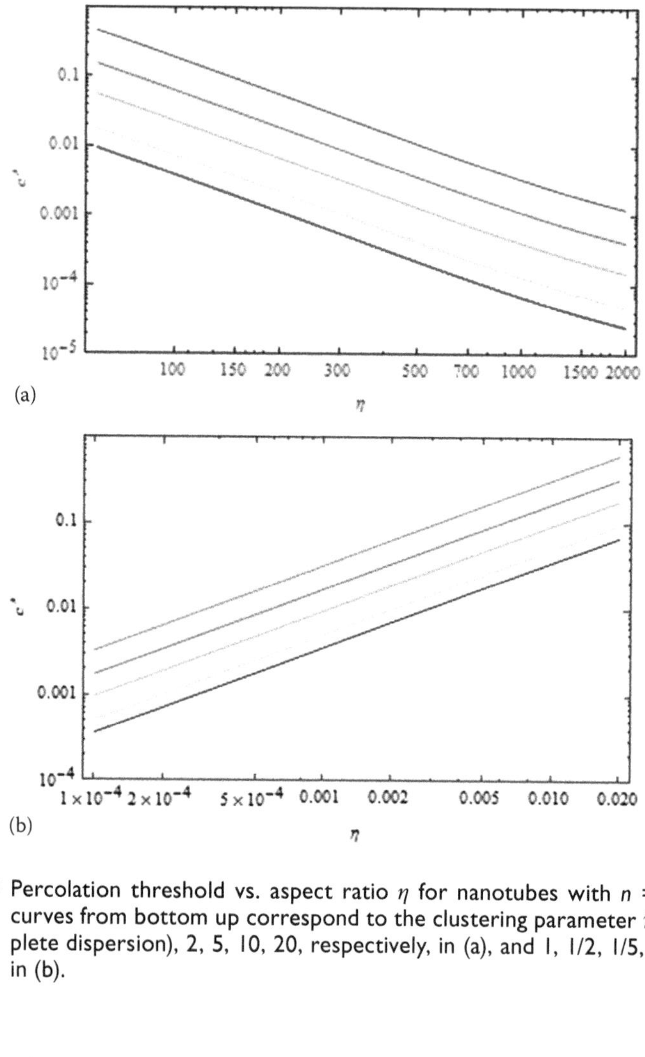

Figure 6.10 Percolation threshold vs. aspect ratio η for nanotubes with $n = 10^7$. The curves from bottom up correspond to the clustering parameter $\tilde{\eta} = 1$ (complete dispersion), 2, 5, 10, 20, respectively, in (a), and 1, 1/2, 1/5, 1/10, 1/20 in (b).

and (6.15) for nanotubes and nanoplatelets, respectively. To account for the clustering effect, the formulas (6.14–15) are further generalized into

$$\boxed{c^* = W_1(\tilde{\eta})\eta^{-1.81}} \tag{6.39}$$

$$\boxed{c^* = W_2(\tilde{\eta})\eta} \tag{6.40}$$

for nanotunes and nanoplatelets, respectively, in the specified ranges of the aspect and contrast ratios, e.g. those given in Subsection 6.1.2.

6.2.2 Asymptotic results

Extreme Contrast Ratios with n → ∞ and 0

In application of electrical nanocomposites, the contrast ratio n of the electrical conductivity is typically in the range 10^{10}–10^{18} that is many orders of magnitude beyond the common values of the aspect ratio. By letting n approach infinitely large, an asymptotic lower bound is obtained from (6.38) as (Xu & Jie, 2014b)

$$\lim_{n\to\infty} K_-^{EB}\Big/K_0 = 1 + \frac{c_1}{3}\left[\frac{2}{S_{11}(\eta) - \frac{c_1}{6}S_{11}(\tilde{\eta})\left(\frac{1}{1-2S_{11}(\eta)} + 3\right)} + \frac{1}{1 - 2S_{11}(\eta) + \frac{c_1}{3}(2S_{11}(\tilde{\eta}) - 1)\left(\frac{2}{S_{11}(\eta)} - 3\right)}\right] \tag{6.41}$$

Similarly, when the contrast ratio n is zero, an upper bound is obtained as (Xu & Jie, 2014b)

$$K_+^{EB}\Big/K_0 = 1 + \frac{c_1}{3}\left[\frac{2}{-1 + S_{11}(\eta) - \frac{c_1}{6}S_{11}(\tilde{\eta})\left(\frac{1}{S_{11}(\eta)} + 3\right)} + \frac{1}{-2S_{11}(\eta) + \frac{c_1}{3}(2S_{11}(\tilde{\eta}) - 1)\left(\frac{4}{1 - S_{11}(\eta)} - 3\right)}\right] \tag{6.42}$$

In Figure 6.11, the asymptotic results (6.41–6.42) are shown versus the volume fraction of nanoparticles with the aspect ratio being 1000 and 1/1000, respectively, in (a) and (b). The steep vertical rise of the curves indicates clearly the occurrence of percolation. With increase of the clustering parameter $\tilde{\eta}$ from 1, the percolation threshold increases in case of nanotubes as shown in Figure 6.11a. A similar phenomenon is observed in (b) in case of nanoplatelets with the aspect ratio being 1/1000, while in this case $\tilde{\eta}$ is decreased from 1.

Complete Clustering with $\tilde{\eta}$ → ∞ and 0

Complete clustering of nanotubes corresponds to $\tilde{\eta} \to \infty$. To check whether it is consistent that no percolation occurs, we take a double limit by further letting the contrast ratio approach infinitely large, as follows (Xu & Jie, 2014b):

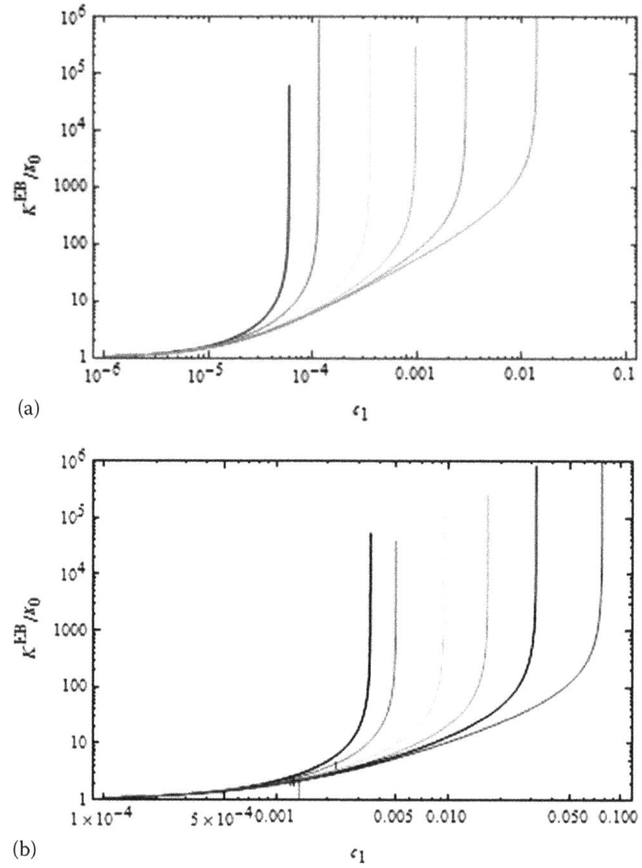

Figure 6.11 Effective conductivity of the asymptotic result (6.39–6.40) vs. volume fraction of nanoparticles with $n \to \infty$ and $\eta = 1000$ (a) and 1/1000 (b). The curves from left to right correspond to clustering parameter $\tilde{\eta} = 1$ (complete dispersion), 2, 5, 10, 20, 50, respectively, in (a), and 1, 1/2, 1/5, 1/10, 1/20, 1/50 in (b).

$$\lim_{n\to\infty}\left(\lim_{\tilde{\eta}\to\infty}\left(\frac{K_{-}^{EB}}{K_0}\right)\right)=$$
$$\frac{\left[\begin{array}{l}18\eta^3\mathrm{gic}^3(\eta)+9\eta^2\sqrt{\eta^2-1}\left(c_1(\eta^2-1)+2(\eta^2+2)\right)\mathrm{gic}^2(\eta)+3\eta(\eta^2-1)\left(c_1^2(\eta^2-1)^2-c_1(\eta^4-8\eta^2+7)+12\eta^2+6\right)\mathrm{gic}(\eta)\\ \quad -(\eta^2-1)^{\frac{3}{2}}\left(c_1^2(\eta^2-4)(\eta^2-1)^2+3c_1(\eta^4-5\eta^2+4)-18\eta^2\right)\end{array}\right]}{3\left(\sqrt{\eta^2-1}+\eta\cdot\mathrm{gic}(\eta)\right)\left[(\eta^2-1)\left(6\eta^2+c_1(\eta^4-5\eta^2+4)\right)+3\eta\sqrt{\eta^2-1}\left(c_1(\eta^2-1)+2(\eta^2+1)\right)\mathrm{gic}(\eta)+6\eta^2\mathrm{gic}^2(\eta)\right]} \tag{6.43}$$

Similarly, in case of complete clustering of nanoplatelets with $\tilde{\eta}\to 0$, the double asymptotic limit is obtained as (Xu & Jie, 2014b)

$$\lim_{n\to\infty}\left(\lim_{\tilde{\eta}\to 0}\left(\frac{K_{-}^{EB}}{K_0}\right)\right)=$$
$$\frac{9\eta^3\mathrm{gic}^3(\eta)+9\eta^2\sqrt{\eta^2-1}(2\eta^2+1)\mathrm{gic}^2(\eta)-3\eta(\eta^2-1)\left(4c_1^2(\eta^2-1)^2-3(\eta^2+2)\eta^2\right)\mathrm{gic}(\eta)+(\eta^2-1)^{\frac{3}{2}}\left(4c_1^2(\eta^2-4)(\eta^2-1)^2+9\eta^4\right)}{3\eta\left(\eta\sqrt{\eta^2-1}+\mathrm{gic}(\eta)\right)\left[(\eta^2-1)\left(3\eta^2+c_1(\eta^4-5\eta^2+4)\right)-3\eta\sqrt{\eta^2-1}\left(c_1(\eta^2-1)-(\eta^2+1)\right)\mathrm{gic}(\eta)+3\eta^2\mathrm{gic}^2(\eta)\right]} \tag{6.44}$$

The asymptotic results (6.43) and (6.44) are plotted in Figure 6.12. Compared with Figure 6.11, there is no steep rise of conductivity in the curves of Figure 6.12. The conductivity values shown in Figure 6.12, though large, remain finite even when the volume fraction $c_1 = 1$, which is incomparable to the infinitely large contrast ratio n. The both figures verify that, in the case of complete clustering, no percolation occurs for any fillers with a finite aspect ratio.

It is finally noted that it is straightforward to extend the previously discussed formulation accounting for the clustering effect to elasticity of composites. The detail is left as an exercise.

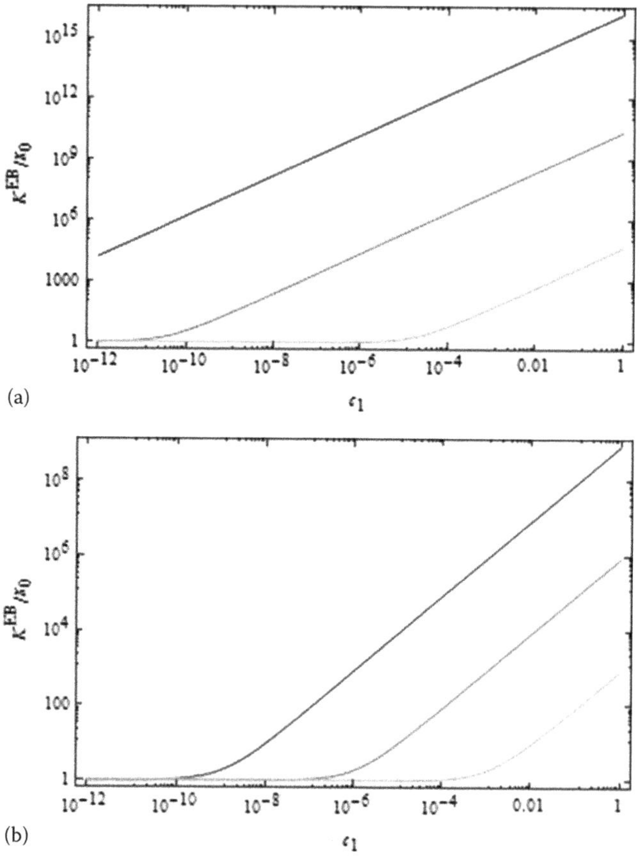

Figure 6.12 Effective conductivity of complete clustering vs. volume fraction of nanotubes (a) and nanoplatelets (b) with $n \to \infty$. The curves from left to right correspond to aspect ratio $\eta = 10^9, 10^6, 10^3$ (a) and $10^{-9}, 10^{-6}, 10^{-3}$ (b), respectively.

Part II

Computational analysis of scale-coupling problems

Chapter 7

Green-function-based variational principles for scale-coupling problems

In a finite body composite, typically there are three scales involved, i.e. a characteristic length scale of BVP size, a characteristic length scale of microstructure, and a length scale of deformation wave. When any two of the three scales are not well separated (i.e. Type-I or Type-II scale-coupling defined in Chapter 1), we say it is a *scale-coupling BVP*, e.g., a boundary layer with the thickness comparable to the microstructure size, a concrete beam or a rockfill dam with the width dimension a few times of the aggregate or rock size, a thin film with the thickness comparable to the crystalline size, and most of elastodynamics BVPs, just name a few. A major challenge confronting theoretical and computational mechanics is that in a scale-coupling BVP the classical homogenization approach becomes inapplicable. In engineering design and computation another relevant concern is, while a fine mesh of finite elements is needed to achieve high numerical accuracy, a certain mesh size threshold exists below which the material constants of these finite elements become nondeterministic when the mesh size scale and the length scale of microstructure are not well separated. In other words, conventional practice of assigning homogenized material constants to the finite elements of a BVP becomes invalid when the implicit assumption of scale separation is violated. To tackle such a scale-coupling BVP, new theory of scale-coupling mechanics is desired.

In this chapter, the Green-function-based variational principles presented in Chapter 3 are generalized from an infinite body composite to a finite body composite involving scale-coupling effects. Relevant studies of a finite body random composite were presented in Hori and Munasighe (1999), Luciano and Willis (2005), and Xu (2009). In Section 7.1 a scheme is first described to decompose a BVP into a slow-scale reference problem and a fast-scale fluctuation problem. In Section 7.2 the Green-function-based variational principles are formulated on scale-coupling problems, in the both deterministic and stochastic versions. Finally, in Section 7.3 the variational principles are applied onto a multiphase composite to demonstrate the size effect of RVE (i.e. Type-II scale-coupling effect).

7.1 DECOMPOSITION OF A BOUNDARY VALUE PROBLEM

7.1.1 Principle of superposition

The governing equations for a boundary value problem of a random heterogeneous composite are written as

$$\nabla \cdot \boldsymbol{\sigma}(x,\vartheta)+\boldsymbol{f}(\boldsymbol{x})=0 \quad \text{in } D \tag{7.1}$$

$$\boldsymbol{\sigma}(\boldsymbol{x},\vartheta)=\boldsymbol{L}(\boldsymbol{x},\vartheta)\boldsymbol{\varepsilon}(\boldsymbol{x},\vartheta) \tag{7.2}$$

$$\boldsymbol{\varepsilon}(\boldsymbol{x},\vartheta)=\nabla^s\boldsymbol{u}(\boldsymbol{x},\vartheta) \tag{7.3}$$

with the boundary conditions

$$\boldsymbol{u}(\boldsymbol{x},\vartheta)=\tilde{\boldsymbol{u}}(x) \quad \text{on } \partial D_u \tag{7.4}$$

$$\boldsymbol{\sigma}(\boldsymbol{x},\vartheta)\cdot\boldsymbol{n}=\tilde{\boldsymbol{t}}(\boldsymbol{x}) \quad \text{on } \partial D_t \tag{7.5}$$

As the attention is focused on scale-coupling effects between random microstructure and a finite body, the body force and boundary conditions are assumed deterministic. It should be noted that uncertainty of microstructure overwhelmingly presents in engineering practices since measurement of material properties at fine scales, in (7.1–7.5) the elastic moduli $\boldsymbol{L}$, is always incomplete and subject to error.

By applying the principle of superposition, the stochastic BVP (7.1–7.5) is decomposed into two sub-problems, a reference problem and a fluctuation problem. The reference problem is deterministic and characterized with a deterministic reference tensor L_0, often conveniently chosen to be uniformly constant, and a body force identical to that of the original BVP, i.e.

$$\nabla \cdot \boldsymbol{\sigma}_0(\boldsymbol{x})+\boldsymbol{f}(\boldsymbol{x})=0 \quad \text{in } D \tag{7.6}$$

$$\boldsymbol{\sigma}_0(\boldsymbol{x})=\boldsymbol{L}_0\boldsymbol{\varepsilon}_0(\boldsymbol{x}) \tag{7.7}$$

$$\boldsymbol{\varepsilon}_0(\boldsymbol{x},\vartheta)=\nabla^s\boldsymbol{u}_0(\boldsymbol{x},\vartheta) \tag{7.8}$$

with the boundary conditions

$$\boldsymbol{u}_0(\boldsymbol{x},\vartheta) = \tilde{\boldsymbol{u}}_0(\boldsymbol{x}) \quad \text{on } \partial D_u \tag{7.9}$$

$$\boldsymbol{\sigma}_0(\boldsymbol{x},\vartheta)\cdot\boldsymbol{n} = \tilde{\boldsymbol{t}}_0(\boldsymbol{x}) \quad \text{on } \partial D_t \tag{7.10}$$

The fluctuation problem, characterized with the same reference moduli $\boldsymbol{L}_0$, is subjected to a "body force" due to divergence of a stress polarization $\boldsymbol{p}$ and the complementary boundary conditions, i.e

$$\nabla\cdot\boldsymbol{\sigma}'(\boldsymbol{x},\vartheta) + \nabla\cdot\boldsymbol{p}(\boldsymbol{x},\vartheta) = 0 \quad \text{in } D \tag{7.11}$$

$$\boldsymbol{\sigma}'(\boldsymbol{x},\vartheta) = \boldsymbol{L}_0\boldsymbol{\varepsilon}'(\boldsymbol{x},\vartheta) \tag{7.12}$$

$$\boldsymbol{\varepsilon}'(\boldsymbol{x},\vartheta) = \nabla^s\boldsymbol{u}'(\boldsymbol{x},\vartheta) \tag{7.13}$$

with boundary conditions

$$\boldsymbol{u}'(\boldsymbol{x},\vartheta) = \tilde{\boldsymbol{u}}'(\boldsymbol{x}) \quad \text{on } \partial D_u \tag{7.14}$$

$$\boldsymbol{\sigma}'(\boldsymbol{x},\vartheta)\cdot\eta = \tilde{\boldsymbol{t}}'(x) \quad \text{on } \partial D_t \tag{7.15}$$

For the sake of brevity, the spatial argument $\boldsymbol{x}$ is dropped henceforth throughout this chapter, unless it is necessary. In the equilibrium equation (7.11), the stochastic "body force" is induced by divergence of the stress polarization

$$\boldsymbol{p}(\vartheta) = (\boldsymbol{L}(\vartheta) - \boldsymbol{L}_0)\,\boldsymbol{\varepsilon}(\vartheta) \tag{7.16}$$

which couples the fluctuation problem to the reference problem by introducing the total strain $\boldsymbol{\varepsilon}(\vartheta)$ into the fluctuation problem. The displacement, strain, and displacement boundary condition of the original BVP are, respectively, superposition of the reference terms and fluctuation terms, i.e.

$$\boldsymbol{u}(\vartheta) = \boldsymbol{u}_0 + \boldsymbol{u}'(\vartheta) \quad \text{in } D \tag{7.17}$$

$$\boldsymbol{\varepsilon}(\vartheta) = \boldsymbol{\varepsilon}_0 + \boldsymbol{\varepsilon}'(\vartheta) \quad \text{in } D \tag{7.18}$$

$$\tilde{\boldsymbol{u}}(\vartheta) = \tilde{\boldsymbol{u}}_0 + \tilde{\boldsymbol{u}}'(\vartheta) \qquad \text{on } \partial D_u \tag{7.19}$$

while the stress, in addition, contains the stress polarization

$$\boldsymbol{\sigma}(\vartheta) = \boldsymbol{\sigma}_0 + \boldsymbol{\sigma}'(\vartheta) + \boldsymbol{p}(\vartheta) \tag{7.20}$$

Denote $\Delta\tilde{\boldsymbol{t}}$ the traction difference on ∂D_t between the original BVP and the reference BVP, i.e.

$$\Delta\tilde{\boldsymbol{t}} = \tilde{\boldsymbol{t}} - \tilde{\boldsymbol{t}}_0 \qquad \text{on } \partial D_t \tag{7.21}$$

Consistency of boundary conditions thus requires

$$\boldsymbol{p}(\vartheta)\cdot\boldsymbol{n} = \Delta\tilde{\boldsymbol{t}} - \tilde{\boldsymbol{t}}' \qquad \text{on } \partial D_t \tag{7.22}$$

The decomposition scheme shown in Figure 7.1 can be verified from a variational perspective. The variations of the potential energy for the original BVP, the reference BVP, and the fluctuation BVP are given, respectively, with concise notation as

$$\delta\mathcal{U} = \langle\boldsymbol{\sigma},\delta\boldsymbol{\varepsilon}\rangle - \langle\boldsymbol{f},\delta\boldsymbol{u}\rangle - (\tilde{\boldsymbol{t}},\delta\boldsymbol{u}) \tag{7.23}$$

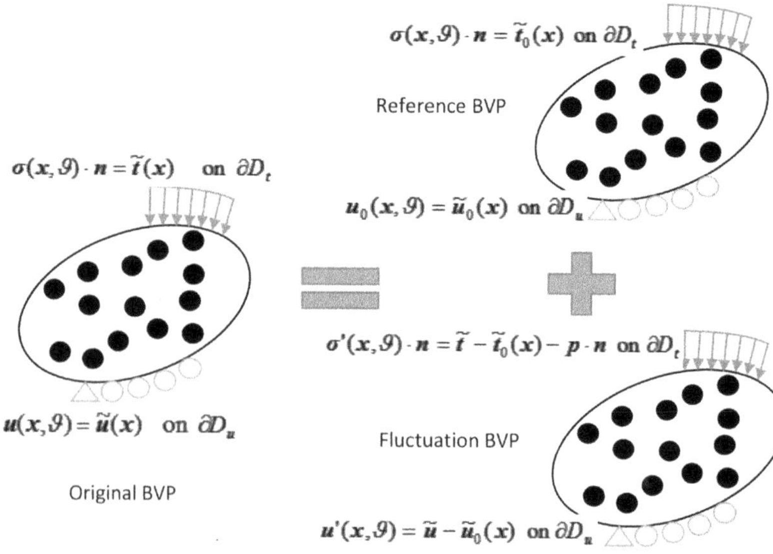

Figure 7.1 Schematic for decomposition of a boundary value problem.

$$\delta\mathcal{U}_0 = \langle \boldsymbol{\sigma}_0, \delta\boldsymbol{\varepsilon}_0 \rangle - \langle \boldsymbol{f}, \delta\boldsymbol{u}_0 \rangle - (\tilde{\boldsymbol{t}}_0, \delta\boldsymbol{u}_0) \tag{7.24}$$

$$\delta\mathcal{U}' = \langle \boldsymbol{\sigma}', \delta\boldsymbol{\varepsilon}' \rangle + \langle \boldsymbol{p}, \delta\boldsymbol{\varepsilon}' \rangle - (\tilde{\boldsymbol{t}}', \delta\boldsymbol{u}') \tag{7.25}$$

By employing (7.18) and (7.20), the first term on the right-hand side of (7.23) is rewritten as

$$\langle \boldsymbol{\sigma}, \delta\boldsymbol{\varepsilon} \rangle = \langle \boldsymbol{\sigma}_0, \delta\boldsymbol{\varepsilon}_0 \rangle + \langle \boldsymbol{\sigma}', \delta\boldsymbol{\varepsilon}' \rangle + \langle \boldsymbol{p}, \delta\boldsymbol{\varepsilon}' \rangle + \langle \boldsymbol{\sigma}_0, \delta\boldsymbol{\varepsilon}' \rangle + \langle \boldsymbol{\sigma}' + \boldsymbol{p}, \delta\boldsymbol{\varepsilon}_0 \rangle \tag{7.26}$$

By applying Gauss' divergence theorem, it follows that

$$\langle \boldsymbol{\sigma}_0, \delta\boldsymbol{\varepsilon}' \rangle = \langle \boldsymbol{f}, \delta\boldsymbol{u}' \rangle + (\tilde{\boldsymbol{t}}_0, \delta\boldsymbol{u}') \tag{7.27}$$

$$\langle \boldsymbol{\sigma}' + \boldsymbol{p}, \delta\boldsymbol{\varepsilon}_0 \rangle = (\tilde{\boldsymbol{t}}', \delta\boldsymbol{u}_0) \tag{7.28}$$

Substituting of (7.26–7.28) into (7.23) shows that $\delta\mathcal{U} = \delta\mathcal{U}_0 + \delta\mathcal{U}'$, which inversely verifies the decomposition scheme.

7.1.2 Stress polarization

In the fast-scale fluctuation BVP, the displacement is expressed in terms of the Green function as

$$u_i'(\boldsymbol{x}) = \int_D G_{ik}(\boldsymbol{x}, \boldsymbol{x}') \nabla p_{kl,l}(\boldsymbol{x}') d\boldsymbol{x}' + \int_{\partial D_t} G_{ik}(\boldsymbol{x}, \boldsymbol{x}') \tilde{t}_k'(\boldsymbol{x}') d\boldsymbol{x}' \tag{7.29}$$

To satisfy the boundary condition (7.14), the Green function in (7.29) is imposed with the following two constrains

$$\int_D G_{ik}(\boldsymbol{x}, \boldsymbol{x}') \nabla p_{kl,l}(\boldsymbol{x}') d\boldsymbol{x}' + \int_{\partial D_t} G_{ik}(\boldsymbol{x}, \boldsymbol{x}') t_k(\boldsymbol{x}') d\boldsymbol{x}' = \tilde{u}_i'(\boldsymbol{x}) \ \forall \boldsymbol{x} \text{ on } \partial D_u \tag{7.30}$$

$$\int_{\partial D_u} G_{ik}(\boldsymbol{x}, \boldsymbol{x}') t_k(\boldsymbol{x}') d\boldsymbol{x}' = 0 \quad \forall \boldsymbol{x} \text{ on } \partial D_u \tag{7.31}$$

for any stress polarization $\boldsymbol{p}$ and traction $\boldsymbol{t}$. By using (7.22), (7.31), and Gauss' divergence theorem, (7.29) becomes

$$u_i'(\boldsymbol{x}) = -\int_D \frac{\partial G_{ik}(\boldsymbol{x},\boldsymbol{x}')}{\partial x_l'} p_{kl}(\boldsymbol{x}')\,d\boldsymbol{x}' + \int_{\partial D_t} G_{ik}(\boldsymbol{x},\boldsymbol{x}')\Delta\tilde{t}_k(\boldsymbol{x}')\,d\boldsymbol{x}' \tag{7.32}$$

which, by employing the definition of the strain (7.13), yields

$$\varepsilon_{ij}'(\boldsymbol{x}) = -\int_D \Gamma_{ijkl}(\boldsymbol{x},\boldsymbol{x}')p_{kl}(\boldsymbol{x}')\,d\boldsymbol{x}' + \int_{\partial D_t} \Lambda_{ijk}(\boldsymbol{x},\boldsymbol{x}')\Delta\tilde{\boldsymbol{t}}_k(\boldsymbol{x}')\,d\boldsymbol{x}' \tag{7.33}$$

with two derivatives of the Green function

$$\Lambda_{ijk}(\boldsymbol{x},\boldsymbol{x}') = \frac{1}{2}\left(\frac{\partial G_{ik}(\boldsymbol{x},\boldsymbol{x}')}{\partial x_j} + \frac{\partial G_{jk}(\boldsymbol{x},\boldsymbol{x}')}{\partial x_i}\right) \tag{7.34}$$

$$\Gamma_{ijkl}(\boldsymbol{x},\boldsymbol{x}') = \frac{1}{2}\left(\frac{\partial G_{ik}(\boldsymbol{x},\boldsymbol{x}')}{\partial x_l'\partial x_j} + \frac{\partial G_{jk}(\boldsymbol{x},\boldsymbol{x}')}{\partial x_l'\partial x_i}\right) \tag{7.35}$$

For sake of analytical simplicity, usually the free space modified Green function is further symmetrized according to (3.22) that satisfies minor symmetries, i.e. $\Gamma_{ijkl} = \Gamma_{jikl} = \Gamma_{ijlk} = \Gamma_{jilk}$. To be consistent with the traction boundary condition (7.15), following (7.33) the derivatives of the Green function, $\boldsymbol{\Lambda}$ and $\boldsymbol{\Gamma}$, should satisfy the following constraint

$$\left[\boldsymbol{L}_0\left(-\int_D \boldsymbol{\Gamma}(\boldsymbol{x},\boldsymbol{x}')\boldsymbol{p}(\boldsymbol{x}')\,d\boldsymbol{x}' + \int_{\partial D_t} \boldsymbol{\Lambda}(\boldsymbol{x},\boldsymbol{x}')\boldsymbol{t}(\boldsymbol{x}')\,d\boldsymbol{x}'\right)\right]\cdot\boldsymbol{n} = \tilde{\boldsymbol{t}}'(\boldsymbol{x}) \qquad \forall\boldsymbol{x} \text{ on } \partial D_t \tag{7.36}$$

for any $\boldsymbol{p}$ and $\boldsymbol{t}$.

Equations (7.33) and (7.16) immediately lead to the exact solution of the polarization function as

$$\boldsymbol{p} = \left[(\boldsymbol{L}-\boldsymbol{L}_0)^{-1} + \boldsymbol{\Gamma}\right]^{-1}(\boldsymbol{\varepsilon}_0 + \boldsymbol{\Lambda}\Delta\tilde{\boldsymbol{t}}) \tag{7.37}$$

which is a generalization of (3.28) in classical micromechanics.

7.2 VARIATIONAL PRINCIPLES FOR A FINITE BODY RANDOM COMPOSITE

7.2.1 Odd-order variational principle

By choosing any trial polarization p that satisfies the traction consistency condition (7.22), the fluctuation strain, fluctuation stress, total strain, and total stress are accordingly generated as

$$\boldsymbol{\varepsilon}' = -\boldsymbol{\Gamma}\boldsymbol{p} + \boldsymbol{\Lambda}\Delta\tilde{t} \tag{7.38}$$

$$\boldsymbol{\sigma}' = -\boldsymbol{L}_0\boldsymbol{\Gamma}\boldsymbol{p} + \boldsymbol{L}_0\boldsymbol{\Lambda}\Delta\tilde{t} \tag{7.39}$$

$$\boldsymbol{\varepsilon} = \boldsymbol{\varepsilon}_0 - \boldsymbol{\Gamma}\boldsymbol{p} + \boldsymbol{\Lambda}\Delta\tilde{t} \tag{7.40}$$

$$\boldsymbol{\sigma} = \boldsymbol{L}_0\boldsymbol{\varepsilon}_0 - \boldsymbol{L}_0\boldsymbol{\Gamma}\boldsymbol{p} + \boldsymbol{L}_0\boldsymbol{\Lambda}\Delta\tilde{t} + \boldsymbol{p} \tag{7.41}$$

which are compatible with the displacement and traction boundary conditions, and by employing Gauss' divergence theorem, satisfy the following inner product relations

$$\langle \boldsymbol{\sigma}_0, \boldsymbol{\varepsilon}' \rangle = \langle \boldsymbol{f}, \boldsymbol{u}' \rangle + (\tilde{\boldsymbol{t}}_0, \boldsymbol{u}') + (\boldsymbol{t}_0, \tilde{\boldsymbol{u}}') \tag{7.42}$$

$$\langle \boldsymbol{\sigma}', \boldsymbol{\varepsilon}' \rangle = -\langle \boldsymbol{p}, \boldsymbol{\varepsilon}' \rangle + (\Delta\tilde{\boldsymbol{t}}, \boldsymbol{u}') + (\boldsymbol{t}' + \boldsymbol{p}\cdot\boldsymbol{n}, \tilde{\boldsymbol{u}}') \tag{7.43}$$

$$\langle \boldsymbol{\sigma}', \boldsymbol{\varepsilon}_0 \rangle = -\langle \boldsymbol{p}, \boldsymbol{\varepsilon}_0 \rangle + (\Delta\tilde{\boldsymbol{t}}, \boldsymbol{u}_0) + (\boldsymbol{t}' + \boldsymbol{p}\cdot\boldsymbol{n}, \tilde{\boldsymbol{u}}_0) \tag{7.44}$$

Note that in (7.42–7.44) the surface integrations are performed over either ∂D_t or ∂D_u due to the prescribed traction or displacement boundary condition.

By applying the minimum potential energy principle to the original BVP, the potential energy $\mathcal{U}$ is bounded from above

$$\mathcal{U} \le \mathcal{U}^+(\boldsymbol{p}) = \Delta\mathcal{U}^+(\boldsymbol{p}) + \mathcal{U}_0 \tag{7.45}$$

where the trail function $\boldsymbol{p}$ satisfies (7.22), $\mathcal{U}$ denotes the exact potential energy, and $\Delta\mathcal{U}^+$ the difference between the upper bound $\mathcal{U}^+$ and $\mathcal{U}_0$, the potential energy of the reference BVP. The bound $\mathcal{U}^+$ and the potential energy of the reference problem $\mathcal{U}_0$ are expressed, respectively, as

$$\mathcal{U}^+ = \frac{1}{2}\langle \boldsymbol{\varepsilon}, \boldsymbol{L}_0\boldsymbol{\varepsilon}\rangle + \frac{1}{2}\langle \boldsymbol{\varepsilon}, (\boldsymbol{L}-\boldsymbol{L}_0)\boldsymbol{\varepsilon}\rangle - \langle \boldsymbol{f}, \boldsymbol{u}\rangle - (\tilde{\boldsymbol{t}}, \boldsymbol{u}) \tag{7.46}$$

$$\mathcal{U}_0 = \frac{1}{2}\langle \boldsymbol{\varepsilon}_0, \boldsymbol{L}_0\boldsymbol{\varepsilon}_0\rangle - \langle \boldsymbol{f}, \boldsymbol{u}_0\rangle - (\tilde{\boldsymbol{t}}_0, \boldsymbol{u}_0) \tag{7.47}$$

By substituting (7.40) into (7.46–7.47) and taking (7.42–7.44) into account, it follows

$$\begin{aligned}\Delta\mathcal{U}^+(\boldsymbol{p},\boldsymbol{L}_0) = &\frac{1}{2}\langle \boldsymbol{p}, \boldsymbol{\Gamma}\boldsymbol{p}\rangle + \frac{1}{2}\langle \boldsymbol{\varepsilon}_0, (\boldsymbol{L}-\boldsymbol{L}_0)\boldsymbol{\varepsilon}_0\rangle + \frac{1}{2}\langle \boldsymbol{\Gamma}\boldsymbol{p} - \boldsymbol{\Lambda}\Delta\tilde{\boldsymbol{t}}, (\boldsymbol{L}-\boldsymbol{L}_0)(\boldsymbol{\Gamma}\boldsymbol{p} - \boldsymbol{\Lambda}\Delta\tilde{\boldsymbol{t}})\rangle \\ &- \langle \boldsymbol{\varepsilon}_0, (\boldsymbol{L}-\boldsymbol{L}_0)(\boldsymbol{\Gamma}\boldsymbol{p} - \boldsymbol{\Lambda}\Delta\tilde{\boldsymbol{t}})\rangle - \frac{1}{2}\langle \boldsymbol{\Lambda}\Delta\tilde{\boldsymbol{t}}, \boldsymbol{p}\rangle - (\Delta\tilde{\boldsymbol{t}}, \boldsymbol{u}_0) + (\boldsymbol{t}_0, \tilde{\boldsymbol{u}}') - \frac{1}{2}(\Delta\tilde{\boldsymbol{t}}, \boldsymbol{u}) \\ &+ \frac{1}{2}(\boldsymbol{t}' + \boldsymbol{p}\cdot\boldsymbol{n}, \tilde{\boldsymbol{u}}')\end{aligned} \tag{7.48}$$

By letting

$$\Delta\boldsymbol{\varepsilon} = \left((\boldsymbol{L}-\boldsymbol{L}_0)^{-1} + \boldsymbol{\Gamma}\right)\boldsymbol{p} - \boldsymbol{\Lambda}\Delta\tilde{\boldsymbol{t}} - \boldsymbol{\varepsilon}_0 \tag{7.49}$$

(7.48) reduces to

$$\begin{aligned}\Delta\mathcal{U}^+(\boldsymbol{p},\boldsymbol{L}_0) = &-\frac{1}{2}\langle \boldsymbol{p}, \boldsymbol{\Gamma}\boldsymbol{p}\rangle - \frac{1}{2}\langle \boldsymbol{p}, (\boldsymbol{L}-\boldsymbol{L}_0)^{-1}\boldsymbol{p}\rangle + \langle \boldsymbol{\varepsilon}_0, \boldsymbol{p}\rangle + \frac{1}{2}\langle \boldsymbol{\Lambda}\Delta\tilde{\boldsymbol{t}}, \boldsymbol{p}\rangle - (\Delta\tilde{\boldsymbol{t}}, \boldsymbol{u}_0) \\ &+ (\boldsymbol{t}_0, \tilde{\boldsymbol{u}}') - \frac{1}{2}(\Delta\tilde{\boldsymbol{t}}, \boldsymbol{u}') + \frac{1}{2}(\boldsymbol{t}' + \boldsymbol{p}\cdot\boldsymbol{n}, \tilde{\boldsymbol{u}}') + \frac{1}{2}\langle \Delta\boldsymbol{\varepsilon}, (\boldsymbol{L}-\boldsymbol{L}_0)\Delta\boldsymbol{\varepsilon}\rangle\end{aligned} \tag{7.50}$$

The complementary potential energy $\mathcal{U}^c$ is similarly bounded from above by invoking the minimum complementary energy principle, i.e.

$$\mathcal{U}^c \leq \mathcal{U}^{c+} = \Delta\mathcal{U}^{c+} + \mathcal{U}_0^c \tag{7.51}$$

with

$$\mathcal{U}^{c+} = \frac{1}{2}\langle \boldsymbol{\sigma}, \boldsymbol{M}_0\boldsymbol{\sigma}\rangle + \frac{1}{2}\langle \boldsymbol{\sigma}, (\boldsymbol{M}-\boldsymbol{M}_0)\boldsymbol{\sigma}\rangle - (\boldsymbol{t}, \tilde{\boldsymbol{u}}) \tag{7.52}$$

$$\mathcal{U}_0^c = \frac{1}{2}\langle \boldsymbol{\sigma}_0, \boldsymbol{M}_0\boldsymbol{\sigma}_0\rangle - (\boldsymbol{t}_0, \tilde{\boldsymbol{u}}_0) \tag{7.53}$$

where $\boldsymbol{M}$ and $\boldsymbol{M}_0$ are the compliance tensors of the original and reference BVPs, respectively. By substituting the self-equilibrated trial stress function (7.41) into (7.51–7.53), it follows that

$$\Delta\mathcal{U}^{c+}(\boldsymbol{p},\boldsymbol{L}_0) = \frac{1}{2}\langle \boldsymbol{p}, \boldsymbol{\Gamma}\boldsymbol{p}\rangle + \frac{1}{2}\langle \boldsymbol{p}, (\boldsymbol{L}-\boldsymbol{L}_0)^{-1}\boldsymbol{p}\rangle - \langle \boldsymbol{\varepsilon}_0, \boldsymbol{p}\rangle - \frac{1}{2}\langle \boldsymbol{\Lambda}\Delta\tilde{\boldsymbol{t}}, \boldsymbol{p}\rangle + (\Delta\tilde{\boldsymbol{t}}, \boldsymbol{u}_0)$$
$$- (\boldsymbol{t}_0, \tilde{\boldsymbol{u}}') + \frac{1}{2}(\Delta\tilde{\boldsymbol{t}}, \boldsymbol{u}') - \frac{1}{2}(\boldsymbol{t}' + \boldsymbol{p}\cdot\boldsymbol{n}, \tilde{\boldsymbol{u}}') + \frac{1}{2}\langle \boldsymbol{L}_0\Delta\boldsymbol{\varepsilon}, (\boldsymbol{M}-\boldsymbol{M}_0)\boldsymbol{L}_0\Delta\boldsymbol{\varepsilon}\rangle \tag{7.54}$$

The first three terms on the right-hand side of (7.50) and (7.54) correspond to those of the classical Hashin-Shtrikman functional (3.47). The fourth terms represent the extra work between the polarization and $\boldsymbol{\Lambda}\Delta\tilde{\boldsymbol{t}}$. The fifth and sixth terms are independent of $\boldsymbol{p}$, while the last terms vanish when $\boldsymbol{p}$ is taken to be the exact solution (7.37). The analytically intractable terms are the seventh and the eighth, and to get rid of these two terms we arrive at the following statement.

To make the potential energy bounds of a stochastic BVP (7.50) and (7.54) analytically tractable, the decomposition of the BVP shall be made in such a way that both the displacement and traction boundary conditions of the reference BVP are identical to those of the original BVP.

With $\tilde{\boldsymbol{u}}' = 0 \;\, \partial D_u$, $\Delta\tilde{\boldsymbol{t}} = 0 \;\, \partial D_t$, (7.50) and (7.54) therefore reduce, respectively, to

$$\Delta\mathcal{U}^{+}(\boldsymbol{p},\boldsymbol{L}_0) = -\frac{1}{2}\langle \boldsymbol{p}, \boldsymbol{\Gamma}\boldsymbol{p}\rangle - \frac{1}{2}\langle \boldsymbol{p}, (\boldsymbol{L}-\boldsymbol{L}_0)^{-1}\boldsymbol{p}\rangle + \langle \boldsymbol{\varepsilon}_0, \boldsymbol{p}\rangle + \frac{1}{2}\langle \Delta\boldsymbol{\varepsilon}, (\boldsymbol{L}-\boldsymbol{L}_0)\Delta\boldsymbol{\varepsilon}\rangle \tag{7.55}$$

$$\Delta\mathcal{U}^{c+}(\boldsymbol{p},\boldsymbol{L}_0) = \frac{1}{2}\langle \boldsymbol{p}, \boldsymbol{\Gamma}\boldsymbol{p}\rangle + \frac{1}{2}\langle \boldsymbol{p}, (\boldsymbol{L}-\boldsymbol{L}_0)^{-1}\boldsymbol{p}\rangle - \langle \boldsymbol{\varepsilon}_0, \boldsymbol{p}\rangle + \frac{1}{2}\langle \boldsymbol{L}_0\Delta\boldsymbol{\varepsilon}, (\boldsymbol{M}-\boldsymbol{M}_0)\boldsymbol{L}_0\Delta\boldsymbol{\varepsilon}\rangle \tag{7.56}$$

When $\boldsymbol{p}$ is taken to be exact, i.e.

$$\boldsymbol{p} = \left[(\boldsymbol{L}-\boldsymbol{L}_0)^{-1} + \boldsymbol{\Gamma}\right]^{-1}\boldsymbol{\varepsilon}_0 \tag{7.57}$$

we obtain $\Delta\boldsymbol{\varepsilon} = 0$, and thereby a complementary relation,

$$\Delta\mathcal{U}^{+} = -\Delta\mathcal{U}^{c+} \tag{7.58}$$

which, by further taking the complementary relation $\mathcal{U} = -\mathcal{U}^{c}$ (3.39) into account, yields

$$-\Delta\mathcal{U}^{c+} \leq -\Delta\mathcal{U}^{c} = \Delta\mathcal{U} \leq \Delta\mathcal{U}^{+} \tag{7.59}$$

with $\Delta\mathcal{U} = \mathcal{U} - \mathcal{U}_0$.

It should be noted that in deriving (7.59), it is necessary to enforce the consistency condition

$$\boldsymbol{p}\cdot\boldsymbol{n} = -\tilde{\boldsymbol{t}}' \text{ on } \partial D_t$$

to satisfy the traction condition $\Delta\tilde{\boldsymbol{t}} = 0 \;\; \partial D_t$. Otherwise in (7.55–7.56) there will be intractable terms related to $\Delta\tilde{\boldsymbol{t}}$ as shown in (7.50) and (7.54). In this decomposition, the traction of the fluctuation BVP, $\tilde{\boldsymbol{t}}'$, is not specified. A convenient choice is to simply set

$$\tilde{\boldsymbol{t}}' = 0 \tag{7.60}$$

which leads to a traction free boundary condition for the stress polarization, i.e.

$$\boldsymbol{p}\cdot\boldsymbol{n} = 0 \text{ on } \partial D_t \tag{7.61}$$

A summary of this exposition is formally stated as

The potential energy of a stochastic boundary value problem (7.1–7.5) is bounded from above and below

$$\mathcal{U}_0 - \Delta\mathcal{U}^{c+} = \mathcal{U}^{-} \leq \mathcal{U} \leq \mathcal{U}^{+} = \Delta\mathcal{U}^{+} + \mathcal{U}_0 \tag{7.62}$$

with

$$\boxed{\begin{aligned}\mathcal{U}^{+}(\boldsymbol{p},\boldsymbol{L}_0) = \mathcal{U}_0 + \frac{1}{2}\langle\boldsymbol{\varepsilon}_0,(\boldsymbol{L}-\boldsymbol{L}_0)\boldsymbol{\varepsilon}_0\rangle + \frac{1}{2}\langle\boldsymbol{p},\boldsymbol{\Gamma}\boldsymbol{p}\rangle + \frac{1}{2}\langle\boldsymbol{\Gamma}\boldsymbol{p},(\boldsymbol{L}-\boldsymbol{L}_0)\,\boldsymbol{\Gamma}\boldsymbol{p}\rangle \\ - \langle\boldsymbol{\Gamma}\boldsymbol{p},(\boldsymbol{L}-\boldsymbol{L}_0)\boldsymbol{\varepsilon}_0\rangle\end{aligned}} \tag{7.63}$$

$$\boxed{\begin{aligned}\mathcal{U}^-(\boldsymbol{p},\boldsymbol{L}_0) &= \mathcal{U}_0 - \frac{1}{2}\langle \boldsymbol{\varepsilon}_0, \boldsymbol{L}_0(\boldsymbol{M}-\boldsymbol{M}_0)\boldsymbol{L}_0\boldsymbol{\varepsilon}_0\rangle - \frac{1}{2}\langle \boldsymbol{p}, \boldsymbol{M}\boldsymbol{p}\rangle - \langle \boldsymbol{L}_0\boldsymbol{\varepsilon}_0, (\boldsymbol{M}-\boldsymbol{M}_0)\boldsymbol{p}\rangle \\ &- \frac{1}{2}\langle \boldsymbol{\Gamma p}, \boldsymbol{L}_0(\boldsymbol{M}-\boldsymbol{M}_0)\boldsymbol{L}_0\boldsymbol{\Gamma p}\rangle - \frac{1}{2}\langle \boldsymbol{p}, \boldsymbol{\Gamma p}\rangle + \langle \boldsymbol{\varepsilon}_0, \boldsymbol{L}_0(\boldsymbol{M}-\boldsymbol{M}_0)\boldsymbol{L}_0\boldsymbol{\Gamma p}\rangle + \langle \boldsymbol{L}_0\boldsymbol{M}\boldsymbol{p}, \boldsymbol{\Gamma p}\rangle\end{aligned}}$$

(7.64)

where the Green function satisfies the following boundary conditions

$$\int_D \boldsymbol{G}(\boldsymbol{x},\boldsymbol{x}')\nabla\cdot\boldsymbol{p}(\boldsymbol{x}')d\boldsymbol{x}' + \int_{\partial D_t} \boldsymbol{G}(\boldsymbol{x},\boldsymbol{x}')\boldsymbol{t}(\boldsymbol{x}')d\boldsymbol{x}' = 0 \quad \forall \boldsymbol{x} \text{ on } \partial D_u \tag{7.65}$$

$$\int_{\partial D_u} \boldsymbol{G}(\boldsymbol{x},\boldsymbol{x}')\boldsymbol{t}(\boldsymbol{x}')d\boldsymbol{x}' = 0 \quad \forall \boldsymbol{x} \text{ on } \partial D_u \tag{7.66}$$

$$\left[\boldsymbol{L}_0\left(-\int_{\boldsymbol{D}} \boldsymbol{\Gamma}(\boldsymbol{x},\boldsymbol{x}')\boldsymbol{p}(\boldsymbol{x}')\mathrm{d}\boldsymbol{x}' + \int_{\partial D_t} \boldsymbol{\Lambda}(\boldsymbol{x},\boldsymbol{x}')\boldsymbol{t}(\boldsymbol{x}')d\boldsymbol{x}'\right)\right]\cdot\boldsymbol{n} = 0 \quad \forall \boldsymbol{x} \text{ on } \partial D_t$$

(7.67)

for any reference moduli $\boldsymbol{L}_0$, traction $\boldsymbol{t}$, and stress polarization $\boldsymbol{p}$ satisfying

$$\boldsymbol{p}\cdot\boldsymbol{n} = 0 \text{ on } \partial D_t.$$

As the bounds (7.62–7.64) are formulated onto a sample of a finite body random heterogeneous solid, by applying ensemble averaging to (7.62–7.64), it yields the following stochastic version.

The probabilistically average potential energy of a stochastic boundary value problem (7.1–7.5) is bounded from above and below

$$\overline{\mathcal{U}^-} \le \overline{\mathcal{U}} \le \overline{\mathcal{U}^+} \tag{7.68}$$

with the expressions of the two bounds given in (7.63–7.64).

Higher probabilistic moments of the potential energy can be similarly bounded from above and below. By taking a functional variation of the upper or lower bound energy to minimize or maximize the functional, it results in the lowest upper bound or the highest lower bound in the function space constructed with an appropriately chosen trial polarization. Similar to what is described in Section 3.1, as the highest order correlation of the underlying microstructure involved in the two bounds (7.63–7.64) is always odd, the variational approach is called the *odd-order variational principle.*

7.2.2 Even-order Hashin-Shtrikman variational principle

Similar to (3.47), define the Hashin-Shtrikman functional of a finite body composite as

$$\boxed{\mathcal{H}(\boldsymbol{p},\boldsymbol{L}_0)=\mathcal{U}_0-\frac{1}{2}\langle\boldsymbol{p},\boldsymbol{\Gamma}\boldsymbol{p}\rangle-\frac{1}{2}\langle\boldsymbol{p},(\boldsymbol{L}-\boldsymbol{L}_0)^{-1}\boldsymbol{p}\rangle+\langle\boldsymbol{\varepsilon}_0,\boldsymbol{p}\rangle} \tag{7.69}$$

By noticing that the last term of (7.55) is positive when $\boldsymbol{L}-\boldsymbol{L}_0>0$ (i.e. positive definite), and the last term of (7.56) is positive when $\boldsymbol{M}-\boldsymbol{M}_0>0$ or equivalently $\boldsymbol{L}-\boldsymbol{L}_0<0$, the following inequalities are obtained:

The potential energy of a stochastic boundary value problem (7.1–7.5) is bounded from above and below

$$\mathcal{H}^-(\boldsymbol{p},\boldsymbol{L}_0)\le\mathcal{U}^-(\boldsymbol{p},\boldsymbol{L}_0)\le\mathcal{U}\le\mathcal{U}^+(\boldsymbol{p},\boldsymbol{L}_0)\le\mathcal{H}^+(\boldsymbol{p},\boldsymbol{L}_0) \tag{7.70}$$

for any reference moduli $\boldsymbol{L}_0$, *and any traction free stress polarization* $\mathbf{p}$, *where* $\mathcal{H}^+$ *and* $\mathcal{H}^-$ *correspond to the functional (7.69) subject to* $\boldsymbol{L}-\boldsymbol{L}_0\le 0$ *and* $\boldsymbol{L}-\boldsymbol{L}_0\ge 0$, *respectively, and* $\mathcal{U}^+,\mathcal{U}^-$ *are given in (7.63–7.64).*

Specifically, when the stress polarization $\boldsymbol{p}$ is chosen to be the exact solution (7.57), all the five terms in (7.70) become equal. Similar to (7.68), the stochastic version of (7.70) is given as follows:

The probabilistically average potential energy of a stochastic boundary value problem (7.1–7.5) is bounded from above and below

$$\overline{\mathcal{H}^-(\boldsymbol{p},\boldsymbol{L}_0)}\le\overline{\mathcal{U}^-(\boldsymbol{p},\boldsymbol{L}_0)}\le\overline{\mathcal{U}}\le\overline{\mathcal{U}^+(\boldsymbol{p},\boldsymbol{L}_0)}\le\overline{\mathcal{H}^+(\boldsymbol{p},\boldsymbol{L}_0)} \tag{7.71}$$

with the expressions given in (7.63–7.64) and (7.69).

Higher probabilistic moments of the potential energy can be similarly bounded from above and below. Compared with the infinite body variational principles formulated in Chapter 3, there are two distinctive features of the scale-coupling variational principles here:

i. There are boundary constraints imposed on the stress polarization; and
ii. There are boundary constraints specified upon the Green function as well.

Effects of stochasticity, diminishing in a wide-scale separation, become essential in a scale-coupling problem, and the stochastic variational principles provide a way to quantify such uncertainty. Identical to the infinite body Hashin-Shtrikman principle formulated in Chapter 3, the scale-coupling HS principle is of even order about the highest correlation of the microstructure involved. With sufficient widening of scale separation, both the odd- and even-order scale-coupling variational principles reduce to their scale-separation counterparts where the details of boundary layers are neglected and simple boundary conditions are imposed. In the next section, such a degree of sufficiency to quantify scale separation, more specifically of Type-II, is particularly investigated on the size effect of RVE.

7.3 MINIMUM SIZE OF REPRESENTATIVE VOLUME ELEMENT

7.3.1 Series representation of the stress polarization

In a boundary value problem of an N-phase random elastic composite, like (4.7), the random field of the elastic moduli is represented as

$$\boldsymbol{L}(\boldsymbol{x},\vartheta)=\sum_{n=1}^{N}\chi_n(\boldsymbol{x},\vartheta)\boldsymbol{L}_n$$

with the indicator function

$$\chi_n(x,\vartheta)=\begin{cases}1 & x\in D_n\\ 0 & x\notin D_n\end{cases}$$

From (7.16) and (7.57), we have

$$\boldsymbol{\varepsilon}=\boldsymbol{\varepsilon}_0-\boldsymbol{\Gamma}\boldsymbol{p} \tag{7.72}$$

Substitution of (4.7) and (7.72) into (7.16) yields a series representation for the stress polarization

$$\boldsymbol{p}=\sum_{r=1}^{N}(\boldsymbol{L}_r-\boldsymbol{L}_0)\chi_r\boldsymbol{\varepsilon}_0-\sum_{r=1}^{N}(\boldsymbol{L}_r-\boldsymbol{L}_0)\chi_r\boldsymbol{\Gamma}\sum_{s=1}^{N}(\boldsymbol{L}_s-\boldsymbol{L}_0)\chi_s\boldsymbol{\varepsilon}_0+\cdots \tag{7.73}$$

which is the scale-coupling version of (3.28). The expansion terms in Equation (7.73) form a function space for the trial polarization to apply the odd- and even-order variational principles formulated previously, which leads to an optimal kernel corresponding to a truncation of the series (7.73). For instance, the first-, and second-order trial polarizations are given, respectively, as

$$\boldsymbol{p}(\boldsymbol{x},\vartheta)=\sum_{r=1}^{N}\chi_r(\boldsymbol{x},\vartheta)\boldsymbol{p}_r(\boldsymbol{x}) \tag{7.74}$$

$$\boldsymbol{p}(\boldsymbol{x},\vartheta)=\sum_{r=1}^{N}\chi_r(\boldsymbol{x},\vartheta)\boldsymbol{p}_r^{(1)}(\boldsymbol{x})+\sum_{r=1}^{N}\chi_r(\boldsymbol{x},\vartheta)\boldsymbol{p}_r^{(2)}(\boldsymbol{x})\boldsymbol{\Gamma}\sum_{s=1}^{N}\chi_s(\boldsymbol{x},\vartheta)\boldsymbol{L}_s(\boldsymbol{x}) \tag{7.75}$$

The first-order approximation (7.74) is interpreted as a local relation between a stress polarization $\boldsymbol{p}_r$ in Phase-r and the corresponding random field $\chi_r(\boldsymbol{x},\vartheta)$, while a higher order approximation takes nonlocal effect into account by using a Green-function-based kernel. It should be noted that in a homogenization problem the first-order approximation (4.11) in Chapter 4 is a reduction of (7.74) to a piecewise constant polarization field, while in a scale-coupling BVP instead of being constant, the value of a polarization function varies within each individual phase.

7.3.2 First-order trial function

To solve $\boldsymbol{p}_r(\boldsymbol{x})$, we apply the stochastic HS variational principle, i.e.

$$\frac{\delta\overline{\mathcal{H}}}{\delta\boldsymbol{p}_r}=0 \tag{7.76}$$

By substituting (7.74) into (7.69) and taking a functional derivative about p_r, it yields

$$\begin{aligned}\frac{\delta\overline{\mathcal{H}}}{\delta\boldsymbol{p}_r(\boldsymbol{x})}&=-(\boldsymbol{L}_r-\boldsymbol{L}_0)^{-1}\sum_{s=1}^{N}\boldsymbol{p}_s(\boldsymbol{x})\overline{\chi_r(\boldsymbol{x},\vartheta)\chi_s(\boldsymbol{x},\vartheta)}\\&\quad-\frac{1}{2}\int_D\boldsymbol{\Gamma}(\boldsymbol{x},\boldsymbol{x}')\sum_{s=1}^{N}\boldsymbol{p}_s(\boldsymbol{x}')\overline{\chi_r(\boldsymbol{x},\vartheta)\chi_s(\boldsymbol{x}',\vartheta)}d\boldsymbol{x}'+\boldsymbol{\varepsilon}_0(\boldsymbol{x})\overline{\chi_r(\boldsymbol{x},\vartheta)}\\&=-\int_D(\boldsymbol{L}_r-\boldsymbol{L}_0)^{-1}\boldsymbol{p}_r(\boldsymbol{x})c_r(\boldsymbol{x})d\boldsymbol{x}-\frac{1}{2}\int_D\boldsymbol{\Gamma}(\boldsymbol{x},\boldsymbol{x}')\sum_{s=1}^{N}\boldsymbol{p}_s(\boldsymbol{x}')c_{rs}(\boldsymbol{x},\boldsymbol{x}')d\boldsymbol{x}'\\&\quad+c_r(\boldsymbol{x})\boldsymbol{\varepsilon}_0(\boldsymbol{x})\end{aligned} \tag{7.77}$$

and

$$\begin{aligned}\frac{\delta\bar{\mathcal{H}}}{\delta\boldsymbol{p}_r(\boldsymbol{x}')}&=-\frac{1}{2}\int_D\boldsymbol{\Gamma}(\boldsymbol{x},\boldsymbol{x}')\sum_{s=1}^{N}\overline{\boldsymbol{p}_s(\boldsymbol{x},\vartheta)\chi_r(\boldsymbol{x}',\vartheta)\chi_s(\boldsymbol{x},\vartheta)}d\boldsymbol{x}\\&=-\frac{1}{2}\int_D\boldsymbol{\Gamma}(\boldsymbol{x},\boldsymbol{x}')\sum_{s=1}^{N}\boldsymbol{p}_s(\boldsymbol{x}')c_{rs}(\boldsymbol{x},\boldsymbol{x}')d\boldsymbol{x}'\end{aligned}\tag{7.78}$$

Since

$$\frac{\delta\bar{\mathcal{H}}}{\delta\boldsymbol{p}_r}=\frac{\delta\bar{\mathcal{H}}}{\delta\boldsymbol{p}_r(\boldsymbol{x})}+\frac{\delta\bar{\mathcal{H}}}{\delta p_r(\boldsymbol{x}')}\tag{7.79}$$

with (7.76–7.78) we obtain the following N equations

$$(\boldsymbol{L}_r-\boldsymbol{L}_0)^{-1}\boldsymbol{p}_r(\boldsymbol{x})c_r(\boldsymbol{x})+\int_D\boldsymbol{\Gamma}(\boldsymbol{x},\boldsymbol{x}')\sum_{s=1}^{N}\boldsymbol{p}_s(\boldsymbol{x}')c_{rs}(\boldsymbol{x},\boldsymbol{x}')d\boldsymbol{x}'=c_r(\boldsymbol{x})\boldsymbol{\varepsilon}_0(\boldsymbol{x})\tag{7.80}$$

with r = 1, 2, …, N. In a statistically homogeneous composite, by using the correlation coefficient (4.10), (7.80) reduces to

$$(\boldsymbol{L}_r-\boldsymbol{L}_0)^{-1}\boldsymbol{p}_r(\boldsymbol{x})+\int_D\boldsymbol{\Gamma}(\boldsymbol{x},\boldsymbol{x}')\sum_{s=1}^{N}\boldsymbol{p}_s(\boldsymbol{x}')\big(\rho_{rs}(\boldsymbol{x}-\boldsymbol{x}')(\delta_{rs}-c_s)+c_s\big)d\boldsymbol{x}'=\boldsymbol{\varepsilon}_0(\boldsymbol{x})\tag{7.81}$$

with r = 1, 2, …, N, which is the scale-coupling version of (4.13).

To incorporate the free traction condition (7.61), (7.80) is reformulated into a weak form by employing a Lagrange multiplier $\boldsymbol{u}_r(x)$

$$\begin{aligned}&\int_D(\boldsymbol{L}_r-\boldsymbol{L}_0)^{-1}\boldsymbol{p}_r(\boldsymbol{x})\,c_r(\boldsymbol{x})\delta\boldsymbol{p}_r(\boldsymbol{x})d\boldsymbol{x}\\&+\int_D\int_D\boldsymbol{\Gamma}(\boldsymbol{x},\boldsymbol{x}')\delta\boldsymbol{p}_r(\boldsymbol{x})\sum_{s=1}^{N}\boldsymbol{p}_s(\boldsymbol{x}')c_{rs}(\boldsymbol{x},\boldsymbol{x}')d\boldsymbol{x}'d\boldsymbol{x}\\&+\int_{D_t}\partial\delta\big[\boldsymbol{u}_r(\boldsymbol{x})\big(\boldsymbol{p}_r(\boldsymbol{x})\cdot\boldsymbol{n}\big)\big]\,d\boldsymbol{x}=\int_D\delta\boldsymbol{p}_r(\boldsymbol{x})c_r(\boldsymbol{x})\boldsymbol{\varepsilon}_0(\boldsymbol{x})\,d\boldsymbol{x}\end{aligned}\tag{7.82}$$

that is equivalently

$$\int_D (\boldsymbol{L}_r - \boldsymbol{L}_0)^{-1} \boldsymbol{p}_r(\boldsymbol{x})\ c_r(\boldsymbol{x}) \delta \boldsymbol{p}_r(\boldsymbol{x}) d\boldsymbol{x} + \int_D \int_D \boldsymbol{\Gamma}(\boldsymbol{x}, \boldsymbol{x}') \delta \boldsymbol{p}_r(\boldsymbol{x})$$
$$\sum_{s=1}^{N} \boldsymbol{p}_s(\boldsymbol{x}') c_{rs}(\boldsymbol{x}, \boldsymbol{x}') d\boldsymbol{x}' d\boldsymbol{x} + \int_{D_t} \partial \delta \boldsymbol{u}_r(\boldsymbol{x}) \big(\boldsymbol{p}_r(\boldsymbol{x}) \cdot \boldsymbol{n} \big) d\boldsymbol{x} \tag{7.83}$$
$$+ \int_{D_t} \partial \boldsymbol{u}_r(\boldsymbol{x}) (\delta \boldsymbol{p}_r(\boldsymbol{x}) \cdot \boldsymbol{n})\ d\boldsymbol{x} = \int_D \delta \boldsymbol{p}_r(\boldsymbol{x}) c_r(\boldsymbol{x}) \boldsymbol{\varepsilon}_0(\boldsymbol{x}) d\boldsymbol{x}$$

with $r = 1, 2, \ldots, N$. The computational detail to solve (7.83) is presented in Chapter 8.

When the first-order trial function is applied to the potential energy functional (7.63) or (7.64), it results in N equations involving the triple correlation function of microstructure, which reduce to (4.78) on a two-phase composite with the scales well separated.

7.3.3 Minimum RVE size in 3D

In a homogenization problem with the scales well separated, (7.81) becomes (4.13) and all the bounds are analytically obtained in Chapter 4. When the size of a RVE domain is finite comparable to the correlation length of heterogeneities, the results obtained from (7.81) lead to the so-called size-dependent HS bounds. The gap between the size-dependent HS bounds and the classical HS bounds demonstrates the size effect of RVE. In the following an assessment made in Xu and Chen (2009) is presented.

As the size of an RVE is compared with the correlation length of microstructure, by taking the correlation length as a variable in the size quantification, we can conveniently use the free space modified Green function $\boldsymbol{\Gamma}^{\infty}$ (A2.6) that is analytically available. In a statistically homogeneous composite, the volume fractions present in (7.80) are all constant, and (7.80) is rewritten in Fourier space as

$$c_r (\boldsymbol{L}_r - \boldsymbol{L}_0)^{-1} \hat{\boldsymbol{p}}_r(\boldsymbol{\xi}) + \sum_{s=1}^{N} \left(\hat{\boldsymbol{\Gamma}}^{\infty} * \hat{c}_{rs} \right)(\boldsymbol{\xi}) \hat{\boldsymbol{p}}_s(\boldsymbol{\xi}) = c_r \hat{\varepsilon}_0(\boldsymbol{\xi}) \tag{7.84}$$

where $r = 1, 2, \ldots, N$, and the symbol $*$ indicates the convolution operator. In a homogenization problem the reference strain $\boldsymbol{\varepsilon}_0$ is taken to be spatially constant, and therefore $\hat{\boldsymbol{\varepsilon}}_0(\boldsymbol{\xi}) = 0$ when $\boldsymbol{\xi} \neq 0$. By Cramer's rule the homogeneous system of linear equations (7.84) results in a trivial solution $\hat{\boldsymbol{p}}_r(\boldsymbol{\xi}) = 0$ when $\boldsymbol{\xi} \neq 0$. Given the constant $\boldsymbol{p}_r$ for all $r = 1, 2, \ldots, N$ by employing (4.10), (7.84) reduces to

$$(\boldsymbol{L}_r - \boldsymbol{L}_0)^{-1}\hat{\boldsymbol{p}}_r(\xi) + \sum_{s=1}^{N}\left(\hat{\boldsymbol{\Gamma}}^{\infty} * \hat{\rho}_{rs}\right)(\xi)\hat{\boldsymbol{p}}_s(\xi)(\boldsymbol{\delta}_{rs} - c_s) = \hat{\boldsymbol{\varepsilon}}_0(\xi) \tag{7.85}$$

which has the real space expression

$$(\boldsymbol{L}_r - \boldsymbol{L}_0)^{-1}\boldsymbol{p}_r + \sum_{s=1}^{N}\boldsymbol{S}_{rs}\boldsymbol{L}_0^{-1}\boldsymbol{p}_s(\delta_{rs} - c_s) = \boldsymbol{\varepsilon}_0 \tag{7.86}$$

with

$$\boldsymbol{S}_{rs}\boldsymbol{L}_0^{-1} = \int_{Y}\boldsymbol{\Gamma}^{\infty}(\boldsymbol{x})\rho_{rs}(\boldsymbol{x})d\boldsymbol{x} \tag{7.87}$$

To obtain a closed-form solution for (7.86), we assume the spatial distribution of each phase is statistically isotropic (i.e. $\rho_{rs} = \rho_0$) and (7.86) becomes

$$(\boldsymbol{L}_r - \boldsymbol{L}_0)^{-1}\boldsymbol{p}_r + \boldsymbol{S}_0^{\rho}\boldsymbol{L}_0^{-1}\sum_{s=1}^{N}\boldsymbol{p}_s(\delta_{rs} - c_s) = \boldsymbol{\varepsilon}_0 \tag{7.88}$$

with

$$\boldsymbol{S}_0^{\rho}\boldsymbol{L}_0^{-1} = \int_{Y}\boldsymbol{\Gamma}^{\infty}(\boldsymbol{x})\rho_0(\boldsymbol{x})d\boldsymbol{x} \tag{7.89}$$

that is written in a form similar to Property (5.11). Note that in (7.89) the size-dependent Eshelby tensor $\boldsymbol{S}_0^{\rho}$ is numerically obtained, which converges to the spherical Eshelby tensor $\boldsymbol{S}_0$ when the RVE size becomes infinitely large compared to the correlation length. Similar to (4.21), (7.88) has the solution (4.22)

$$\boldsymbol{p}_r = (\boldsymbol{L}_r - \boldsymbol{L}_0)\tilde{\boldsymbol{A}}_r\boldsymbol{\varepsilon}_0 \qquad r = 1,2,\ldots, N$$

with

$$\tilde{\boldsymbol{A}}_r = \boldsymbol{A}_r\left(\sum_{s=1}^{N}c_s\boldsymbol{A}_s\right)^{-1} \quad \text{and}$$

$$\boldsymbol{A}_r = \left[\boldsymbol{I} + \boldsymbol{S}_0^{\rho}(\boldsymbol{L}_0^{-1}\boldsymbol{L}_r - \boldsymbol{I})\right]^{-1} \tag{7.90}$$

Similar to the HS bound (4.26), the size-dependent Hashin-Shtrikman bound $\boldsymbol{L}_\rho^{(2)}$ is given as

$$\boldsymbol{L}_\rho^{(2)} = \sum_{r=1}^{N} c_r \boldsymbol{L}_r \tilde{\boldsymbol{A}}_r \tag{7.91}$$

which corresponds to the upper or lower bound when $\boldsymbol{L}_r \leq \boldsymbol{L}_0$ or $\boldsymbol{L}_r \geq \boldsymbol{L}_0$ for all r = 1, 2,...,N. When applied to a two-phase random composite, by choosing $\boldsymbol{L}_0 = \boldsymbol{L}_2$, (7.91) reduces to

$$\boldsymbol{L}_\rho^{(2)} = \boldsymbol{L}_2 + \boldsymbol{c}_1\left[(\boldsymbol{L}_1 - \boldsymbol{L}_2)^{-1} + \boldsymbol{c}_2\ \boldsymbol{S}_0^{\rho}\boldsymbol{L}_2^{-1}\right]^{-1} \tag{7.92}$$

as the upper bound when $\boldsymbol{L}_1 < \boldsymbol{L}_2$ or the lower bound when $\boldsymbol{L}_1 > \boldsymbol{L}_2$. By applying Hill's notation to (7.92), with $S_0^{\rho} \sim \left(3\kappa_S^{\rho}, 2\mu_S^{\rho}\right)$, it follows that

$$\kappa_\rho^{(2)} = \kappa_2 + \frac{c_1}{(\kappa_1 - \kappa_2)^{-1} + \boldsymbol{c}_2 \dfrac{\kappa_S^{\rho}}{\kappa_2}} \tag{7.93}$$

$$\mu_\rho^{(2)} = \mu_2 + \frac{c_1}{(\mu_1 - \mu_2)^{-1} + c_2 \dfrac{\mu_S^{\rho}}{\mu_2}} \tag{7.94}$$

When the RVE size is infinitely large, substitution of the "moduli" of the spherical Eshelby tensor (4.29–4.30) into (7.93–7.94) results in the classical Hashin-Shtrikman bounds presented in Chapter 4

$$\kappa^{(2)} = (c_1\kappa_1 + c_2\kappa_2) - \frac{c_1c_2(\kappa_1 - \kappa_2)^2}{c_1\kappa_2 + c_2\kappa_1 + \dfrac{4}{3}\mu_2} \tag{7.95}$$

$$\mu^{(2)} = (c_1\mu_1 + c_2\mu_2) - \frac{c_1c_2(\mu_1 - \mu_2)^2}{c_1\mu_2 + c_2\mu_1 + \dfrac{\mu_2(9\kappa_2 + 8\mu_2)}{6(\kappa_2 + 2\mu_2)}} \tag{7.96}$$

The relative difference between the size-dependent HS bounds (7.93–7.94) and the classical HS bounds (7.95–7.96),

$$\frac{\kappa_p^{(2)} - \kappa^{(2)}}{\kappa^{(2)}} \text{ or } \frac{\mu_p^{(2)} - \mu^{(2)}}{\mu^{(2)}} \tag{7.97}$$

demonstrates the size effect of RVE as quantified below.

In a three-dimensional isotropic random medium, the correlation length ℓ_c is defined as (2.6a). Choose a legitimate correlation function

$$\rho_0(\boldsymbol{r}) = \exp\left[-\frac{\left(3\sqrt{\pi/2}\right)^{2/3}}{2}\frac{|\boldsymbol{r}|^2}{\ell_c^2}\right] \tag{7.98}$$

that is positive definite everywhere. Note that the microstructure of a two-phase composite matching the correlation (7.98) can be simulated by using the SRC morphological model described in Chapter 2.

The RVE size L in terms of the correlation length is hereby characterized as $\frac{L}{2\ell_c}$, times of "correlation diameter". Given a particular number of correlation diameters, the correlation (7.98) in a finite-size RVE can be transformed into its Fourier expression. By further employing (A2.6) the modified Green function in Fourier space and Parseval's theorem, the size-dependent Eshelby tensor $\boldsymbol{S}_0^\rho$ in (7.89) is conveniently calculated in Fourier space with fast Fourier transform (FFT).

To obtain an upper bound estimate for the minimum RVE size, we specially choose a solid-void composite and a solid-rigid composite. In Figure 7.2 the solid-void composite demonstrates that the RVE size effect on the bulk modulus is two orders of magnitude stronger than on the shear modulus. The size effect becomes greater with increase of the volume fraction of voids or rigid inclusions. When the Poisson's ratio is changed from 0.33 to 0.2, the RVE size effect slightly decreases in the solid-void case but remains little change in the solid-rigid case, as detailed in Xu and Chen (2009). The minimum numbers of correlation diameters are summarized in Table 7.1 on four cases. Choose, say, 0.1% as the criterion for the maximum allowable error to apply the RVE concept, the threshold of scale separation is found from Table 7.1 to be about 20 correlation diameters.

These results, in terms of the correlation length, can be extended to microstructure that is practically described with physical parameters such as the size of inclusions. For instance, the correlation length of nonoverlapping identical spheres of radius a in a hosting matrix is derived from the Percus-Yevick radial distribution function as

$$\ell_c = \frac{1 - c_1}{(1 + 2c_1)^{2/3}} a \tag{7.99}$$

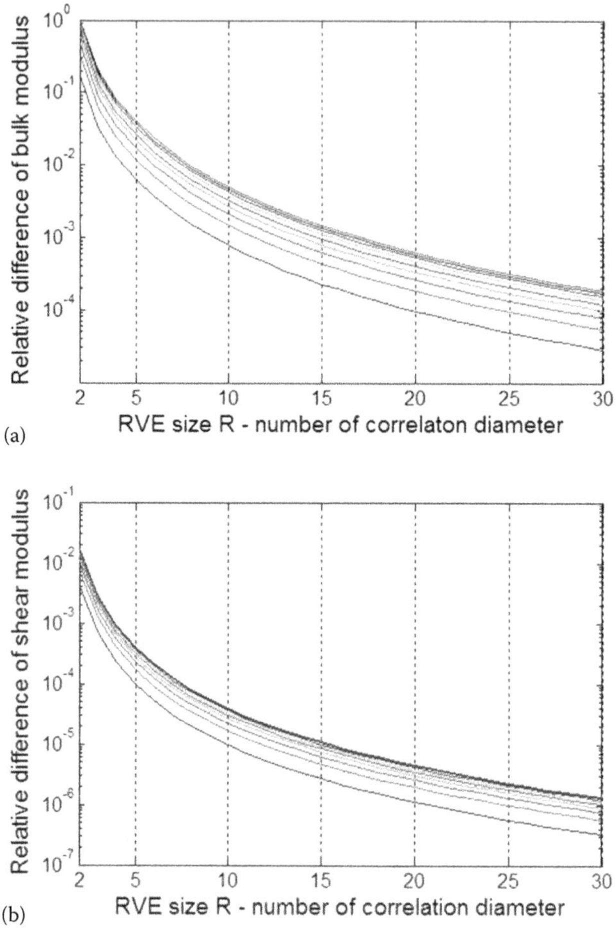

Figure 7.2 3D RVE size effect - relative differences of the bulk modulus (a) and shear modulus (b) vs. RVE size in terms of the number of correlation diameters for a solid-void two-phase composite with Poisson's ratio v = 0.33 (the curves from top to bottom correspond to the void volume fraction from 0.9 down to 0.1). (Reprinted from *Mech. Mater.*, 41, Xu, X.F. and X. Chen, Stochastic homogenization of random multi-phase composites and size quantification of representative volume element, 174–186, Copyright 2009, with permission from Elsevier. Permission conveyed through Copyright Clearance Center, Inc.)

which is graphically illustrated in Figure 7.3. The minimum correlation length is estimated to be about $0.20a$ in the case of random closed packing when the volume fraction of spheres $c_1 \approx 0.64$. In the case of nearly crystal packing with $c_1 \approx 0.74$, the result (7.99) becomes inapplicable because in this highly regular microstructure, randomness almost disappears and the assumption of an isotropic correlation breaks down. Based on the results

Table 7.1 Minimum RVE size in terms of the number of correlation diameters in a 3D composite

	I	*II*	*III*	*IV*
Criteria	*Solid-void* *v = 0.33*	*Solid-void* *v = 0.2*	*Solid-rigid* *v = 0.33*	*Solid-rigid* *v = 0.2*
5%	6	5	5	5
1%	10	8	8	8
0.5%	13	10	10	10
0.1%	22	17	17	17

Source: Reprinted from *Mech. Mater.*, 41, Xu, X.F. and X. Chen, Stochastic homogenization of random multi-phase composites and size quantification of representative volume element, 174–186, Copyright 2009, with permission from Elsevier. Permission conveyed through Copyright Clearance Center, Inc.

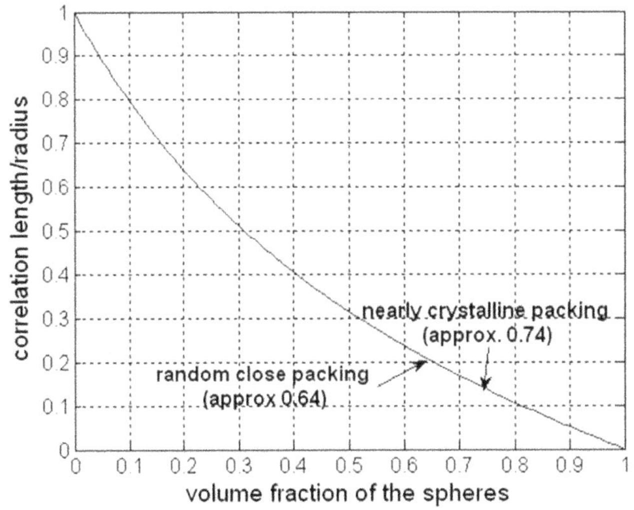

Figure 7.3 Volume fraction of spheres vs. correlation length of randomly dispersed non-overlapping identical spheres. (Reprinted from *Mech. Mater.*, 41, Xu, X.F. and X. Chen, Stochastic homogenization of random multi-phase composites and size quantification of representative volume element, 174–186, Copyright 2009, with permission from Elsevier. Permission conveyed through Copyright Clearance Center, Inc.)

in Table 7.1 and (7.99), the minimum RVE size is recalculated in terms of spherical diameters and listed in Table 7.2. Table 7.2 shows that the minimum RVE size highly depends on the volume fraction of spheres. In the case of a low-volume fraction such as $c_1 = 0.1$, given the criterion 0.1% the minimum RVE size is 28 spherical diameters; in other words, the minimum RVE size contains about 4,000 spheres. At the other end of the spectrum,

Table 7.2 Minimum RVE size in terms of the number of spherical diameters (Xu & Chen, 2009) subject to accuracy criteria of 5%, 1%, 0.5%, and 0.1% for 3D composites

Criteria	$c_1 = 0.1$	$c_1 = 0.2$	$c_1 = 0.3$	$c_1 = 0.4$	$c_1 = 0.5$	$c_1 = 0.6$	$c_1 = 0.64$
5%	8	9	12	15	19	25	29
1%	13	16	20	25	32	42	48
0.5%	16	20	25	31	40	53	60
0.1%	28	35	43	54	70	93	105

Source: Reprinted from *Mech. Mater.*, 41, Xu, X.F. and X. Chen, Stochastic homogenization of random multi-phase composites and size quantification of representative volume element, 174–186, Copyright 2009, with permission from Elsevier. Permission conveyed through Copyright Clearance Center, Inc.

i.e., the jammed state with the maximum $c_1 \approx 0.64$, an RVE size of 105 spherical diameters is required, which contains about 1.4 million spheres!

Another interesting result of (7.99) is the correlation length is close to the radius of spheres when the volume fraction of spheres becomes small in a dilute case. This result indicates that the correlation diameter of polycrystalline is approximately of the average grain size. Based on the 0.1% criterion, the minimum RVE size is about 20 times of crystalline size, which is close to relevant experimental and numerical results (Cho & Chasiotis, 2007; Liu, 2005; Ren & Zheng, 2002).

It is further noted that the minimum RVE size investigated by Drugan and Willis (1996) is focused on the nonlocal effect due to strain fluctuation, while the scale-coupling effect due to a finite-size RVE such as fluctuation of volume fractions is not considered. The size effect described here, related to $\int \rho(r)r^2\, dr$, the second moment of the 1D correlation coefficient, proves to be much stronger than the effect due to strain fluctuation that depends on the first moment $\int \rho(r)r\, dr$. In other words, even subject to a constant strain eliminating the nonlocal effect, the effective elastic moduli of a finite-size RVE varies from one sample to another and the random variation can be neglected only when the RVE size increases to a certain level such as the size threshold quantified previously.

7.3.4 Minimum RVE size in 2D

With respect to a random composite symmetric about the axis-3 normal to the plane of isotropy, the size-dependent Eshelby tensor $\boldsymbol{S}_0^\rho$ and the HS bound $\boldsymbol{L}_\rho^{(2)}$ are transversely isotropic and characterized with five independent constants, with the latter expanded as

$$\boldsymbol{L}_\rho^{(2)} = \begin{bmatrix} L_{\rho,11}^{(2)} & L_{\rho,12}^{(2)} & L_{\rho,13}^{(2)} & 0 & 0 & 0 \\ & L_{\rho,11}^{(2)} & L_{\rho,13}^{(2)} & 0 & 0 & 0 \\ & & L_{\rho,33}^{(2)} & 0 & 0 & 0 \\ & & & L_{\rho,44}^{(2)} & 0 & 0 \\ & \text{Sym} & & & L_{\rho,44}^{(2)} & 0 \\ & & & & & (L_{\rho,11}^{(2)} - L_{\rho,12}^{(2)})/2 \end{bmatrix} \tag{7.100}$$

In a plain strain problem, two constants are needed, i.e., $L_{\rho,11}^{(2)}$ and $L_{\rho,12}^{(2)}$, which are obtained from (7.91) or (7.92). The size-dependent Eshelby tensor $\boldsymbol{S}_0^\rho$ is numerically evaluated via (7.89), which converges to the circular Eshelby tensor $\boldsymbol{S}_0$ (4.44) in case of an infinitely large RVE. The size effect of a two-dimensional problem is similarly quantified based on relative difference of the moduli between a finite-size RVE and an infinitely large RVE, e.g.

$$\frac{L_{\rho,11}^{(2)} - L_{11}^{(2)}}{L_{11}^{(2)}}, \frac{L_{\rho,12}^{(2)} - L_{12}^{(2)}}{L_{12}^{(2)}} \tag{7.101}$$

In a two-dimensional random medium such as a randomly distributed fibrous composite, the correlation length ℓ_c is defined as (2.6b). Similar to the 3D case (7.98), a legitimate exponential correlation function is chosen as

$$\rho_0(\boldsymbol{r}) = \exp\left(-\frac{|\boldsymbol{r}|^2}{\ell_c^2}\right) \tag{7.102}$$

By taking advantage of FFT, the relative difference in (7.101) is conveniently calculated for a number of volume fractions ranging from 0.1 to 0.9 for a solid-void composite and a solid-rigid composite with $\upsilon = 0.33$ and 0.2. Similar to the 3D results, the 2D results indicate that the RVE size effect becomes slightly stronger with increase of the volume fraction of voids or rigid inclusions. In the solid-void case, the RVE size effect increases

Table 7.3 Minimum RVE size in terms of the number of correlation diameters in a 2D composite

	I	II	III	IV
Criteria	*Solid-void* $\nu = 0.33$	*Solid-void* $\nu = 0.2$	*Solid-rigid* $\nu = 0.33$	*Solid-rigid* $\nu = 0.2$
5%	7	6	4	4
1%	15	13	10	10
0.5%	22	19	14	14
0.1%	47	42	30	30

Source: Reprinted from *Mech. Mater.*, 41, Xu, X.F. and X. Chen, Stochastic homogenization of random multi-phase composites and size quantification of representative volume element, 174–186, Copyright 2009, with permission from Elsevier. Permission conveyed through Copyright Clearance Center, Inc.

with rise of the Poisson's ratio, while in the solid-rigid case remains almost same. The minimum RVE sizes in terms of the number of correlation diameters are summarized in Table 7.3. By comparing Table 7.1 with Table 7.3, to reduce one order of magnitude on the relative error, the RVE size need be approximately double and triple in 3D and 2D, respectively. In other words, the RVE size effect is much stronger in 2D than in 3D.

Chapter 8

Multiscale stochastic finite element method and multiphase composites

With the scale-coupling variational principles formulated in Chapter 7, a specifically developed finite element method is implemented to numerically solve the equations with the scales coupled between the slow-scale reference BVP and the fast-scale fluctuation BVP. The whole approach, named as the *multiscale stochastic finite element method* (MSFEM), covers the scale decomposition scheme, the stochastic variational principles, and a finite element formulation. In this chapter the MSFEM is formulated to tackle a discontinuous multiphase composite, and in Chapter 9 a perturbation-type MSFEM is introduced to efficiently solve a continuous random media problem, e.g. foundation settlement in soil mechanics.

Based on the Hashin-Shtrikman variational principle, a finite body random composite was first tackled with finite elements in Luciano and Willis (2005), where the scale-coupling phenomenon was termed as nonlocal interactions. In Xu et al. (2009), a scale-coupling variational principle is rigorous formulated, with a boundary condition of the unknown stress polarization distinctively specified.

As demonstrated in Chapter 7, a boundary value problem of a random multiphase composite is decomposed into a reference BVP and a fluctuation BVP (see Figure 7.1). The reference problem is characterized with a homogeneous material subject to boundary conditions and a body force identical to those of the original BVP. The reference BVP can be conveniently solved with classical finite element method. The solution of the reference problem represents a certain slow-scale component of the final result. With the slow-scale solution serving as an input, more specifically the slow-scale strain and the slow-scale Green function, the fluctuation BVP is solved by applying the odd-order principle (7.63–7.64) or the even-order HS principle (7.69). The special finite element method (FEM) developed to solve the fluctuation BVP is also called *Green-function-based FEM* due to the use of the modified Green function rather than the Green function in the classical FEM. A flow chart is drawn in Figure 8.1 to illustrate the multiscale computational process.

In Section 8.1, the slow-scale reference BVP is solved with the classical finite element method, in which formulation of the classical FEM is briefly reviewed. The fast-scale fluctuation BVP is then tackled in Section 8.2 by

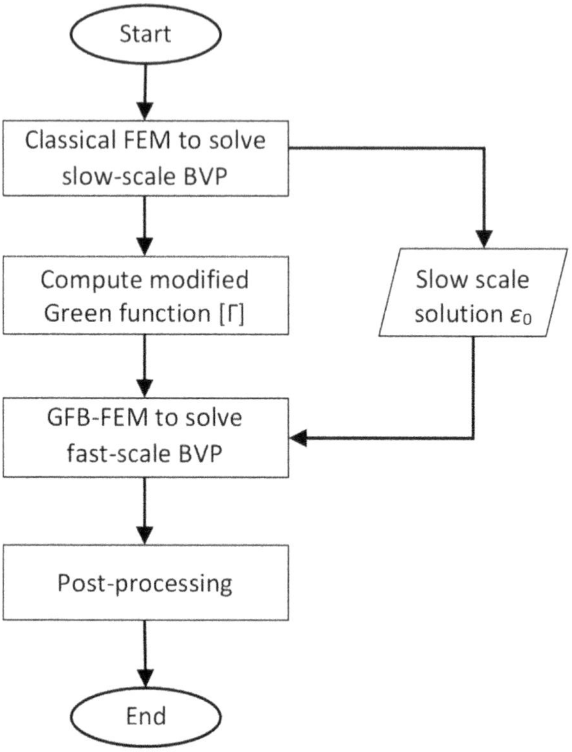

Figure 8.1 Flow chart of multiscale stochastic finite element method.

formulating the novel Green-function-based FEM. Finally, in Section 8.3 two numerical examples are presented to demonstrate unique capacity and algorithmic features of the MSFEM.

8.1 CLASSICAL FINITE ELEMENTS ON THE SLOW-SCALE BVP

The reference BVP is characterized with homogeneous elastic tensor $\boldsymbol{L}_0$, and a body force f identical to that of the original BVP (Figure 7.1). The governing equations (7.6–7.10) are given here, with the boundary conditions adjusted accordingly

$$\nabla \cdot \boldsymbol{\sigma}_0(\boldsymbol{x}) + \boldsymbol{f}(\boldsymbol{x}) = 0 \quad \text{in } D \tag{8.1}$$

$$\boldsymbol{\sigma}_0(\boldsymbol{x}) = \boldsymbol{L}_0 \boldsymbol{\varepsilon}_0(\boldsymbol{x}) \tag{8.2}$$

$$\boldsymbol{\varepsilon}_0(\boldsymbol{x},\vartheta) = \nabla^s \boldsymbol{u}_0(\boldsymbol{x},\vartheta) \tag{8.3}$$

with boundary conditions

$$\boldsymbol{u}_0(\boldsymbol{x},\vartheta) = \tilde{\boldsymbol{u}}(\boldsymbol{x}) \quad \text{on} \quad \partial D_u \tag{8.4}$$

$$\boldsymbol{\sigma}_0(\boldsymbol{x},\vartheta)\cdot \boldsymbol{n} = \tilde{\boldsymbol{t}}(\boldsymbol{x}) \quad \text{on} \quad \partial D_t \tag{8.5}$$

By equating the variation (7.24) to zero, and considering the constitutive and displacement-strain relations (8.2–8.3), it follows that

$$\left\langle \boldsymbol{L}_0 \nabla^s \boldsymbol{u}_0, \delta \nabla^s \boldsymbol{u}_0 \right\rangle = \left\langle \boldsymbol{f}, \delta \boldsymbol{u}_0 \right\rangle + (\tilde{\boldsymbol{t}}, \delta \boldsymbol{u}_0) \tag{8.6}$$

where the test function $\delta \boldsymbol{u}_0$ should satisfy the homogeneous displacement boundary condition.

In finite element discretization, the BVP domain D is meshed into N_e elements with each consisting of $n_p^{(e)}$ elemental nodes, and correspondingly $n_p^{(e)}$ shape functions $N_i(\boldsymbol{x})$, with $i = 1, 2, \ldots, n_p^{(e)}$. A shape function $N_i(\boldsymbol{x})$ of element e vanishes outside of the element, and satisfies the identity condition in that $N_i(\boldsymbol{x}) = 1$ and 0 when $\boldsymbol{x}$ is located right at node i and any other nodes of the element, respectively. In element e denote $U_{0ik}^{(e)}$ the unknown nodal displacement at node i along direction x_k. The displacement at any point within the elemental domain D_e other than the nodes is interpolated via the following equation

$$u_{0k}^{(e)}(\boldsymbol{x}) = \sum_{i=1}^{n_p^{(e)}} U_{0ik}^{(e)} N_i(\boldsymbol{x}) \tag{8.7a}$$

the matrix form of which is written as

$$\{\mathrm{u}_0\}^{(e)} = [\mathrm{N}]\{\mathrm{U}_0\}^{(e)} \tag{8.7b}$$

By substituting (8.7) into (8.3), the elemental strain is obtained as

$$\varepsilon_{0kj}^{(e)}(\boldsymbol{x}) = \frac{1}{2}\sum_{i=1}^{n_p^{(e)}} \left(U_{0ik}^{(e)} \frac{\partial N_i(\boldsymbol{x})}{\partial x_k} + U_{0ij}^{(e)} \frac{\partial N_i(\boldsymbol{x})}{\partial x_j} \right) \tag{8.8a}$$

or

$$\{\varepsilon_0\}^{(e)} = [\mathrm{B}]\{\mathrm{U}_0\}^{(e)} \tag{8.8b}$$

where [B] is the so-called strain-displacement matrix. For example, in a four-node quadrilateral element, (8.7b) and (8.8b) are expressed, respectively, as

$$\begin{Bmatrix} u_{01} \\ u_{02} \end{Bmatrix}^{(e)} = \begin{bmatrix} N_1 & 0 & N_2 & 0 & N_3 & 0 & N_4 & 0 \\ 0 & N_1 & 0 & N_2 & 0 & N_3 & 0 & N_4 \end{bmatrix} \begin{Bmatrix} U_{011}^{(e)} \\ U_{012}^{(e)} \\ \cdots \\ U_{042}^{(e)} \end{Bmatrix} \tag{8.9a}$$

$$\begin{Bmatrix} \varepsilon_{01} \\ \varepsilon_{02} \\ \gamma_{012} \end{Bmatrix}^{(e)} = \begin{bmatrix} \nabla N_{1,1} & 0 & \nabla N_{2,1} & 0 & \nabla N_{3,1} & 0 & \nabla N_{4,1} & 0 \\ 0 & \nabla N_{1,2} & 0 & \nabla N_{2,2} & 0 & \nabla N_{3,2} & 0 & \nabla N_{4,2} \\ \nabla N_{1,2} & \nabla N_{1,1} & \nabla N_{2,2} & \nabla N_{2,1} & \nabla N_{3,2} & \nabla N_{3,1} & \nabla N_{4,2} & \nabla N_{4,1} \end{bmatrix} \begin{Bmatrix} U_{011}^{(e)} \\ U_{012}^{(e)} \\ \cdots \\ U_{042}^{(e)} \end{Bmatrix} \tag{8.9b}$$

where the engineering shear strain $\gamma_{012} = 2\varepsilon_{012}$.

Substitution of (8.7–8.8) into (8.6) yields the following matrix equation

$$[\mathrm{K}_0]\{\mathrm{U}_0\} = \{\mathrm{F}_0\} \tag{8.10}$$

where the stiffness matrix is obtained from the assembled elemental stiffness matrices as

$$\begin{aligned} [\mathrm{K}_0] &= \bigcup_{e=1}^{N_e} [\mathrm{K}_0]^{(e)} \\ &= \bigcup_{e=1}^{N_e} \int_{D_e} [\mathrm{B}]^T [\mathrm{L}_0][\mathrm{B}]\, dV \end{aligned} \tag{8.11}$$

and similarly the force vector is given as

$$\begin{aligned}\{\mathrm{F}_0\} &= \bigcup_{e=1}^{N_e} \{\mathrm{F}_0\}^{(e)} \\ &= \bigcup_{e=1}^{N_e} \left[\int_{D_e} [\mathrm{N}]^T \{\mathrm{f}\} dV + \int_{\partial D_e} [\mathrm{N}]^T \{\tilde{\mathrm{t}}\} dS \right]\end{aligned} \tag{8.12}$$

The symbol $\bigcup$ in (8.11–8.12) denotes assemblage of matrices or vectors, and $[\mathrm{L}_0]$ the matrix of the elastic moduli. The displacement boundary condition is imposed on (8.10) by applying the method of large numbers or explicit specification.

8.2 GREEN-FUNCTION-BASED FINITE ELEMENTS ON THE FAST-SCALE BVP

8.2.1 Formulation

Similar to (7.11–7.15), the governing equations of the fast-scale fluctuation BVP are given here, with the boundary conditions adjusted accordingly

$$\nabla \cdot \boldsymbol{\sigma}'(\boldsymbol{x},\vartheta) + \nabla \cdot \boldsymbol{p}(\boldsymbol{x},\vartheta) = 0 \text{ in } D \tag{8.13}$$

$$\boldsymbol{\sigma}'(\boldsymbol{x},\vartheta) = \boldsymbol{L}_0 \boldsymbol{\varepsilon}'(\boldsymbol{x},\vartheta) \tag{8.14}$$

$$\boldsymbol{\varepsilon}'(\boldsymbol{x},\vartheta) = \nabla^s \boldsymbol{u}'(\boldsymbol{x},\vartheta) \tag{8.15}$$

with boundary conditions

$$\boldsymbol{u}'(\boldsymbol{x},\vartheta) = 0 \quad \text{on } \partial D_u \tag{8.16}$$

$$\boldsymbol{\sigma}'(\boldsymbol{x},\vartheta) \cdot \boldsymbol{n} = 0 \quad \text{on } \partial D_t \tag{8.17}$$

$$\boldsymbol{p} \cdot \boldsymbol{n} = 0 \text{ on } \partial D_t \tag{8.18}$$

As described in Chapter 7, given the highest order correlation of the microstructure being odd or even, the corresponding even or odd variational principle is chosen to solve the fast-scale fluctuation BVP. Hereby the highest order correlation is assumed to be of the second order, and accordingly the HS principle is employed. With the first-order trial function (7.74) to approximate the unknown polarization, we obtain (7.80) shown here:

$$(\boldsymbol{L}_r - \boldsymbol{L}_0)^{-1}\boldsymbol{p}_r(\boldsymbol{x})c_r(\boldsymbol{x}) + \int_D \boldsymbol{\Gamma}(\boldsymbol{x},\boldsymbol{x}')\sum_{s=1}^{N}\boldsymbol{p}_s(\boldsymbol{x}')c_{rs}(\boldsymbol{x},\boldsymbol{x}')d\boldsymbol{x}' = c_r(\boldsymbol{x})\boldsymbol{\varepsilon}_0(\boldsymbol{x}) \tag{8.19}$$

with r = 1, 2,...,N, which is to be solved subject to the constraint (8.18) the traction free boundary condition on ∂D_t. Note that, a prerequisite for application of the HS principle yielding (8.19), is positive or negative semi-definiteness of $\boldsymbol{L}_r - \boldsymbol{L}_0$ for any phase r within the composite; otherwise the solution of (8.19) merely represents an approximation with its physical meaning unclear.

To solve (8.19) using finite elements, (8.19) is reformulated in a weak form as follows

$$\begin{aligned}&\int_D \delta\boldsymbol{p}_r(\boldsymbol{x})(\boldsymbol{L}_r-\boldsymbol{L}_0)^{-1}\boldsymbol{p}_r(\boldsymbol{x})c_r(\boldsymbol{x})d\boldsymbol{x}+\int_D\int_D \delta\boldsymbol{p}_r(\boldsymbol{x})\boldsymbol{\Gamma}(\boldsymbol{x},\boldsymbol{x}')\sum_{s=1}^{N}\boldsymbol{p}_s(\boldsymbol{x}')c_{rs}(\boldsymbol{x},\boldsymbol{x}')d\boldsymbol{x}'d\boldsymbol{x}\\&=\int_D \delta\boldsymbol{p}_r(\boldsymbol{x})c_r(\boldsymbol{x})\boldsymbol{\varepsilon}_0(\boldsymbol{x})d\boldsymbol{x}\end{aligned} \tag{8.20}$$

with r = 1, 2,...,N. In a statistically homogeneous random composite, c_r(x) is constant throughout the BVP domain. In such a case (8.20) reduces to

$$\begin{aligned}&c_r\int_D \delta\boldsymbol{p}_r(\boldsymbol{x})(\boldsymbol{L}_r - \boldsymbol{L}_0)^{-1}\boldsymbol{p}_r(\boldsymbol{x})_r\, d\boldsymbol{x} + \int_D\int_D \delta\boldsymbol{p}_r(\boldsymbol{x})\boldsymbol{\Gamma}(\boldsymbol{x},\boldsymbol{x}')\\&\times\sum_{s=1}^{N}\boldsymbol{p}_s(\boldsymbol{x}')c_{rs}(\boldsymbol{x},\boldsymbol{x}')d\boldsymbol{x}'d\boldsymbol{x} = c_r\int_D \delta\boldsymbol{p}_r(\boldsymbol{x})_r\boldsymbol{\varepsilon}_0(\boldsymbol{x})d\boldsymbol{x}\end{aligned} \tag{8.21}$$

Similar to (8.7), by using the elemental shape functions $N_i(\boldsymbol{x})$, the elemental polarization is represented as

$$\boldsymbol{p}_r^{(e)}(\boldsymbol{x}) = \sum_{i=1}^{n_p^{(e)}}\boldsymbol{P}_{ri}^{(e)}N_i(\boldsymbol{x}) \tag{8.22a}$$

with $\boldsymbol{P}_{ri}^{(e)}$ denoting $\boldsymbol{p}_r$ valued at node i of element e, which in the matrix form is expressed as

$$\{\mathrm{p_r}\}^{(e)} = [\mathrm{N}]\{\mathrm{P_r}\}^{(e)} \tag{8.22b}$$

For example, in a four-node quadrilateral element (8.22b) is expanded as

$$\begin{Bmatrix} p_{r1} \\ p_{r2} \\ p_{r4} \end{Bmatrix}^{(e)} = \begin{bmatrix} N_1 & 0 & 0 & N_2 & 0 & 0 & N_3 & 0 & 0 & N_4 & 0 & 0 \\ 0 & N_1 & 0 & 0 & N_2 & 0 & 0 & N_3 & 0 & 0 & N_4 & 0 \\ 0 & 0 & N_1 & 0 & 0 & N_2 & 0 & 0 & N_3 & 0 & 0 & N_4 \end{bmatrix} \begin{Bmatrix} P_{r11} \\ P_{r12} \\ P_{r14} \\ \cdots \\ P_{r44} \end{Bmatrix}^{(e)}$$

where the subscript 4 denotes the shear component.

The numerical Green function [G] is conveniently obtained by applying the zero displacement boundary condition (8.16) to the inverse of $[\mathrm{K}_0]$ in (8.11), i.e.

$$[\mathrm{G}] = [\tilde{\mathrm{K}}_0]^{-1} \tag{8.23}$$

where the tilde denotes the modification using the method of large numbers to take the zero displacement condition (8.16) into account. A numerical value of the Green function $G_{ik}(\boldsymbol{x}, \boldsymbol{x}')$ with $i,k = 1,2,3$ physically means the resulting displacement at point $\boldsymbol{x}$ along direction i produced by a unit force acting at point $\boldsymbol{x}'$ along direction k, which is approximated via interpolation of the shape functions as

$$G_{ik}(\boldsymbol{x},\boldsymbol{x}') = \sum_{m}^{N_p^{(e)}} \sum_{n}^{N_p^{(e')}} N_m(\boldsymbol{x})[\mathrm{G}]_{IK} N_n(\boldsymbol{x}') \tag{8.24}$$

In (8.24) the subscripts m, n denote the local nodal numbers of two elements containing $\boldsymbol{x}$ and $\boldsymbol{x}'$, respectively, and I, K represent two degrees of freedom corresponding to node m in direction i and node n in direction k, respectively, as shown in Figure 8.2. Substitution of (8.24) into (7.35) yields a numerical approximation of the modified Green function as

$$\Gamma_{ijkl}(\boldsymbol{x},\boldsymbol{x}') = \frac{1}{2}\left[\sum_{m}^{N_p^{(e)}} \sum_{n}^{N_p^{(e')}} \nabla N_{m,j}(\boldsymbol{x})[\mathrm{G}]_{IK} \nabla N_{n,l}(\boldsymbol{x}') + \sum_{m}^{N_p^{(e)}} \sum_{n}^{N_p^{(e')}} \nabla N_{m,i}(\boldsymbol{x})[\mathrm{G}]_{JK} \nabla N_{n,l}(\boldsymbol{x}') \right] \tag{8.25}$$

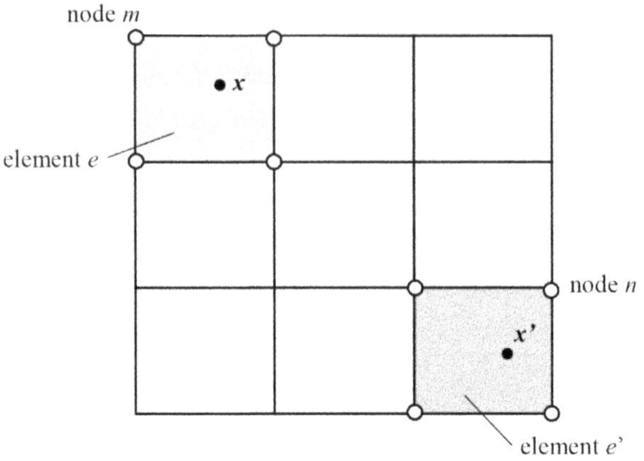

Figure 8.2 Schematic of two elements and the nodes in interpolation of the Green function (8.24).

By substituting (8.22) and (8.25) into (8.21), it results in the following N matrix equations

$$c_r[\Delta \mathrm{M}_r]\{\mathrm{P}_r\}+\sum_{s=1}^{N}[\Gamma_{rs}]\{\mathrm{P}_s\}=c_r\{\mathrm{E}_0\} \quad r=1,2,\ldots,N \tag{8.26}$$

with

$$[\Delta \mathrm{M}_r]=\bigcup_{e=1}^{Ne}\int_{D_e}[\mathrm{N}]^T[\mathrm{L}_r-\mathrm{L}_0]^{-1}[\mathrm{N}]\,dV \tag{8.27}$$

$$[\Gamma_{rs}]=\bigcup_{e=1}^{Ne}\bigcup_{e'=1}^{Ne}\int_{D_e}[\mathrm{N}]^{T}(\boldsymbol{x})\int_{D_{e'}}[\Gamma](\boldsymbol{x},\boldsymbol{x}')[\mathrm{N}](\boldsymbol{x}')c_{rs}(\boldsymbol{x},\boldsymbol{x}')\,d\boldsymbol{x}'\,d\boldsymbol{x} \tag{8.28}$$

$$\{\mathrm{E}_0\}=\bigcup_{e=1}^{Ne}\int_{D_e}[\mathrm{N}]^T\{\varepsilon_0\}^{(e)}\,dV \tag{8.29}$$

Note that in converting (8.21) to the matrix form (8.26), the rank-two polarization and strain tensors in (8.21) are rearranged into rank-one vectors, and accordingly the rank-4 $\boldsymbol{\Gamma}$ tensor in (8.25) is rearranged into $[\Gamma]$ in (8.28) which, in case of the plane strain, is expressed as

$$[\Gamma]=\begin{bmatrix} \Gamma_{1111} & \Gamma_{1122} & \Gamma_{1144} \\ & \Gamma_{2222} & \Gamma_{2244} \\ \text{sym} & & \Gamma_{4444} \end{bmatrix} \tag{8.30}$$

To impose the traction free boundary condition (8.18) onto the unknown polarization $\{P_r\}$, we simply apply the method of large numbers to the left hand side of (8.26). Alternatively, Eq. (8.19) can be solved by using the collocation method, with the modified Green function obtained from (8.25). A comparison of the collocation method and the finite element method on solving (8.19) is provided in Shen and Xu (2010).

It is finally remarked that to achieve maximum accuracy and efficiency, the mesh size is adjusted differently in computation of the slow-scale BVP, the fast-scale BVP, and the Green function. As shown in the numerical examples of Section 8.3, numerical accuracy of the modified Green function plays the most important role in the final result.

8.2.2 Probabilistic evaluation of stress and strain

Based on the stress polarization solved from (8.26), probabilistic moments of the stress and the strain can be numerically evaluated. In an N-Phase composite, by choosing the elastic moduli of the matrix phase, i.e. Phase-N, as the reference moduli (i.e. $\boldsymbol{L}_0 = \boldsymbol{L}_N$), the polarization in the matrix phase

$$\boldsymbol{p}_N = 0 \tag{8.31}$$

The mean strain at any point within the BVP domain equals to the sum of all the conditional mean strains of the N phases, i.e.

$$\bar{\varepsilon}(x) = \sum_{r=1}^{N} c_r \bar{\boldsymbol{\varepsilon}}_r(\boldsymbol{x}) \tag{8.32}$$

with the conditional mean strain

$$\bar{\boldsymbol{\varepsilon}}_r(\boldsymbol{x}) = \operatorname{Exp}\left\langle \boldsymbol{\varepsilon}(\boldsymbol{x}) \middle| \boldsymbol{x} \in D_r \right\rangle \tag{8.33}$$

Ensemble averaging of (7.72) results in

$$\bar{\boldsymbol{\varepsilon}} = \boldsymbol{\varepsilon}_0 - \boldsymbol{\Gamma}\bar{\boldsymbol{p}} \tag{8.34}$$

which, by applying the first-order trial polarization (7.74) and the result (8.31), yields

$$\boxed{\bar{\boldsymbol{\varepsilon}} = \boldsymbol{\varepsilon}_0 - \sum_{r=1}^{N-1} c_r \boldsymbol{\Gamma} \boldsymbol{p}_r} \tag{8.35}$$

By substituting (7.72) and (7.74) into the following conditional expectation

$$\begin{aligned}\operatorname{Exp}\left\langle \boldsymbol{\varepsilon}(\boldsymbol{x},\vartheta) \middle| \boldsymbol{x} \in D_r \right\rangle &= \frac{\operatorname{Exp}\left\langle \boldsymbol{\varepsilon}(\boldsymbol{x},\vartheta)\chi_r(\boldsymbol{x},\vartheta)\right\rangle}{\operatorname{Pro}(\boldsymbol{x} \in D_r)} \\ &= \frac{\operatorname{Exp}\left\langle \boldsymbol{\varepsilon}(\boldsymbol{x},\vartheta)\chi_r(\boldsymbol{x},\vartheta)\right\rangle}{c_r}\end{aligned} \tag{8.36}$$

it yields that

$$\bar{\boldsymbol{\varepsilon}}_r(\boldsymbol{x}) = \boldsymbol{\varepsilon}_0(\boldsymbol{x}) - \frac{1}{c_r}\sum_{s=1}^{N-1}\int_D \boldsymbol{\Gamma}(\boldsymbol{x},\boldsymbol{x}')c_{rs}(\boldsymbol{x},\boldsymbol{x}')\boldsymbol{p}_s(\boldsymbol{x}')dx' \tag{8.37}$$

which, by taking (8.19) into account, leads to the mean strain in Phase-r as

$$\boxed{\bar{\boldsymbol{\varepsilon}}_r = (\boldsymbol{L}_r - \boldsymbol{L}_0)^{-1}\boldsymbol{p}_r} \tag{8.38}$$

with $r = 1, 2, \ldots, N-1$. Similarly, the mean strain in the matrix phase (i.e. Phase-N) is obtained from (8.36) with $r = N$, as

$$\boxed{\bar{\boldsymbol{\varepsilon}}_N = \boldsymbol{\varepsilon}_0 - \frac{1}{c_N}\sum_{r=1}^{N-1}\boldsymbol{\Gamma} c_{Nr}\boldsymbol{p}_r} \tag{8.39}$$

With the traction free condition (8.17), (7.41) becomes

$$\boldsymbol{\sigma} = \boldsymbol{\sigma}_0 - \boldsymbol{L}_0\boldsymbol{\Gamma}\boldsymbol{p} + \boldsymbol{p} \tag{8.40}$$

The mean stress is directly obtained by taking ensemble averaging of (8.40) as

$$\boxed{\bar{\boldsymbol{\sigma}} = \boldsymbol{\sigma}_0 + \sum_{r=1}^{N-1} c_r(\boldsymbol{I} - \boldsymbol{L}_0\boldsymbol{\Gamma})\boldsymbol{p}_r} \tag{8.41}$$

By substitute (7.20) and (8.40) into the conditional expectation

$$\mathrm{Exp}\left\langle \boldsymbol{\sigma}(\boldsymbol{x},\vartheta) \middle| \boldsymbol{x} \in D_r \right\rangle = \frac{\mathrm{Exp}\left\langle \boldsymbol{\sigma}(\boldsymbol{x},\vartheta)\chi_r(\boldsymbol{x},\vartheta)\right\rangle}{\mathrm{Pro}(\boldsymbol{x} \in D_r)} \tag{8.42}$$

the mean stress in Phase-r is derived as

$$\boxed{\begin{aligned}\bar{\boldsymbol{\sigma}}_r &= \frac{\overline{(\boldsymbol{\sigma}_0 - \boldsymbol{L}_0\boldsymbol{\Gamma}\boldsymbol{p} + \boldsymbol{p})\chi_r}}{c_r} \\ &= \boldsymbol{L}_r(\boldsymbol{L}_r - \boldsymbol{L}_0)^{-1}\boldsymbol{p}_r\end{aligned}} \tag{8.43}$$

with r = 1, 2,...,N–1. The mean stress in the matrix, Phase-N, is similarly obtained as

$$\boxed{\bar{\boldsymbol{\sigma}}_N = \boldsymbol{\sigma}_0 - \frac{1}{c_N}\boldsymbol{L}_0\sum_{r=1}^{N-1}\boldsymbol{\Gamma}c_{Nr}\boldsymbol{p}_r} \tag{8.44}$$

By comparing (8.43–8.44) with (8.38–8.39), it is found that the constitutive law of Phase-r remains unchanged between the mean strain and the mean stress, i.e.

$$\bar{\sigma}_r = L_r\bar{\boldsymbol{\varepsilon}}_r \tag{8.45}$$

for all the phases r = 1, 2,...,N.

With (7.72) and (7.74), the variance of the strain is derived as

$$\begin{aligned}&\mathrm{Var}\left(\boldsymbol{\varepsilon}(\boldsymbol{x})\right) \\ &= \overline{\boldsymbol{\varepsilon}^2} - \bar{\boldsymbol{\varepsilon}}^2 \\ &= \overline{(\boldsymbol{\varepsilon}_0 - \boldsymbol{\Gamma}\boldsymbol{p})^2} - \overline{\boldsymbol{\varepsilon}_0 - \boldsymbol{\Gamma}\boldsymbol{p}}^2 \\ &= \sum_{s=1}^{N-1}\sum_{t=1}^{N-1}\int_D\int_D \boldsymbol{\Gamma}(\boldsymbol{x},\boldsymbol{x}')\boldsymbol{p}_s(\boldsymbol{x}')\boldsymbol{\Gamma}(\boldsymbol{x},\boldsymbol{x}'')\boldsymbol{p}_t(\boldsymbol{x}'')\left(c_{st}(\boldsymbol{x}',\boldsymbol{x}'') - c_sc_t\right)d\boldsymbol{x}''d\boldsymbol{x}'\end{aligned} \tag{8.46}$$

By substituting (7.72) and (7.74) into the following conditional expectation

$$\begin{aligned}\overline{\boldsymbol{\varepsilon}_r^2}(x) &= \mathrm{Exp}\left\langle \boldsymbol{\varepsilon}^2(\boldsymbol{x},\vartheta) \middle| \boldsymbol{x} \in D_r \right\rangle \\ &= \frac{\mathrm{Exp}\left\langle \boldsymbol{\varepsilon}^2(\boldsymbol{x},\vartheta)\chi_r(\boldsymbol{x},\vartheta)\right\rangle}{c_r}\end{aligned} \tag{8.47}$$

the variance of the strain in Phase-r, $\mathrm{Var}(\boldsymbol{\varepsilon}_r(\boldsymbol{x}))$, is derived as

$$
\begin{aligned}
&\mathrm{Var}(\boldsymbol{\varepsilon}_r(\boldsymbol{x}))\\
&=\overline{\boldsymbol{\varepsilon}_r^2}-\overline{\boldsymbol{\varepsilon}_r}^2\\
&=\frac{1}{c_r}\sum_{s=1}^{N-1}\sum_{t=1}^{N-1}\int_D\int_D \boldsymbol{\Gamma}(\boldsymbol{x},\boldsymbol{x}')\boldsymbol{p}_s(\boldsymbol{x}')\boldsymbol{\Gamma}(\boldsymbol{x},\boldsymbol{x}'')\boldsymbol{p}_t(\boldsymbol{x}'')c_{rst}(\boldsymbol{x},\boldsymbol{x}',\boldsymbol{x}'')d\boldsymbol{x}''d\boldsymbol{x}'\\
&\quad-\left((\boldsymbol{L}_r-\boldsymbol{L}_0)^{-1}\boldsymbol{p}_r-\boldsymbol{\varepsilon}_0\right)^2
\end{aligned}
\tag{8.48}
$$

with $r = 1, 2,\ldots,N-1$ and c_{rst} denoting the triple correlation function. The variance of the strain in the matrix is similarly obtained as

$$
\begin{aligned}
\mathrm{Var}(\boldsymbol{\varepsilon}_N(\boldsymbol{x}))=&\frac{1}{c_N}\sum_{s=1}^{N-1}\sum_{t=1}^{N-1}\int_D\int_D \boldsymbol{\Gamma}(\boldsymbol{x},\boldsymbol{x}')\boldsymbol{p}_s(\boldsymbol{x}')\boldsymbol{\Gamma}(\boldsymbol{x},\boldsymbol{x}'')\boldsymbol{p}_t(\boldsymbol{x}'')c_{Nst}(\boldsymbol{x},\boldsymbol{x}',\boldsymbol{x}'')d\boldsymbol{x}''d\boldsymbol{x}'\\
&-\left(\frac{1}{c_N}\sum_{r=1}^{N-1}\boldsymbol{\Gamma}c_{Nr}\boldsymbol{p}_r\right)^2
\end{aligned}
\tag{8.49}
$$

Similar to (8.46–8.49), with the detail omitted three variances of the stresses are directly given, respectively, as

$$
\begin{aligned}
\mathrm{Var}(\boldsymbol{\sigma}(\boldsymbol{x}))=&\sum_{s=1}^{N-1}\sum_{t=1}^{N-1}\int_D\int_D \boldsymbol{L}_0\boldsymbol{\Gamma}(\boldsymbol{x},\boldsymbol{x}')\boldsymbol{p}_s(\boldsymbol{x}')\boldsymbol{L}_0\boldsymbol{\Gamma}(\boldsymbol{x},\boldsymbol{x}'')\boldsymbol{p}_t(\boldsymbol{x}'')\left(c_{st}(\boldsymbol{x},\boldsymbol{x}'')-c_sc_t\right)d\boldsymbol{x}''d\boldsymbol{x}'\\
&-\sum_{s=1}^{N-1}\sum_{t=1}^{N-1}\int_D \boldsymbol{p}_s(\boldsymbol{x})\boldsymbol{L}_0\boldsymbol{\Gamma}(\boldsymbol{x},\boldsymbol{x}')\boldsymbol{p}_t(\boldsymbol{x}')\left(c_{st}(\boldsymbol{x},\boldsymbol{x}')-c_sc_t\right)d\boldsymbol{x}'\\
&+\sum_{r=1}^{N-1}c_r\boldsymbol{p}_r^2-\left(\sum_{r=1}^{N-1}c_r\boldsymbol{p}_r\right)^2
\end{aligned}
\tag{8.50}
$$

$$\mathrm{Var}\left(\boldsymbol{\sigma}_r(\boldsymbol{x})\right) = \frac{1}{c_r}\sum_{s=1}^{N-1}\sum_{t=1}^{N-1}\int_D\int_D \boldsymbol{L}_0\boldsymbol{\Gamma}(\boldsymbol{x},\boldsymbol{x}')\boldsymbol{p}_s(\boldsymbol{x}')\boldsymbol{L}_0\boldsymbol{\Gamma}(\boldsymbol{x},\boldsymbol{x}'')\boldsymbol{p}_t(\boldsymbol{x}'')$$
$$c_{rst}(\boldsymbol{x},\boldsymbol{x}',\boldsymbol{x}'')d\boldsymbol{x}''d\boldsymbol{x}' - \left((\boldsymbol{L}_r\boldsymbol{L}_0^{-1} - \boldsymbol{I})^{-1}\boldsymbol{p}_r - \boldsymbol{\sigma}_0\right)^2 \tag{8.51}$$

with $r = 1, 2,\ldots,N-1$

$$\mathrm{Var}\left(\boldsymbol{\sigma}_N(\boldsymbol{x})\right) = \frac{1}{c_N}\sum_{s=1}^{N-1}\sum_{t=1}^{N-1}\int_D\int_D \boldsymbol{L}_0\boldsymbol{\Gamma}(\boldsymbol{x},\boldsymbol{x}')\boldsymbol{p}_s(\boldsymbol{x}')\boldsymbol{L}_0\boldsymbol{\Gamma}(\boldsymbol{x},\boldsymbol{x}'')\boldsymbol{p}_t(\boldsymbol{x}'')$$
$$c_{Nst}(\boldsymbol{x},\boldsymbol{x}',\boldsymbol{x}'')d\boldsymbol{x}''d\boldsymbol{x}' - \left(\frac{\boldsymbol{L}_0}{c_N}\sum_{r=1}^{N-1}\boldsymbol{\Gamma}c_{Nr}\boldsymbol{p}_r\right)^2 \tag{8.52}$$

By comparing (8.48–8.49) with (8.51–8.52), it is noted that the variances of the stress and the strain in each individual phase follow the constitutive law as well.

Higher moments of strains and stresses can be similarly calculated by using the solved polarization functions, which involve higher order correlation functions, and correspondingly more computation. Probabilistic moments of other quantities of interest are derivable in a way similar to those of strains and stresses described above, and a probabilistic evaluation of displacement is especially described in the introduction of the perturbation-type MSFEM in Chapter 9.

8.3 NUMERICAL EXAMPLES

To illustrate the scale-coupling computation and its unconventional outcome, in the following two examples a two-phase random composite is selected to be statistically homogeneous (i.e. the volume fractions and the autocorrelation are independent of location). By choosing the elastic moduli of the hosting matrix as the reference moduli and applying the HS principle, the final solution represents the lower or upper bound of the potential energy, dependent on whether the matrix is softer or stiffer than the inclusions. Since $\boldsymbol{L}_0 = \boldsymbol{L}_2$ it immediately yields the polarization in the matrix $\boldsymbol{p}_2 = 0$. With the slow-scale strain solved following the procedure described in Section 8.1, the fast-scale unknown $\boldsymbol{p}_1$ in the inclusion phase is obtained by solving the two-phase version of (8.19) as

$$c_1(\boldsymbol{L}_1 - \boldsymbol{L}_0)^{-1}\boldsymbol{p}_1(\boldsymbol{x}) + \int_D \boldsymbol{\Gamma}(\boldsymbol{x},\boldsymbol{x}')\boldsymbol{p}_1(\boldsymbol{x}')c_{11}(\boldsymbol{x}-\boldsymbol{x}')d\boldsymbol{x}' = c_1\boldsymbol{\varepsilon}_0(\boldsymbol{x}) \tag{8.53}$$

Similarly, in the two-phase case the matrix equation (8.26) reduces to

$$\left(c_1[\Delta\,\mathrm{M}_1]+[\Gamma_{11}]\right)\{\mathrm{P}_1\}=c_1\{\mathrm{E}_0\} \tag{8.54}$$

which is exactly the matrix form of (8.53). In the following two examples the autocorrelation function is assumed to be exponential i.e.

$$c_{11}(\boldsymbol{x},\boldsymbol{x}')=c_1c_2\rho(\boldsymbol{x},\boldsymbol{x}')+\boldsymbol{c}_1^2 \tag{8.55}$$

with

$$\rho(\boldsymbol{x},\boldsymbol{x}')=\exp\left(-\frac{|\boldsymbol{x}-\boldsymbol{x}'|^2}{\ell_c^2}\right) \tag{8.56}$$

where ℓ_c denotes the correlation length.

8.3.1 A thin plate subject to pure shear

A 10 cm-by-10 cm thin plate containing randomly distributed particles is subjected to pure shear deformation, as shown in Figure 8.3. The reinforcing particle phase has the volume fraction c_1 = 0.4. The Poisson's ratios for the both phases are identically $\nu_1 = \nu_2 = 0.2$, and the Young's moduli are E_1 = 200 GPa and E_2 = 220 GPa for the matrix and the particles, respectively. Since the matrix phase is relatively soft, the numerical result corresponds to the lower bound of the potential energy. An example of microstructure that matches the solution closely is a two-dimensional finite-rank laminate mentioned in Section 4.2 with interlaminate spacing characterized by the correlation length.

With the boundary conditions specified in Figure 8.3, the slow-scale shear strain (i.e. the global shear strain) is directly found to be $\varepsilon_{0,12} = 2.5\ \mu$ cm/cm. By applying Hooke's law, the slow-scale shear stress is calculated as 416.67 MPa. The numerical result for the polarization $p_{1,12}$ at the central cross section x_1 = 5 cm is shown in Figure 8.4a. When the correlation length is infinitely large, the integral term in (8.53) vanishes and it directly leads to $p_{1,12}$ = 41.667 MPa. With decrease of the correlation length, the boundary layer effect emerges as shown in the curves of Figure 8.4a (i.e. fluctuation in a boundary layer with the thickness about its correlation length). When the correlation length decreases toward zero, the boundary layer gradually vanishes, and by employing the analytical result (4.43) we obtain $p_{1,12}$ = 39.987 MPa. In Figure 8.4b the mesh convergence study indicates that the Galerkin FEM achieves a

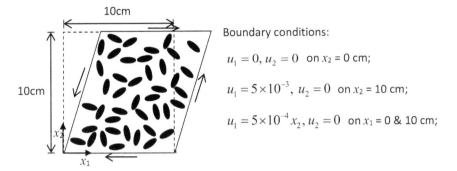

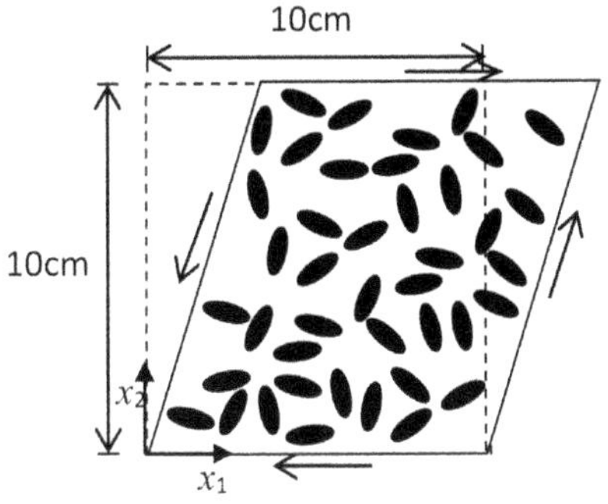

Boundary conditions:

$u_1 = 0, u_2 = 0$ on $x_2 = 0$ cm;

$u_1 = 5\times10^{-3}$, $u_2 = 0$ on $x_2 = 10$ cm;

$u_1 = 5\times10^{-4}x_2, u_2 = 0$ on $x_1 = 0$ & 10 cm;

Figure 8.3 Boundary conditions of a thin plate subject to pure shear. (Reprinted from *Comput. Struct.*, 87, Xu, X.F., X. Chen, and L. Shen, Green-function-based multiscale method for uncertainty quantification of finite body random heterogeneous materials, 1416–1426, Copyright 2009, with permission from Elsevier. Permission conveyed through Copyright Clearance Center, Inc.)

better mesh convergence rate than the collocation method, while the latter has the advantage of saving computation in generating algebraic matrices.

With the unknown stress polarization solved, probabilistic moments of strains and stresses in Section 8.2 can be calculated. The mean stresses of the two phases in case of $\ell_c = 2$ cm are shown in Figure 8.4. The accuracy of the numerical result is confirmed by the mean stress values at the four corners, as theoretically the conditional mean shear strain at the four corners for the both phases is identical to the slow-scale shear strain.

There are two interesting and unconventional observations made from this example, which are not available from a traditional deterministic mono-scale computational analysis.

The first observation is made based on Figure 8.4a in that the mean strain and the mean stress in the reinforcement increase with rise of the correlation length, physically meaning increase of the size of the reinforcing particles. Since in this example the global strain remains fixed, conversely the mean strain and the mean stress in the matrix phase decrease with enlargement of the reinforcing particles. From a strength perspective, the reinforcing particles with a smaller size are desired in that the stress is more evenly distributed between two phases and thereby the interphase stress gradients become alleviated.

The second observation is made based on the result of Figure 8.5. In the reinforcing phase, the mean shear stress has its minimum at the center and increases outwardly until it reaches the maximum at the four corners. The phenomenon becomes completely opposite in the case of the matrix phase. The observation suggests that, in a reinforcing composite such as a concrete block under a pure shear loading, the strength of the reinforcement and the matrix should be emphasized near the boundary and in the center of the domain, respectively.

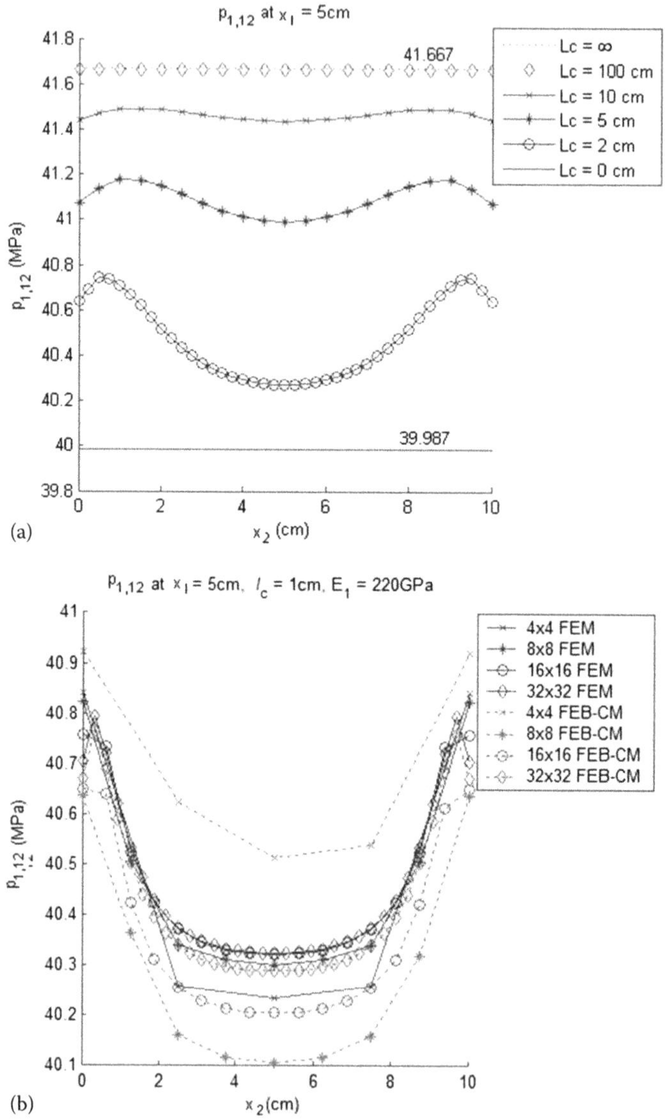

Figure 8.4 (a) Distribution of the polarization $p_{1,12}$ at the central cross section $x_1 = 5$ cm for different correlation lengths. (Adapted from *Comput. Struct.*, 87, Xu, X.F., X. Chen, and L. Shen, Green-function-based multiscale method for uncertainty quantification of finite body random heterogeneous materials, 1416–1426, Copyright 2009.); (b) Mesh convergence of $p_{1,12}$ at the central cross section with $\ell c = 1$ cm using FEM and Collocation method. (Reprinted from *Comput. Mech.*, 45, Shen, L. and X.F. Xu, Multiscale stochastic finite element modeling of random elastic heterogeneous materials, 607–621, Copyright 2010, with permission from Elsevier. Permission conveyed through Copyright Clearance Center, Inc.)

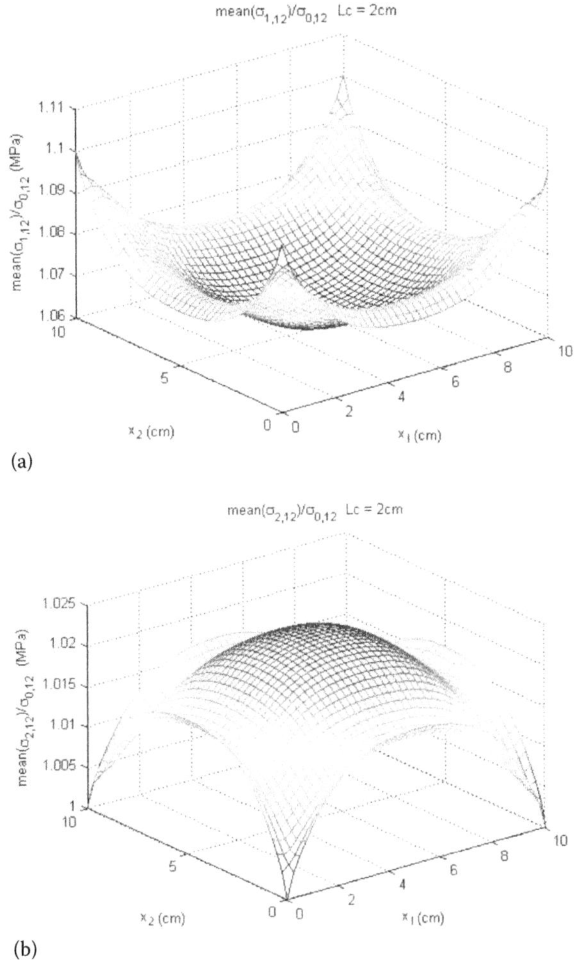

Figure 8.5 (a) Distribution of the mean stress in the particles $\overline{\sigma_{1,12}}/\sigma_{0,12}$ for ℓ_c = 2 cm with mesh 40 × 40; (b) Distribution of the mean stress in the matrix $\sigma_{2,12}/\sigma_{0,12}$ for ℓ_c = 2 cm with mesh 40 × 40. (Reprinted from *Comput. Struct.*, 87, Xu, X.F., X. Chen, and L. Shen, Green-function-based multiscale method for uncertainty quantification of finite body random heterogeneous materials, 1416–1426, Copyright 2009, with permission from Elsevier. Permission conveyed through Copyright Clearance Center, Inc.)

8.3.2 A thin composite bar subject to pure tension

In this example all the material parameters are kept identical to the previous example, except for the volume fractions changed to $c_1 = c_2 = 0.5$. The geometry and the boundary conditions are shown in Figure 8.6. Different from the previous example, the slow-scale BVP in this example demands a classical finite element analysis due to the irregular shape of the domain.

To confirm mesh convergence or sufficiency, the Multiscale Stochastic FEM is run with two mesh sizes, 2.5 mm (8 × 28) and 1.25 mm (16 × 56). Figure 8.7 shows the calculated mean stresses of the two phases, where the correlation length is taken to be 1 mm. The mean stress of the reinforcing phase is certainly higher due to its elastic moduli stiffer than those of the matrix. When the correlation length increases to 2 mm (i.e. increase of the particle size), as shown in Figure 8.8, the mean stress in the reinforcing phase decreases and conversely, the mean stress in the matrix phase increases slightly. This observation is opposite to that in the previous example under a pure shear loading. The reason is in this example there is stress concentration due to the continuously narrowed cross section, and the smaller particles corresponding to $\ell_c = 1$ mm come with a greater scale-coupling effect than those particles corresponding to $\ell_c = 2$ mm. It is expected that, with the particle size continuously decreased, the mean stress in the particles will eventually become less concentrated; in other words, the scale-coupling effect will gradually disappear with decrease of the particle size.

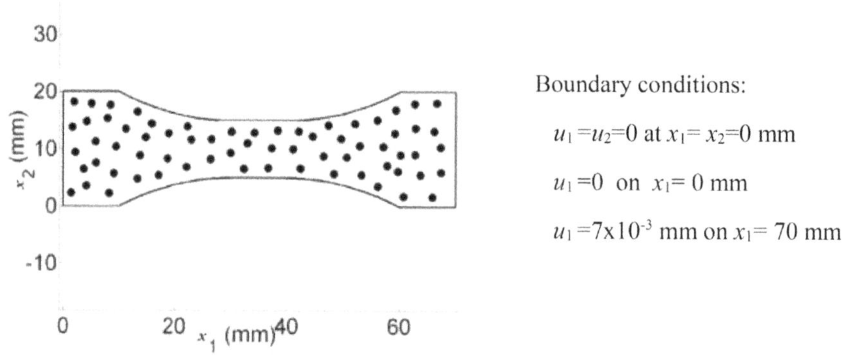

Figure 8.6 Geometry and boundary conditions of a thin composite bar subject to pure tension. (With kind permission from Springer Science+Business Media: *Comput. Mech.*, Multiscale stochastic finite element modeling of random elastic heterogeneous materials, 45, 2010, 607–621, Shen, L. and X.F. Xu. Permission conveyed through Copyright Clearance Center, Inc.)

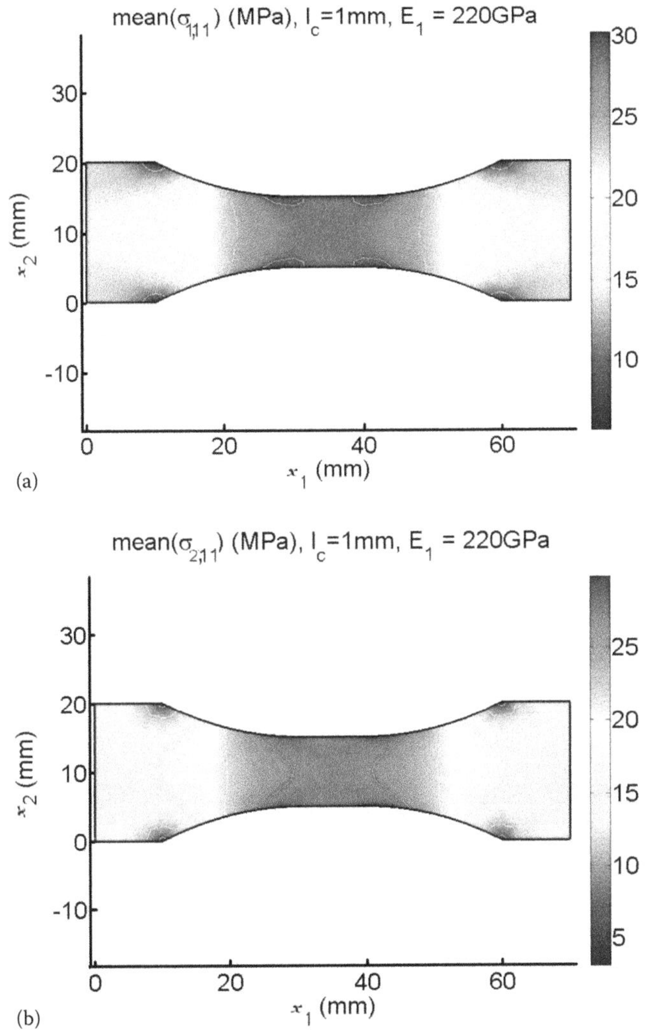

Figure 8.7 The conditional mean stresses for the particle phase (a) and the matrix phase (b) with the correlation length ℓ_c = 1 mm. (With kind permission from Springer Science+Business Media: *Comput. Mech.*, Multiscale stochastic finite element modeling of random elastic heterogeneous materials, 45, 2010, 607–621, Shen, L. and X.F. Xu. Permission conveyed through Copyright Clearance Center, Inc.)

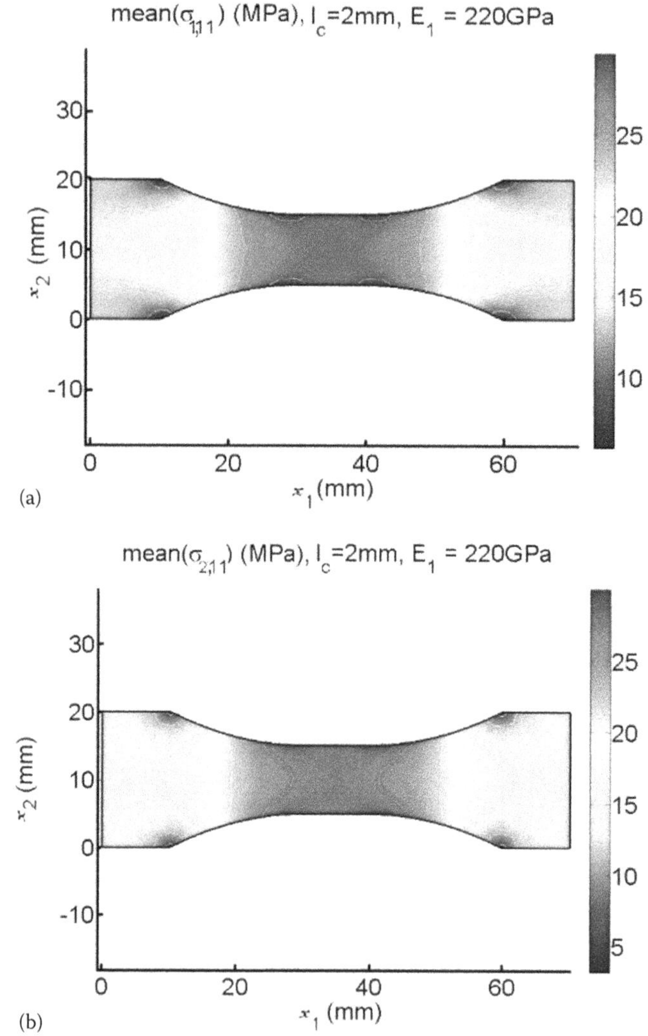

Figure 8.8 The conditional mean stresses for the particle phase (a) and the matrix phase (b) with the correlation length $\ell_c = 2$ mm. (With kind permission from Springer Science+Business Media: *Comput. Mech.*, Multiscale stochastic finite element modeling of random elastic heterogeneous materials, 45, 2010, 607–621, Shen, L. and X.F. Xu. Permission conveyed through Copyright Clearance Center, Inc.)

To optimize microstructure from a strength perspective, we need to look at a certain yield criterion $Y_\sigma = \sigma^e(\boldsymbol{\sigma})$ (e.g. the von Mises effective stress):

$$\sigma^e = \frac{1}{\sqrt{2}}\sqrt{(\sigma_{11}-\sigma_{22})^2+(\sigma_{11})^2+(\sigma_{22})^2+6(\sigma_{12})^2} \tag{8.57}$$

By using a Taylor series the mean effective stress is expanded as

$$\overline{\sigma^e} = \sigma^e(\overline{\boldsymbol{\sigma}}) + \frac{1}{2}\sum_{ij,kl}\frac{\partial^2 Y_\sigma}{\partial\sigma_{ij}\partial\sigma_{kl}}\overline{\sigma_{ij}\sigma_{kl}} + \cdots \tag{8.58}$$

In Xu et al. (2009) the first-order truncation of (8.58) is used. Compared with the slow-scale solution, the mean von Mises stress of the matrix phase becomes smaller near all the boundaries but greater at the center of the section. The mean von Mises stress of the reinforcing phase is about 1.1 times, the ratio between Young's modulus of the two phases, of the slow-scale solution in the most part of the domain, but increases significantly near the boundary of the narrowed section. This observation is consistent with the second observation made in the previous example under pure shear.

8.3.3 Concluding remarks

The above two examples demonstrate that the multiscale stochastic FEM provides certain important insights into stress and strain analysis of a finite body random heterogeneous material, which are not available from a conventional mono-scale computation. According to the MSFEM results, an optimized composite is to emphasize high strength of the reinforcing phase near all the boundaries and of the matrix phase at the center. With respect to an optimal particle size, there is no general rule, and MSFEM computation should be run on each individual case, since the scale-coupling resonance depends on a combination of several factors including geometry, boundary conditions, and interphase contrast of the elastic moduli. A few observations made about the MSFEM are summarized as follows:

- A distinctive feature of the MSFEM is computation of the modified Green function $\boldsymbol{\Gamma}$ (i.e. the second derivative of the Green function), upon which the accuracy of MSFEM computation mostly relies. A numerical experiment conducted in Shen and Xu (2010) shows that higher order finite elements provide better numerical efficiency than linear finite elements;
- The modified Green function $\boldsymbol{\Gamma}$ is singular, and it is found that use of triangular elements may lead to certain problems of oscillating effect and mesh sensitivity. Quadrilateral elements are therefore preferred;

- In MSFEM computation the Galerkin FEM achieves a higher rate of mesh convergence than the collocation method, while the latter has advantage of reduced computation;
- The amount of MSFEM computation increases significantly with decrease of the correlation length, due to a higher resolution demanded in computing the corresponding modified Green function $\boldsymbol{\Gamma}$. Therefore the domain size should be appropriately chosen to perform MSFEM computation efficiently.

Chapter 9

Multiscale stochastic finite element method and continuous random media

To numerically model a finite body continuous random medium, classical deterministic finite element method has been extended to stochastic finite element method (SFEM). A variety of formulations have been proposed to develop the stochastic FEM, including perturbation analysis (Vanmarcke & Grigoriu, 1993; Der Kiureghian & Ke, 1999), weighted integral method (Deodatis & Shinozuka, 1991), spectral analysis (Ghanem & Spanos, 1991), variational analysis (Liu et al., 1988; Hien & Kleiber, 1990; Elishakoff, Ren, & Ke, 1996; Xu, 2012d), and others. In this chapter, by distinguishing a quasiweak form from a weak form in real and random space, a unifying framework of variational formulation is presented to cover both classical displacement-based SFEM (Section 9.1) and Multiscale Stochastic FEM (Section 9.2). The formulation shows that Monte Carlo SFEM, perturbation-type SFEM, and weighted integral SFEM belong to the quasiweak form, while the weak form yields spectral SFEM, pseudospectral SFEM, and MSFEM. In Section 9.3, a perturbation-type MSFEM is specifically formulated to solve continuous random media problems efficiently, especially soils in geotechnical engineering.

9.1 FORMULATION OF STOCHASTIC FEM

9.1.1 Quasiweak form and weak form

Suppose a scalar elliptic partial differential equation is characterized with a continuous random field coefficient $K(\boldsymbol{x}, \vartheta)$ such as thermal conductivity, fluid permeability, and the like. The governing equations of the scalar elliptic BVP are written as

$$\nabla \cdot \left(K(\boldsymbol{x},\vartheta)\nabla u(\boldsymbol{x},\vartheta)\right) + f(\boldsymbol{x}) = 0 \quad in \quad D \tag{9.1}$$

$$u(\boldsymbol{x},\vartheta) = \tilde{u} \qquad on \ \partial D_u \tag{9.2}$$

$$\left(K(\boldsymbol{x},\vartheta)\nabla u(\boldsymbol{x},\vartheta)\right)\cdot \boldsymbol{n} = \tilde{t} \qquad on \ \partial D_t \tag{9.3}$$

where the body force and boundary conditions are assumed deterministic. With regard to an individual sample of the stochastic BVP (9.1–9.3), the variational form of (9.1–9.3) is given as

$$\delta\mathcal{U}(u,\vartheta)=0 \tag{9.4}$$

with the variation

$$\begin{aligned}\delta\mathcal{U}(u,\vartheta)=&\int_D\left(\nabla\delta u(\boldsymbol{x},\vartheta)\cdot K(\boldsymbol{x},\vartheta)\nabla u(\boldsymbol{x},\vartheta)-\delta u(\boldsymbol{x},\vartheta)f(\boldsymbol{x})\right)d\boldsymbol{x}\\&-\int_{\partial D_t}\delta u(\boldsymbol{x},\vartheta)\tilde{t}(\boldsymbol{x})d\boldsymbol{x}\end{aligned} \tag{9.5}$$

where the trial function u satisfies the boundary condition (9.2) and the test function δu satisfies the homogeneous boundary condition.

Since the coefficient $K(\boldsymbol{x}, \vartheta)$ is non-negative at any location in the domain D, and non-vanishing altogether, it leads to positivity of the second variation

$$\delta^2\mathcal{U}(u,\vartheta)=\int_D\nabla\delta u\cdot K(\boldsymbol{x},\vartheta)\nabla\delta u\ d\boldsymbol{x}>0 \tag{9.6}$$

which indicates that the solution of the variational form (9.4–9.5) is sought to minimize the potential energy of each individual BVP sample. The variational form (9.4–9.5) is weak in real space but strong in random space, and therefore is called *quasiweak*. A straightforward application of the quasiweak form (9.4–9.5) is Monte Carlo finite element method, which is universal and robust but often requests a great amount of sampling and computation, especially on problems involving extreme rare events.

By taking ensemble averaging of $\mathcal{U}$ in (9.4–9.5), it yields the first-moment stochastic variational form

$$\delta\overline{\mathcal{U}(u)}=0 \tag{9.7}$$

with

$$\delta\overline{\mathcal{U}(u)}=\int_D\overline{\nabla\delta u(\boldsymbol{x},\vartheta)\cdot K(\boldsymbol{x},\vartheta)\nabla u(\boldsymbol{x},\vartheta)-\delta u(\boldsymbol{x},\vartheta)f(\boldsymbol{x})}d\boldsymbol{x}-\int_{\partial D_t}\overline{\delta u(\boldsymbol{x},\vartheta)\tilde{t}(\boldsymbol{x})}d\boldsymbol{x} \tag{9.8}$$

Similar to (9.6), it follows that $\delta^2\overline{\mathcal{U}}>0$, which indicates that the solution of (9.7–9.8) renders the potential energy of an ensemble of the BVP samples minimum in a probabilistically average sense. The form (9.7–9.8) is called a *weak form*, as it is completely weak in both real space and random space.

It should be noted that, while the physical meaning for the quadratic behavior of the first-moment weak form (9.8) is clear, the extension to a higher moment weak form (e.g. $\delta(\mathcal{U}-\overline{\mathcal{U}})^2=0$) remains a question for further study. A higher moment variational formulation opens a new way to tackle complex dynamic systems on intriguing topics such as bifurcations, metastability, chaotic behaviors, etc.

9.1.2 Finite element formulation

Similar to the deterministic finite elements described in Section 8.1, by meshing the domain D into N_e elements, and using $n_p^{(e)}$ shape functions $N_i(\boldsymbol{x})$, i = 1, 2, …,$n_p^{(e)}$, in each element, the elemental random field of displacement is represented as

$$u^{(e)}(\boldsymbol{x},\vartheta)=\sum_{i=1}^{n_p^{(e)}}U_i^{(e)}(\vartheta)N_i(\boldsymbol{x}) \tag{9.9}$$

with the matrix form

$$u^{(e)}(\boldsymbol{x},\vartheta)=\{\mathrm{N}\}^T(\boldsymbol{x})\{\mathrm{U}\}^{(e)}(\vartheta) \tag{9.10}$$

The gradient of u is directly derived from (9.9–9.10) as

$$\nabla u^{(e)}(x,\vartheta)=\sum_{\mathrm{i}=1}^{n_p^{(e)}}U_i^{(e)}(\vartheta)\nabla N_i(\mathrm{x}) \tag{9.11}$$

$$\{\nabla\,\mathrm{u}\}^{(e)}(\boldsymbol{x},\vartheta)=[\mathrm{B}](\boldsymbol{x})\{\mathrm{U}\}^{(e)}(\vartheta) \tag{9.12}$$

In a four-node quadrilateral element (9.10) and (9.12) are explicitly expressed, respectively, as

$$u^{(e)}=\begin{Bmatrix}N_1 & N_2 & N_3 & N_4\end{Bmatrix}\begin{Bmatrix}U_1^{(e)}\\ U_2^{(e)}\\ U_3^{(e)}\\ U_4^{(e)}\end{Bmatrix}$$

$$\begin{Bmatrix} \nabla u_1 \\ \nabla u_2 \end{Bmatrix}^{(e)} = \begin{bmatrix} \nabla N_{1,1} & \nabla N_{2,1} & \nabla N_{3,1} & \nabla N_{4,1} \\ \nabla N_{1,2} & \nabla N_{2,2} & \nabla N_{3,2} & \nabla N_{4,2} \end{bmatrix} \begin{Bmatrix} U_1^{(e)} \\ U_2^{(e)} \\ U_3^{(e)} \\ U_4^{(e)} \end{Bmatrix}$$

Substitution of (9.9–9.12) into the quasiweak form (9.4–9.5) yields the matrix equation of the latter as

$$[\mathrm{K}](\vartheta)\{\mathrm{U}\}(\vartheta) = \{\mathrm{F}\} \tag{9.13}$$

with the random stiffness matrix

$$[\mathrm{K}](\vartheta) = \bigcup_{e=1}^{N_e} \int_{D_e} [\mathrm{B}]^T(\boldsymbol{x}) K(\boldsymbol{x},\vartheta)[\mathrm{B}](\boldsymbol{x})\, dV \tag{9.14}$$

and the deterministic force vector

$$\{\mathrm{F}\} = \bigcup_{e=1}^{N_e} \left[\int_{D_e} \{\mathrm{N}\}^T \{\mathrm{f}\} dV + \int_{\partial D_e} \{\mathrm{N}\}^T \{\tilde{\mathrm{t}}\} dS \right] \tag{9.15}$$

The matrix equation of the weak form (9.7–9.8) is similarly obtained as

$$\overline{\delta\{\mathrm{U}\}^T(\vartheta)[\mathrm{K}](\vartheta)\{\mathrm{U}\}(\vartheta)} = \overline{\delta\{\mathrm{U}\}^T(\vartheta)\{\mathrm{F}\}} \tag{9.16}$$

Distinguished from the quasiweak form (9.13), on the left-hand side of the weak form (9.16) the variation $\delta\{\mathrm{U}\}^T(\vartheta)$ becomes correlated with [K] and {U} due to application of ensemble averaging, and cannot be simply cancel out with the other one on the right-hand side.

Based on the above discussion of the quasiweak form and the weak form, the stochastic FEM is next formulated by decomposing a random field input into a number of random variables.

9.1.3 Quasiweak form: Perturbation-type SFEM

Suppose a random field coefficient $K(\boldsymbol{x}, \vartheta)$ is a function of N_K random variables $\boldsymbol{\eta} = \{\eta_1, \eta_2, \cdots, \eta_{N_K}\}$, i.e.

$$K(\boldsymbol{x},\vartheta) = g_K(\boldsymbol{\eta};\boldsymbol{x}).$$

Without loss of generality it is assumed that the random variables $\boldsymbol{\eta}$ are uncorrelated and normalized with zero-mean and unit variance, i.e., $\overline{\eta_i(\vartheta)\eta_j(\vartheta)} = \delta_{ij}$. By employing a Taylor series expansion, the coefficient $K(\boldsymbol{x}, \vartheta)$ is represented as

$$K(\boldsymbol{x},\vartheta) = K_0(\boldsymbol{x}) + \sum_{m=1}^{N_K} K_m(\boldsymbol{x})\eta_m(\vartheta) + \sum_{m,n=1}^{N_K} K_{mn}(\boldsymbol{x})\eta_m(\vartheta)\eta_n(\vartheta) + \cdots \quad (9.17)$$

with

$$\begin{aligned} K_0(\boldsymbol{x}) &= g_K(0;\boldsymbol{x}) \\ K_m(\boldsymbol{x}) &= \left.\frac{\partial g_K}{\partial \eta_m}\right|_0 \\ K_{mn}(\boldsymbol{x}) &= \left.\frac{\partial^2 g_K}{\partial \eta_m \partial \eta_n}\right|_0 \end{aligned} \quad (9.18a, b, c)$$

When $K(\boldsymbol{x}, \vartheta)$ is a Gaussian random field, only the first two terms of (9.17) remain which are obtained through the Karhunen-Loeve (K-L) expansion or the Fourier series expansion described in Section 2.1.

Following (9.17), the random stiffness matrix (9.14) is expanded in the Taylor series as

$$[\mathrm{K}](\vartheta) = [\mathrm{K}_0] + \sum_{m=1}^{N_K} [\mathrm{K}_\mathrm{m}]\eta_m + \sum_{m,n=1}^{N_K} [\mathrm{K}_\mathrm{mn}]\eta_m\eta_n + \cdots \quad (9.19)$$

with the deterministic stiffness matrices

$$[\mathrm{K}_0] = \bigcup_{e=1}^{N_e} \int_{D_e} [\mathrm{B}]^T(\boldsymbol{x}) K_0(\boldsymbol{x}) [\mathrm{B}](\boldsymbol{x})\ dV \quad (9.20)$$

$$[\mathrm{K}_\mathrm{m}] = \bigcup_{e=1}^{N_e} \int_{D_e} [\mathrm{B}]^T(\boldsymbol{x}) K_m(\boldsymbol{x}) [\mathrm{B}](\boldsymbol{x})\ dV \quad (9.21)$$

$$[\mathrm{K}_\mathrm{mn}] = \bigcup_{e=1}^{N_e} \int_{D_e} [\mathrm{B}]^T(\boldsymbol{x}) K_{mn}(\boldsymbol{x}) [\mathrm{B}](\boldsymbol{x})\ dV \quad (9.22)$$

Similarly, the unknown displacement vector is expanded as

$$\{U\}(\vartheta) = \{U_0\} + \sum_{m=1}^{N_K} \{U_m\}\eta_m + \sum_{m,n=1}^{N_K} \{U_{mn}\}\eta_m\eta_n + \cdots \tag{9.23}$$

By substituting (9.19) and (9.23) into the quasiweak form (9.13), and matching all the terms containing specific random variables, it results in a series of matrix equations

$$\begin{aligned}
&[K_0]\{U_0\} = \{F\} \\
&[K_0]\{U_m\} + [K_m]\{U_0\} = 0 \\
&[K_m]\{U_n\} + [K_0]\{U_{mn}\} + [K_{mn}]\{U_0\} = 0 \\
&[K_m]\{U_{nl}\} + [K_{mn}]\{U_l\} + [K_{ml}]\{U_n\} + [K_{nl}]\{U_m\} + [K_n]\{U_{ml}\} + [K_l]\{U_{mn}\} = 0 \\
&\cdots
\end{aligned} \tag{9.24a–d}$$

where m, n, l = 1, 2, …, N_K. The result (9.24) corresponds to the perturbation-type stochastic FEM. It can be similarly shown that the weighted integral SFEM is associated with the quasiweak form as well.

To ensure convergence and accuracy, the perturbation-type SFEM demands that the partial derivatives of the coefficient $K(\boldsymbol{x}, \vartheta)$ in (9.18) should decay sufficiently fast with increase of the order in the Taylor series expansion (e.g. the coefficient of variation $\max\left[\frac{1}{K_0(\boldsymbol{x})}\sqrt{\sum_{n=1}^{N_K} K_n^2(\boldsymbol{x})}\right]$ being about 10% or less in most practical examples).

9.1.4 Weak form: Spectral and pseudospectral SFEM

By substituting (9.19) and (9.23) into the weak form (9.16), it yields a series of matrix equations

$$\begin{aligned}
&[K_0]\{U_0\} + \sum_{m=1}^{N_K} [K_m]\{U_m\} + \sum_{m=1}^{N_K} ([K_0]\{U_{mm}\} + [K_{mm}]\{U_0\}) + \cdots = \{F\} \\
&[K_0]\{U_m\} + [K_m]\{U_0\} = 0 \\
&\sum_{m=1}^{N_K} \overline{\eta_m\eta_n\eta_l}([K_0]\{U_m\} + [K_m]\{U_0\}) + \cdots = \delta_{nl}\{F\} \\
&\cdots
\end{aligned} \tag{9.25a–c}$$

It turns out that the weak form (9.25) corresponds to the so-called stochastic Galerkin formulation by using a set of random basis functions obtained from the Taylor series (9.17). Compared with the quasiweak form (9.24), the weak form (9.25) based on the basis functions of a Taylor series offers no particular advantage, in that to ensure convergence and accuracy it is also subjected to the constraint imposed on small fluctuation of $K(\boldsymbol{x}, \vartheta)$.

However, the convergence rate of the weak form (9.25) can be significantly improved by using the Gram-Schmidt process to transform a set of Taylor basis functions into an orthogonal set $Q_m(\boldsymbol{\eta})$ with

$$\overline{Q_m(\boldsymbol{\eta})Q_n(\boldsymbol{\eta})} = \overline{Q_n^2(\boldsymbol{\eta})}\delta_{mn}.$$

When random variables $\boldsymbol{\eta}$ are Gaussian, the orthogonal set $Q_m(\boldsymbol{\eta})$ becomes specifically the Hermite polynomials $H_m(\boldsymbol{\eta})$. By replacing a set of Taylor basis functions with the corresponding orthogonal basis functions, the resulting series expansion in stochastic analysis is called a *polynomial chaos expansion* (PCE). In a PCE, the Taylor expansion (9.17) is rewritten as

$$K(\boldsymbol{x},\vartheta) = \sum_{m=0}^{M_K} K_m(\boldsymbol{x})Q_m\left(\boldsymbol{\eta}(\vartheta)\right) \tag{9.26}$$

with $Q_0 = 1$. The number of PCE terms M_K depends on N_K and n_{pc} the order of a PCE. In the case of the Hermite polynomial chaos, there is a specific formula

$$M_K = \frac{(N_K + n_{pc})!}{N_K!n_{pc}!} - 1 \tag{9.27}$$

By applying a PCE, the stiffness matrix is represented as

$$[\mathrm{K}](\vartheta) = \sum_{m=0}^{M_K} [\mathrm{K}_\mathrm{m}]Q_m\left(\boldsymbol{\eta}(\vartheta)\right) \tag{9.28}$$

where the deterministic matrices $[\mathrm{K}_\mathrm{m}]$ are given in (9.21). Similarly the unknown displacement vector is expanded as

$$\{\mathrm{U}\}(\vartheta) = \sum_{m=0}^{M_K} \{\mathrm{U}_\mathrm{m}\}\mathrm{Q}_m\left(\boldsymbol{\eta}(\vartheta)\right) \tag{9.29}$$

And finally the weak form (9.25) is simplified into

$$\sum_{n,l=0}^{M_K} \overline{Q_m Q_n Q_l}[\mathrm{K_n}]\{\mathrm{U_l}\} = \overline{Q_m}\{\mathrm{F}\} \tag{9.30}$$

with $m = 0, 1, 2, \ldots, M_K$. When a set of orthogonal basis functions $Q_m(\boldsymbol{\eta})$ are chosen to be multivariate orthogonal polynomials about the random variables $\boldsymbol{\eta}$, the weak form (9.30) corresponds to the so-called *spectral SFEM*. To a low-dimensional stochastic problem, the spectral SFEM offers a great advantage over the perturbation-type SFEM since the constraint of small fluctuation imposed on the latter is removed.

Major computation of the spectral SFEM is spent on solving the coupled equations in (9.30) resulting from many nonvanishing coefficients $\overline{Q_m Q_n Q_l}$. To decouple these equations, a special set of basis functions, denoted as $\mathcal{L}_k$, are alternatively constructed by employing the Lagrange interpolation polynomials as follows:

$$\mathcal{L}_k(\boldsymbol{\eta}) = \ell_{1,k_1}\ell_{2,k_2}\cdots\ell_{n,k_n} \quad k = (k_1, k_2, \cdots, k_n) \tag{9.31}$$

$$\ell_{i,k_i}(\eta_i) = \prod_{j_i=1, k_i \neq j_i}^{r_i-1} \frac{\eta_i - \eta^*_{i,j_i}}{\eta^*_{i,k_i} - \eta^*_{i,j_i}} \tag{9.32}$$

where $\left(\eta^*_{i,1}, \eta^*_{i,2}, \cdots, \eta^*_{i,r_i}\right)$ is a set of r_i data points sampled for random variable η_i. Since the Lagrange interpolation polynomials hold not only orthogonality $\overline{\mathcal{L}_m(\boldsymbol{\eta}^*)\mathcal{L}_n(\boldsymbol{\eta}^*)} = \delta_{mn}$ but also the decoupling property $\overline{\mathcal{L}_m(\boldsymbol{\eta}^*)\mathcal{L}_n(\boldsymbol{\eta}^*)\mathcal{L}_l(\boldsymbol{\eta}^*)} = \delta_{mn}\delta_{ml}$, the equations in (9.30) are completely decoupled by using such sampling points. With the unknown {U} solved at the sampling points, its probabilistic moments are evaluated by using a certain efficient quadrature rule (e.g. sparse grids for smooth functions). This formulation results in the so-called *pseudospectral SFEM* characterized with a feature of collocation-type simplicity.

A major issue facing the *spectral SFEM* and the *pseudospectral SFEM* is exponential increase of the unknown vectors with increase of a domain size and the order of a PCE. As noted in Section 2.1, when a domain size is significantly larger than the correlation length, the Karhunen-Loeve expansion becomes the Fourier series expansion with a big number of the expansion terms N_K. The order of a PEC, n_{pc}, depends on the degree of non-Gaussianity of a random field; i.e. the more a random field is deviated from Gaussianity the greater its PCE order n_{pc} is in need. As indicated in (9.27), a combination of two big numbers N_K and n_{pc} results in a prohibitively huge number of equations. Such a curse-of-dimensionality issue is not uncommon in random media problems. To overcome this high-dimensionality issue, a scheme of random field based orthogonal expansion is introduced next.

9.2 RANDOM FIELD BASED ORTHOGONAL EXPANSION

9.2.1 New orthogonal expansion

As shown in (9.27), the high dimensionality issue of the *spectral SFEM* and *pseudospectral SFEM* arises due to the use of a random variable based representation. To circumvent the issue, a random field based (RFB) orthogonal expansion is proposed to replace (9.26) as

$$K(\boldsymbol{x},\vartheta) = \sum_{m=0}^{\tilde{M}_K} K_m(\boldsymbol{x}) Q_m\left(\tilde{Y}(\boldsymbol{x},\vartheta)\right) \tag{9.33}$$

with

$$\overline{Q_m(\tilde{Y}_1) Q_n(\tilde{Y}_2)} = \rho_n(\tilde{Y}_1,\tilde{Y}_2)\delta_{mn} \tag{9.34}$$

where $\tilde{Y}_i = \tilde{Y}(\boldsymbol{x}_i,\vartheta)$ denotes an underlying normalized random field with zero-mean and unit variance, and $\rho_n(\tilde{Y}_1,\tilde{Y}_2)$ the autocorrelation of Q_n at two points $\boldsymbol{x}_1$ and $\boldsymbol{x}_2$.

When the underlying normalized random field is Gaussian $\tilde{Z}$ with its autocorrelation ρ, the n-point correlation is analytically available following the generalized Mehler's formula (Slepian, 1972):

$$\boxed{\overline{\tilde{Z}_{b_1}(\boldsymbol{x}_1,\vartheta)\cdots\tilde{Z}_{b_n}(\boldsymbol{x}_n,\vartheta)} = b_1!\cdots b_n! \sum_{v_{12}=0}^{\infty}\cdots\sum_{v_{n-1,n}=0}^{\infty} \delta_{b_1 d_1}\cdots\delta_{b_n d_n} \prod_{j<k} \frac{\rho^{v_{jk}}(\boldsymbol{x}_j,\boldsymbol{x}_k)}{v_{jk}!}} \tag{9.35}$$

where $d_k = \sum_{j\neq k} v_{jk}$, $v_{jk} = v_{kj}$, $\delta_{b_k d_k} = \begin{cases} 1 & b_k = d_k \\ 0 & b_k \neq d_k \end{cases}$. The auto- and triple-point correlations are specifically given as

$$\boxed{\overline{H_m(\tilde{Z}_1)H_n(\tilde{Z}_2)} = n!\rho^n(\tilde{Z}_1,\tilde{Z}_2)\delta_{mn}} \tag{9.36}$$

$$\boxed{\begin{aligned} c_{mnl}(x,x',x'') &= \overline{H_m(x,\vartheta)H_n(x',\vartheta)H_l(x'',\vartheta)} \\ &= \frac{m!n!l!}{i_1!i_2!i_3!}\rho^{i_1}(x-x')\rho^{i_2}(x-x'')\rho^{i_3}(x'-x'') \\ i_1 &= \frac{m+n-l}{2};i_2=\frac{m+l-n}{2};i_3=\frac{n+l-m}{2} \end{aligned}} \tag{9.37}$$

where i_1, i_2, i_3 are nonnegative integers.

Compared with (9.26–9.27), by using an RFB orthogonal expansion, the number of expansion terms in (9.33) is exponentially reduced to the order of the PCE or nonlinearity, i.e.

$$\tilde{M}_K = n_{pc}$$

For example, a lognormal random field $K(\boldsymbol{x},\vartheta)$ is represented by a series of the Hermite polynomials about an underlying normalized Gaussian random field $\tilde{Z}$, with the random orthogonal basis functions given as

$$\begin{aligned} H_0 = 1, H_1(\boldsymbol{x},\vartheta) = \tilde{Z}(\boldsymbol{x},\vartheta), H_2(\boldsymbol{x},\vartheta) = \tilde{Z}^2(\boldsymbol{x},\vartheta)-1, \\ H_3(\boldsymbol{x},\vartheta) = \tilde{Z}^3(\boldsymbol{x},\vartheta) - 3\tilde{Z}(\boldsymbol{x},\vartheta),\cdots \end{aligned} \tag{9.38}$$

In a random multiphase composite, the random field is characterized with a piecewise constant discrete probability density function, of which the RFB orthogonal expansion is presented in the form (4.7).

An RFB orthogonal expansion offers a great advantage over a random variable based representation (9.26) in terms of dimensionality. However, an RFB orthogonal expansion (9.33) is incompatible with the classical displacement-based finite elements as explained below. By substituting (9.33) into (9.14), it follows that

$$[\mathrm{K}](\vartheta) = \sum_{m=0}^{\tilde{M}_K}[\mathrm{K}_\mathrm{m}](\vartheta) \tag{9.39a}$$

$$[\mathrm{K}_\mathrm{m}](\vartheta) = \bigcup_{e=1}^{N_e}\int_{D_e}[\mathrm{B}]^T(x)K_m(\boldsymbol{x})Q_m(\boldsymbol{x},\vartheta)[\mathrm{B}](\boldsymbol{x})\,dV \tag{9.39b}$$

which cannot be further calculated without decomposing the random field $Q_m(\boldsymbol{x},\vartheta)$ into random variables. It is therefore concluded that an RFB orthogonal expansion is inapplicable to the classical displacement-based finite element method. We show next that, the Green-function-based FEM, which tackles a BVP of a random multiphase composite perfectly in Chapter 8, fits an RFB expansion as well.

9.2.2 RFB orthogonal expansion on a continuous random medium

Identical to the BVP decomposed in Section 7.1, a stochastic elliptic BVP (9.1–9.3) is decomposed to obtain two bounds of the potential energy as follows

$$\boxed{\begin{aligned}\mathcal{U}^{+}(\boldsymbol{p},K_0)=\mathcal{U}_0+\frac{1}{2}\left\langle\nabla u_0,(K-K_0)\nabla u_0\right\rangle+\frac{1}{2}\left\langle\boldsymbol{p},\boldsymbol{\Gamma p}\right\rangle+\frac{1}{2}\left\langle\boldsymbol{\Gamma p},(K-K_0)\boldsymbol{\Gamma p}\right\rangle\\ -\left\langle\boldsymbol{\Gamma p},(K-K_0)\nabla u_0\right\rangle\end{aligned}} \tag{9.40}$$

$$\boxed{\begin{aligned}\mathcal{U}^{-}(\boldsymbol{p},K_0)=\mathcal{U}_0-\frac{1}{2}\left\langle\nabla u_0,K_0\left(\frac{1}{a}-\frac{1}{a_0}\right)K_0\nabla u_0\right\rangle-\frac{1}{2}\left\langle\boldsymbol{p},\left(\frac{\boldsymbol{p}}{K}\right)\right\rangle\\ -\left\langle K_0\nabla u_0,\left(\frac{1}{K}-\frac{1}{K_0}\right)\boldsymbol{p}\right\rangle-\frac{1}{2}\left\langle\boldsymbol{\Gamma p},K_0\left(\frac{1}{K}-\frac{1}{K_0}\right)K_0\boldsymbol{\Gamma p}\right\rangle\\ -\frac{1}{2}\left\langle\boldsymbol{p},\boldsymbol{\Gamma p}\right\rangle+\left\langle\nabla u_0,K_0\left(\frac{1}{K}-\frac{1}{K_0}\right)K_0\boldsymbol{\Gamma p}\right\rangle+\left\langle\frac{K_0\boldsymbol{p}}{K},\boldsymbol{\Gamma p}\right\rangle\end{aligned}} \tag{9.41}$$

which are exactly the conductivity version of (7.63–7.64). When the ratio between a domain size and a correlation length is sufficiently large (i.e. the scales are well separated), the above bounds reduces exactly to the bounds (3.55–3.56) of scale separation.

Like (7.73) the solution of the polarization is expanded as an infinite series

$$\begin{aligned}\boldsymbol{p}&=(K-K_0)\nabla u_0-(K-K_0)\boldsymbol{\Gamma}(K-K_0)\nabla u_0\\&\quad+(K-K_0)\boldsymbol{\Gamma}(K-K_0)\boldsymbol{\Gamma}(K-K_0)\nabla u_0+\cdots\end{aligned} \tag{9.42}$$

Given (9.33) an RFB orthogonal expansion of the random field K, the first- and second-order trial functions are written for the unknown polarization, respectively, as

$$\boldsymbol{p}(\boldsymbol{x},\vartheta)=\sum_{m=0}^{\tilde{M}_K}\boldsymbol{p}_m(\boldsymbol{x})Q_m\left(\tilde{Y}(\boldsymbol{x},\vartheta)\right) \tag{9.43}$$

$$\boldsymbol{p}(\boldsymbol{x},\vartheta)=\sum_{m=0}^{\tilde{M}_K}\boldsymbol{p}_m(\boldsymbol{x})Q_m\left(\tilde{Y}(\boldsymbol{x},\vartheta)\right)+\sum_{m,n=0}^{\tilde{M}_K}\boldsymbol{p}_{mn}(\boldsymbol{x})Q_m\left(\tilde{Y}(\boldsymbol{x},\vartheta)\right) \int_D \boldsymbol{\Gamma}(\boldsymbol{x},\boldsymbol{x}')Q_n\left(\tilde{Y}(\boldsymbol{x}',\vartheta)\right)d\boldsymbol{x}' \tag{9.44}$$

Substitution of a trial function into (9.40) or (9.41) and subsequent minimization lead to a solution for the unknown polarization. The Green-function-based formulation (9.40) or (9.41) suits perfectly such an RFB orthogonal expansion, and thereby circumvents the curse of dimensionality. Below we provide an example of random media homogenization to demonstrate the advantage of an RFB expansion over a random variable based representation. Another example of an RFB expansion on random media is detailed in (Xu, 2011b).

In a flow transport problem, the coefficient of the elliptic equation is termed as *permeability*. We choose a 2D lognormal random field

$$K(\boldsymbol{x},\vartheta)=\exp\left(\sigma_Z\tilde{Z}(\boldsymbol{x},\vartheta)\right) \tag{9.45}$$

with $\tilde{Z}$ a normalized Gaussian random field, and σ_Z the standard deviation. The 2D lognormal random field (9.45) has a well-known closed form solution

$$K^e=\exp(\overline{\ln K}) \tag{9.46}$$

for the effective permeability (Matheron, 1967). The lognormal random field (9.45) is represented in an RFB orthogonal expansion as

$$K(\boldsymbol{x},\vartheta)=\sum_{m=0}^{\tilde{M}_K}K_m H_m\left(\tilde{Z}(\boldsymbol{x},\vartheta)\right) \tag{9.47}$$

with each K_m calculated by applying (2.25) as

$$K_m = \frac{1}{\sqrt{2\pi m\,!}}\int_{-\infty}^{+\infty} \exp(\sigma_Z z) H_m(z) e^{-\frac{z^2}{2}}\, dz \\ = \frac{\sigma_Z^n}{m\,!}\exp\left(\frac{\sigma_Z^2}{2}\right) \tag{9.48}$$

To simplify the trial function, let each expansion term of the first-order trial function (9.43) be a constant, i.e.

$$p_i(\boldsymbol{x},\vartheta) = \sum_{n=0}^{\tilde{M}_K} p_{i,n} H_n(\boldsymbol{x},\vartheta) \tag{9.49}$$

where the subscript i denotes a component of the vector $\boldsymbol{p}$ along direction-i. Note that, since K_0 in (9.48) refers to the first term of the RFB orthogonal expansion (9.47), hereby we especially denote K_r the reference permeability. In an infinitely large two-dimensional domain, the Green function and the modified Green function are analytically available, respectively, as (A2.8) and (A2.12).

Let the lognormal random medium be subject to a constant flow gradient $\nabla_1 u_0$ along direction-1. By substituting (9.49) and (A2.12) into (9.41) or (3.55), and taking (9.36–9.37) into account, the upper bound of the potential is finally obtained in terms of p_1 as

$$U^+(p_1,K_r) = \frac{1}{2}K_0(\nabla_1 u_0)^2 + \frac{1}{4K_r}\sum_{n=1}^{\tilde{M}_K} n!p_{1,n}^2 + \frac{1}{2K_r^2}\sum_{m,n,l=1}^{\tilde{M}_K} n!d_{mnl}K_m p_{1,n}p_{1,l} \\ - \frac{\nabla_1 u_0}{2K_r}\sum_{n=1}^{\tilde{M}_K} n!K_n p_{1,n} \tag{9.50}$$

with

$$d_{mnl} = \frac{1}{4\pi}\int_0^{2\pi}\int_0^{\infty}\int_0^{\infty}\frac{\cos 2\varphi}{rs}c_{mnl}(r,s,\varphi)\,dr\,ds\,d\varphi \tag{9.51}$$

where the triple correlation $c_{mnl}(r,s,\varphi)$, with its meaning similar to c_{111} in (4.102), is calculated via (9.37). Suppose the autocorrelation function is exponential with $\ell_c = 1$

$$\rho(\boldsymbol{x}) = \exp\left(-\frac{|\boldsymbol{x}|^2}{\ell_c^2}\right) \tag{9.52}$$

Extremization of (9.50) about $p_{1,n}$ yields the following matrix equation:

$$[\mathrm{K}]\{\mathrm{P}_1\} = \{\mathrm{RHS}\} \tag{9.53}$$

where $\{\mathrm{P}_1\}$ is the vector consisting of unknown $p_{1,n}$, and the components of the matrix

$$[\mathrm{K}]_{nl} = 1 + \frac{2}{K_r} \sum_{m=1}^{\tilde{M}_K} d_{mnl} K_m \qquad n,l = 1,2,\cdots,\tilde{M}_K \tag{9.54}$$

The advantage of an RFB expansion is demonstrated by the order of a PCE or nonlinearity $\tilde{M}_K$ that can be chosen as a big number. In this example, the nonlinearity or PCE order is chosen to be 9, since the value of K_9 calculated from (9.48) becomes sufficiently small, close to zero. By substituting the solved $p_{1,n}$ into (9.50), the upper bound of the effective permeability is obtained, as shown in Figure 9.1. The calculated upper bound using the spatial statistics up to the triple correlation is 1.0841 when the standard deviation has a value of 1, 8.41% higher than the exact solution. The figure also shows independence of the upper bound estimate from the reference permeability. When the standard deviation of the underlying Gaussian random field in (9.47) decreases toward zero, as verified in Figure 9.2, the upper bound estimate converges to the exact solution.

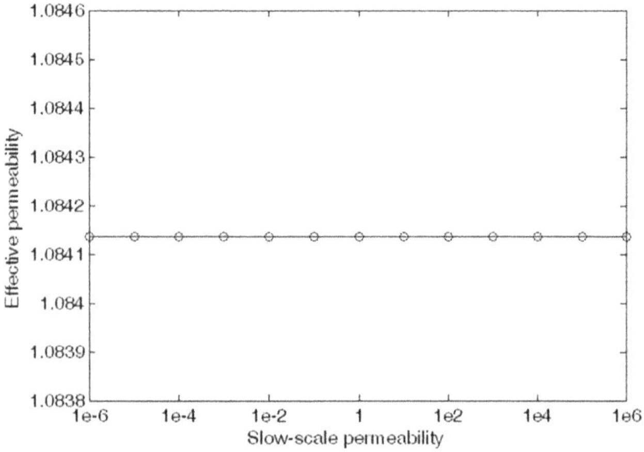

Figure 9.1 Calculated upper bound effective permeability of the lognormal random medium, which is independent of the slow-scale reference permeability. (Republished with permission of Research Publishing Services, from Xu, X.F., *Convolved orthogonal expansions on dynamics and random media problems*, Proc. of the 5th Asian-Pacific Symposium on Structural Reliability and its Applications, P262, 2012e.)

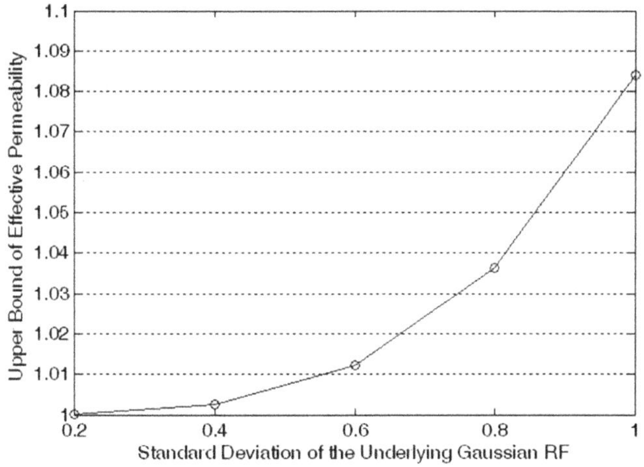

Figure 9.2 Convergence of the upper bound estimate with decrease of standard deviation of the underlying Gaussian random field. (Republished with permission of Research Publishing Services, from Xu, X.F., *Convolved orthogonal expansions on dynamics and random media problems*, Proc. of the 5th Asian-Pacific Symposium on Structural Reliability and its Applications, P262, 2012e.)

A concluding remark is that in a scale-coupling problem, when the domain size is finite, the modified Green function (A2.12) and the potential energy (9.50) demand finite element computation, which is exactly the application of the multiscale stochastic FEM in a random media problem. In Section 9.3, a fast-computing procedure of the MSFEM is introduced to deal with such computation, which is especially useful in geotechnical engineering.

9.2.3 RFB orthogonal expansion on stochastic dynamics

The idea of an RFB orthogonal expansion can be similarly applied to stochastic dynamics in which an orthogonal expansion is made about an underlying one-dimensional stochastic process, instead of a two-dimensional random field in random media. Below, two oscillators are presented, one linear and the other nonlinear, to illustrate application of an RFB orthogonal expansion on stochastic dynamics.

Linear Oscillator

A linear oscillator with its natural frequency ω_n and damping ratio ζ is written as

$$\ddot{u}_0 + 2\zeta\omega_n\dot{u}_0 + \omega_n^2 u_0 = f \tag{9.55}$$

$$u_0(0) = \dot{u}_0(0) = 0 \tag{9.56}$$

The random excitation f is assumed to have an RFB orthogonal expansion similar to (9.47)

$$f(t,\vartheta) = \sum_{n=0}^{n_{pc}} f_n(t) H_n\left(\tilde{Z}(t,\vartheta)\right) \tag{9.57}$$

the response u_0 is directly obtained in terms of G the Green function or the so-called transfer function in dynamics, as

$$u_0(t,\vartheta) = G * \sum_{n=0}^{n_{pc}} f_n(t) H_n\left(\tilde{Z}(t,\vartheta)\right) \tag{9.58}$$

with

$$G(t) = \frac{1}{\omega_n \sqrt{1-\zeta^2}} e^{-\zeta\omega_n t} \sin\left(\omega_n \sqrt{1-\zeta^2}\, t\right) \tag{9.59}$$

$$\hat{G}(\omega) = \frac{1}{\omega_n^2 - \omega^2 + \sqrt{-1}2\zeta\omega\omega_n} \tag{9.60}$$

where the asterisk denotes the convolution operator. All the statistics about the response (9.58) can be calculated by using the generalized Mehler's formula (9.35), e.g. the first three correlations of the nonstationary output are calculated as

$$\bar{u}_0(t) = \int_0^t G(t-\tau) f_0(\tau)\, d\tau \tag{9.61a}$$

$$R_2(t_1,t_2) = \int_0^{t_2}\int_0^{t_1} G(t_1-\tau_1)G(t_2-\tau_2) \sum_{n=0}^{n_{pc}} n!\, \rho^n(\tau_1-\tau_2) f_n(\tau_1) f_n(\tau_2)\, d\tau_1 d\tau_2 \tag{9.61b}$$

$$R_3(t_1,t_2,t_3)=\int_0^{t_3}\int_0^{t_2}\int_0^{t_1} G(t_1-\tau_1)G(t_2-\tau_2)G(t_3-\tau_3)$$
$$\sum_{i,j,k=0}^{n_{pc}} c_{ijk}(\tau_1-\tau_2,\tau_1-\tau_3,\tau_2-\tau_3)\, d\tau_1 d\tau_2 d\tau_3 \tag{9.61c}$$

with c_{ijk} given in (9.37).

When the excitation f is stationary, (9.58) reduces to

$$u_0(t,\vartheta)=G*\sum_{n=0}^{N} f_n H_n(t,\vartheta) \tag{9.62}$$

with f_n being constants. An example to calculate the response of a lognormal random excitation is provided in Xu (2011b).

Nonlinear Oscillator

Next we consider a Duffing oscillator

$$\ddot{u}+2\zeta\omega_n\dot{u}+\omega_n^2(u+\varepsilon u^3)=f \tag{9.63}$$

with initial conditions

$$u(0)=\dot{u}(0)=0 \tag{9.64}$$

Similar to the decomposition of a heterogeneous BVP formulated in Section 7.1, the nonlinear oscillator (9.63) is decomposed into a linear reference oscillator (9.55–9.56), and a nonlinear fluctuation oscillator below

$$\ddot{u}'+2\zeta\omega_n\dot{u}'+\omega_n^2 u'=-\varepsilon\omega_n^2(u_0+u')^3 \tag{9.65}$$

$$u'(0)=\dot{u}'(0)=0 \tag{9.66}$$

with the total displacement and the total velocity being superposition of those of the two suboscillators

$$u=u_0+u' \tag{9.67}$$

$$\dot{u} = \dot{u}_0 + \dot{u}' \tag{9.68}$$

Suppose the excitation f is a stationary Gaussian process characterized with its power spectral density $\hat{C}_f$, and it yields that the response of the reference linear oscillator (9.55–9.56) is a stationary Gaussian process as well, i.e.

$$u_0 = G * f \tag{9.69}$$

which is rewritten in terms of a normalized Gaussian process $\tilde{Z}$ as

$$u_0 = \sigma_0 \tilde{Z} \tag{9.70}$$

with $\tilde{Z}$ characterized with zero-mean, unit variance and a power spectral density $\hat{C}_Z$

$$\hat{C}_Z(\omega) = \frac{\hat{C}_f}{\sigma_0^2} \left| \hat{G}(\omega) \right|^2 \tag{9.71}$$

$$\sigma_0^2 = \frac{1}{2\pi} \int_{-\infty}^{\infty} \left| \frac{1}{\omega_n^2 - \omega^2 + \sqrt{-1} 2\zeta\omega\omega_n} \right|^2 \hat{C}_f(\omega)\, d\omega \tag{9.72}$$

By substituting the linear output of (9.65) into u' on the right-hand side reiteratively, the stationary output of the nonlinear fluctuation oscillator (9.65–9.66) is obtained in an infinite series as

$$u' = -\sigma_0 \beta G * \tilde{Z}^3 + 3\sigma_0 \beta G * \tilde{Z}^2 (\beta G * \tilde{Z}^3) - \cdots \tag{9.73}$$

where a nonlinear parameter

$$\beta = \varepsilon \sigma_0^2 \omega_n^2.$$

A close look at (9.73) tells that in each term the order of $\tilde{Z}$ is always odd, and therefore the mean of the response

$$\bar{u} = 0.$$

By using the random basis functions H_n in (9.38), the total response as a sum of (9.70) and (9.73) is rewritten as

$$u/\sigma_0 = H_1 - \beta G*(H_3+3H_1)+3\beta G*\left[(H_2+1)\left(\beta G*(H_3+3H_1)\right)\right] + O(\beta^3 G*G*G) \tag{9.74}$$

which in the frequency domain is expressed as

$$\hat{u}/\sigma_0 = \hat{H}_1 - \beta\hat{G}(\hat{H}_3+3\hat{H}_1)+3\beta\hat{G}\left[\left(\hat{H}_2+2\pi\delta(0)\right)*\left(\beta\hat{G}(\hat{H}_3+3\hat{H}_1)\right)\right]+O\left((\beta\hat{G})^3\right) \tag{9.75}$$

The stationary PSD of the response is thereby obtained as

$$\begin{aligned}\hat{C}_u &= \hat{u}\hat{u}^* \\ &= \sigma_0^2\hat{C}_Z\left[1-3\beta(\hat{G}+\hat{G}^*)+3\beta^2\left(3\left|\hat{G}\right|^2+2\left|\hat{G}\right|^2\frac{\hat{C}_Z*\hat{C}_Z*\hat{C}_Z}{\hat{C}_Z}+3(\hat{G}^2+\hat{G}^{*2})\right.\right. \\ &\left.\left.+6\hat{G}(\hat{G}*\hat{C}_Z*\hat{C}_Z)+6\hat{G}^*(\hat{G}^**\hat{C}_Z*\hat{C}_Z)+3(\hat{G}+\hat{G}^*)\int_{-\infty}^{\infty}\hat{G}(\omega)\hat{C}_Z(\omega)\,d\omega\right)\right] \\ &+O\left(\left(\beta\left|\hat{G}\right|\right)^3\right)\end{aligned} \tag{9.76}$$

with $\hat{C}_Z = \left|\hat{H}_1\right|^2$. To justify the perturbation approach, a condition of weak nonlinearity is required in that the perturbation factor $\beta\left|\hat{G}\right| << 1$, i.e. $\dfrac{\varepsilon\sigma_0^2}{\sqrt{\left[1-(\omega/\omega_n)^2\right]^2+(2\zeta\omega/\omega_n)^2}} << 1$. In the low-frequency range $\omega << \omega_n$, it yields the condition $\varepsilon\sigma_0^2 << 1$. In the high-frequency range $\omega >> \omega_n$, the condition becomes much relaxed as $\varepsilon\sigma_0^2(\omega_n/\omega)^2 << 1$. In the intermediate-frequency range, especially near resonance when $\omega \sim \omega_n$, the weak nonlinearity condition becomes damping controlled, i.e.

$$\boxed{\frac{\varepsilon\sigma_0^2}{2\zeta} << 1} \tag{9.77}$$

Overall the applicability of the perturbation approach to all the frequency ranges is controlled by the damping controlled factor $\dfrac{\varepsilon\sigma_0^2}{2\zeta}$ in (9.77) (Xu & Stefanou, 2012b).

When the random excitation f is a white noise with its intensity I, from (9.71–9.72) the PSD of $\tilde{Z}$ is obtained as

$$\hat{C}_Z(\omega) = \frac{I}{\sigma_0^2}\left|\hat{G}(\omega)\right|^2$$

with its variance

$$\sigma_0^2 = \frac{1}{2\pi} I \int_{-\infty}^{\infty} \left| \frac{1}{\omega_n^2 - \omega^2 + \sqrt{-1}2\zeta\omega\omega_n} \right|^2 d\omega = \frac{I}{4\zeta\omega_n^3} \tag{9.78}$$

By substituting (9.78) into (9.77), the damping controlled factor becomes (Xu & Stefanou, 2012b)

$$\boxed{\frac{\varepsilon I}{8\zeta^2\omega_n^3} << 1} \tag{9.79}$$

Suppose the damping ratio $\zeta = 0.02$ and the natural frequency $\omega_n = 1$. The response PSD is approximated with the first two terms on the right-hand side of (9.76). With the intensity of the white noise fixed at $I = 1$, Figure 9.3 shows that the PSD shifts to the upper right with increase of the nonlinearity parameter ε from 0 ($\frac{\varepsilon I}{8\zeta^2\omega_n^3} = 0$), to 0.00025 ($\frac{\varepsilon I}{8\zeta^2\omega_n^3} = 0.078125$) and to 0.0005 ($\frac{\varepsilon I}{8\zeta^2\omega_n^3} = 0.15625$). When the nonlinearity parameter ε is fixed at 0.00025,

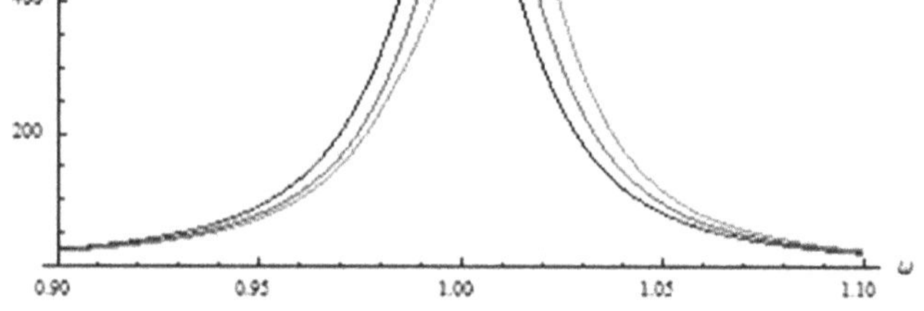

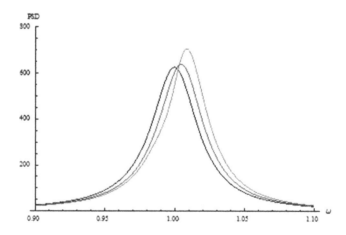

Figure 9.3 The PSD of the displacement (from left to right $\mathcal{E}$ = 0, 0.00025, and 0.0005, respectively) with $\zeta = 0.02$, $\omega_n = 1$, and I = 1. (From Xu, X.F. and G. Stefanou, *Int. J. Uncertainty Quantification*, 2: 383–395, 2012b. With permission from Begell House Inc. Publishers.)

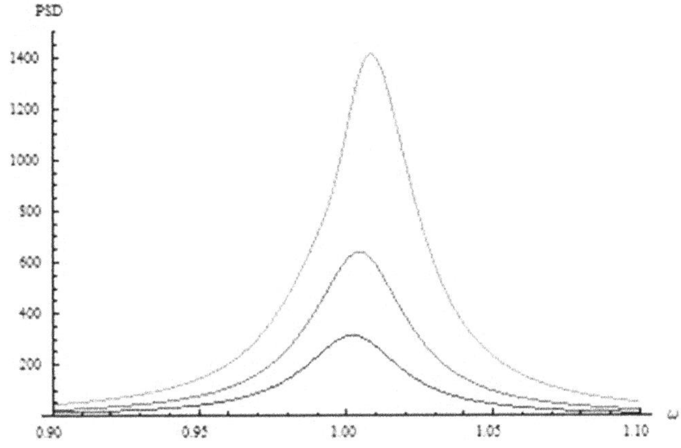

Figure 9.4 The PSD of the displacement (from bottom to top I = 0.5, 1, and 2, respectively) with $\zeta = 0.02$, $\omega_n = 1$, and $\varepsilon = 0.00025$. (From Xu, X.F. and G. Stefanou, *Int. J. Uncertainty Quantification*, 2: 383–395, 2012b. With permission from Begell House Inc. Publishers.)

Figure 9.4 shows a shift of the PSD to the upper right with increase of the intensity I from 0.5 ($\frac{\varepsilon I}{8\zeta^2\omega_n^3} = 0.039063$) to 1 ($\frac{\varepsilon I}{8\zeta^2\omega_n^3} = 0.078125$), and to 2 ($\frac{\varepsilon I}{8\zeta^2\omega_n^3} = 0.15625$). The results reflect exactly the well-known phenomenon concerning the natural frequency of a single-well Duffing oscillator.

As it is analytically known that

$$\int_{-\infty}^{\infty} \frac{2\left(\omega_n^2 - \omega^2\right)}{\left(\omega_n^2 - \omega^2\right)^2 + (2\zeta\omega\omega_n)^2} \left| \frac{1}{\omega_n^2 - \omega^2 + \sqrt{-1}2\zeta\omega\omega_n} \right|^2 d\omega = \frac{\pi}{2\zeta\omega_n^5} \tag{9.80}$$

the variance calculated from integration of the first two terms of (9.76) is simply obtained as

$$\sigma^2 = \sigma_0^2\left(1 - 3\varepsilon\sigma_0^2\right) \tag{9.81}$$

which is identical to the result obtained using other approaches (e.g. Lin, 1976). In addition to serving as verification to an RFB orthogonal expansion, the above example demonstrates the potential of such a new expansion in nonlinear dynamics.

9.3 FAST-COMPUTING PROCEDURE OF THE MULTISCALE STOCHASTIC FEM

9.3.1 Formulation of the perturbation-type MSFEM

In Chapter 8 and Section 9.2, the MSFEM is formulated based on the Green-function-based variational principles. In a random medium characterized with a continuous probability density function (e.g. a soil in geotechnical engineering), the degree of random fluctuation is usually limited, or more specifically, the coefficient of variation C_V of a soil property is usually less than 1. In such a case, a special perturbation-type MSFEM is developed to provide a fast-computing efficiency (Xu, 2015). A unique advantage of the perturbation-type MSFEM is its enabling of computation performed on locations of interest only, thereby saving a lot of computation resources and time.

The governing equations and boundary conditions of a finite stochastic geotechnical body are identical to (7.1–7.5). Following exactly the decomposition made in Chapter 8, the stochastic BVP is decomposed into a reference BVP (8.1–8.5) and a fluctuation BVP (8.13–8.18), the latter of which in this section should be more correctly renamed as a perturbation BVP due to the requirement of limited fluctuation. The deterministic slow-scale reference BVP (8.1–8.5) is solved by using the classical FEM. To solve the fast-scale BVP (8.13–8.18), instead of applying the GFB variational principles in Chapter 7 to solve the unknown stress polarization, we directly adopt the perturbation approach to truncate the infinite series solution by taking advantage of the limited fluctuation. Following (7.29) and the traction free boundary condition (7.60), the perturbation displacement is given in terms of the Green function as

$$u_i'(\boldsymbol{x}) = -\int_D G_{ik}(\boldsymbol{x},\boldsymbol{x}')\nabla p_{kl,l}(\boldsymbol{x}')d\boldsymbol{x}' \tag{9.82}$$

which, by using Gauss' divergence theorem, becomes

$$u_i'(\boldsymbol{x}) = -\int_D (\nabla G)_{ikl}(\boldsymbol{x},\boldsymbol{x}')p_{kl}(\boldsymbol{x}')d\boldsymbol{x}' \tag{9.83}$$

with

$$(\nabla G)_{ikl}(\boldsymbol{x},\boldsymbol{x}') = \frac{\partial G_{ik}(\boldsymbol{x},\boldsymbol{x}')}{\partial x_l'} \tag{9.84}$$

By substituting (7.16) and (7.18) into (9.83), it follows that

$$\boldsymbol{u}'(\boldsymbol{x}) = -\int_D \nabla \boldsymbol{G}(\boldsymbol{x},\boldsymbol{x}')(\boldsymbol{L}(\boldsymbol{x}') - \boldsymbol{L}_0)\left(\boldsymbol{\varepsilon}_0(\boldsymbol{x}') + \boldsymbol{\varepsilon}'(\boldsymbol{x}')\right)d\boldsymbol{x}'$$

the derivative of which yields

$$\boldsymbol{\varepsilon}'(\boldsymbol{x}) = -\int_D \boldsymbol{\Gamma}(\boldsymbol{x},\boldsymbol{x}')\left(\boldsymbol{L}(\boldsymbol{x}') - \boldsymbol{L}_0\right)\left(\boldsymbol{\varepsilon}_0(\boldsymbol{x}') + \boldsymbol{\varepsilon}'(\boldsymbol{x}')\right)d\boldsymbol{x}' \tag{9.85}$$

By substituting the right-hand side of (9.85) into the term $\boldsymbol{\varepsilon}'(\boldsymbol{x})$ on the right-hand side iteratively, it yields an infinite series solution for the perturbation displacement

$$\boldsymbol{u}' = -\nabla \boldsymbol{G} \Delta \boldsymbol{L} \boldsymbol{\varepsilon}_0 + \nabla \boldsymbol{G}\, \Delta \boldsymbol{L}(\boldsymbol{\Gamma} \Delta \boldsymbol{L} \boldsymbol{\varepsilon}_0) - \cdots\cdots \tag{9.86}$$

The Green function and the modified Green function are computed using the finite element method, with the formulas given in (8.24–8.25). According to the definition (9.84), the formula to compute the function $\nabla \boldsymbol{G}$ is given as

$$(\nabla \mathrm{G})_{ikl}(\boldsymbol{x},\boldsymbol{x}') = \sum_m^{N_p^{(e)}} \sum_n^{N_p^{(e')}} \mathrm{N}_m(\boldsymbol{x})[\mathrm{G}]_{IK} \nabla \mathrm{N}_{n,l}(\boldsymbol{x}') \tag{9.87}$$

where the subscripts I,K denote two degrees of freedom corresponding to node m in direction i and node n in direction k, respectively.

It is noted that in homogenization of heterogeneous materials a series solution similar to (9.86) was given early in Kröner (1972). With the proof given in Chapter 7 on decomposition of a heterogeneous BVP (7.1–7.5), the series solution (9.86) extends the perturbation method to finite size heterogeneous BVPs subject to general boundary conditions.

9.3.2 Convergence of the perturbation-type MSFEM

In deterministic computation of a periodic heterogeneous medium, convergence of the series solution (9.86) has been shown in Fokin(1982); Eyre and Milton (1999); Michel, Moulinec, and Suquet (1999); and others. In stochastic homogenization of a random medium, a convergence study was conducted by Xu and Graham-Brady (2006) using the Fourier-Galerkin method. Below, the convergence of the series solution (9.86) in stochastic computation of a geotechnical soil is demonstrated.

For the sake of simplicity, suppose a geotechnical soil is characterized with the Lame's constant λ and the shear modulus μ as two correlated random fields, i.e.

$$\lambda(\boldsymbol{x},\vartheta)=\bar{\lambda}(\boldsymbol{x})+\sigma_\lambda\tilde{Z}(\boldsymbol{x},\vartheta) \tag{9.88}$$

$$\mu(\boldsymbol{x},\vartheta)=\bar{\mu}(\boldsymbol{x})+\sigma_\mu\tilde{Z}(\boldsymbol{x},\vartheta) \tag{9.89}$$

where $\tilde{Z}$ denotes an underlying zero mean unit variance Gaussian random field, and σ_λ, σ_μ the standard deviations. The elastic moduli tensor of the soil is therefore expressed in terms of $\tilde{Z}$ as

$$\boldsymbol{L}(\boldsymbol{x},\vartheta)=\bar{\boldsymbol{L}}+\boldsymbol{L}_\sigma\tilde{Z}(\boldsymbol{x},\vartheta) \tag{9.90a}$$

$$L_{ijkl}(\boldsymbol{x},\vartheta)=\left[\bar{\lambda}\delta_{ij}\delta_{kl}+\bar{\mu}(\delta_{ik}\delta_{jl}+\delta_{il}\delta_{jk})\right]+\left[\sigma_\lambda\delta_{ij}\delta_{kl}+\sigma_\mu(\delta_{ik}\delta_{jl}+\delta_{il}\delta_{jk})\right]\tilde{Z}(\boldsymbol{x},\vartheta) \tag{9.90b}$$

or in a matrix form

$$[\mathrm{L}](\boldsymbol{x},\vartheta)=[\bar{\mathrm{L}}]+[\mathrm{L}_\sigma]\tilde{Z}(\boldsymbol{x},\vartheta) \tag{9.91}$$

with

$$[\bar{\mathrm{L}}]=\begin{bmatrix} 2\bar{\mu}+\bar{\lambda} & \bar{\lambda} & \bar{\lambda} & 0 & 0 & 0 \\ \bar{\lambda} & 2\bar{\mu}+\bar{\lambda} & \bar{\lambda} & 0 & 0 & 0 \\ \bar{\lambda} & \bar{\lambda} & 2\bar{\mu}+\bar{\lambda} & 0 & 0 & 0 \\ 0 & 0 & 0 & \bar{\mu} & 0 & 0 \\ 0 & 0 & 0 & 0 & \bar{\mu} & 0 \\ 0 & 0 & 0 & 0 & 0 & \bar{\mu} \end{bmatrix} \tag{9.92}$$

$$[\mathrm{L}_\sigma]=\begin{bmatrix} 2\sigma_\mu+\sigma_\lambda & \sigma_\lambda & \sigma_\lambda & 0 & 0 & 0 \\ \sigma_\lambda & 2\sigma_\mu+\sigma_\lambda & \sigma_\lambda & 0 & 0 & 0 \\ \sigma_\lambda & \sigma_\lambda & 2\sigma_\mu+\sigma_\lambda & 0 & 0 & 0 \\ 0 & 0 & 0 & \sigma_\mu & 0 & 0 \\ 0 & 0 & 0 & 0 & \sigma_\mu & 0 \\ 0 & 0 & 0 & 0 & 0 & \sigma_\mu \end{bmatrix} \tag{9.93}$$

where the use of the engineering shear strain is adopted.

The probabilistic information about a response can be evaluated from (9.86) via probabilistic moments, e.g. the mean and the mean square values of the perturbation displacement are given, respectively, as

$$\overline{\boldsymbol{u}'} = -\overline{\nabla \boldsymbol{G}(\boldsymbol{L}-\boldsymbol{L}_0)\boldsymbol{\varepsilon}_0} + \overline{\nabla \boldsymbol{G}(\boldsymbol{L}-\boldsymbol{L}_0)\boldsymbol{\Gamma}(\boldsymbol{L}-\boldsymbol{L}_0)\boldsymbol{\varepsilon}_0} - \cdots \tag{9.94}$$

$$\begin{aligned}\overline{(\boldsymbol{u}')^2} &= \overline{\left(\nabla \boldsymbol{G}(\boldsymbol{L}-\boldsymbol{L}_0)\boldsymbol{\varepsilon}_0\right)\left(\nabla \boldsymbol{G}(\boldsymbol{L}-\boldsymbol{L}_0)\boldsymbol{\varepsilon}_0\right)} \\ &\quad -2\overline{\left(\nabla \boldsymbol{G}(\boldsymbol{L}-\boldsymbol{L}_0)\boldsymbol{\varepsilon}_0\right)\left(\nabla \boldsymbol{G}(\boldsymbol{L}-\boldsymbol{L}_0)\boldsymbol{\Gamma}(\boldsymbol{L}-\boldsymbol{L}_0)\boldsymbol{\varepsilon}_0\right)} + \cdots\end{aligned} \tag{9.95}$$

Choose the reference moduli of the slow-scale BVP (8.1–8.5) as $\boldsymbol{L}_0 = \bar{\boldsymbol{L}}$. By taking (9.90) into account, the first term on the right-hand side of (9.94) vanishes, and the second term becomes

$$\nabla \boldsymbol{G}\overline{(\boldsymbol{L}-\bar{\boldsymbol{L}})\boldsymbol{\Gamma}(\boldsymbol{L}-\bar{\boldsymbol{L}})}\boldsymbol{\varepsilon}_0 = \nabla \boldsymbol{G}\boldsymbol{L}_\sigma \boldsymbol{\Gamma}\boldsymbol{L}_\sigma \rho_Z \boldsymbol{\varepsilon}_0 \tag{9.96}$$

where ρ_Z denotes the correlation coefficient of the underlying normalized random field $\tilde{Z}$. The nth term of the series (9.86) is similarly given as

$$\nabla \boldsymbol{G}\overline{(\boldsymbol{L}-\bar{\boldsymbol{L}})\left(\boldsymbol{\Gamma}(\boldsymbol{L}-\bar{\boldsymbol{L}})\right)^{n-1}}\boldsymbol{\varepsilon}_0 = \nabla \boldsymbol{G}\boldsymbol{L}_\sigma (\boldsymbol{\Gamma}\boldsymbol{L}_\sigma)^{n-1} R_n \boldsymbol{\varepsilon}_0 \tag{9.97}$$

where $n = 2, 3, \ldots$, and R_n denotes the n-point correlation function of the random field $\tilde{Z}$, i.e. $R_n(\boldsymbol{x}_1, \boldsymbol{x}_2, \cdots, \boldsymbol{x}_n) = \overline{\tilde{Z}(\boldsymbol{x}_1)\tilde{Z}(\boldsymbol{x}_2)\cdots\tilde{Z}(\boldsymbol{x}_n)}$. The right-hand side of (9.97) is bounded from above by replacing R_n with the least upper bound $B(n)$ that is a function of n, i.e.

$$\nabla \boldsymbol{G}\boldsymbol{L}_\sigma (\boldsymbol{\Gamma}\boldsymbol{L}_\sigma)^{n-1} R_n \boldsymbol{\varepsilon}_0 \le B(n)\nabla \boldsymbol{G}\boldsymbol{L}_\sigma (\boldsymbol{\Gamma}\boldsymbol{L}_\sigma)^{n-1}\boldsymbol{\varepsilon}_0 \tag{9.98}$$

The maximal of the n-point correlation R_n is $\overline{\tilde{Z}^n} = (n-1)!!$ when all the n points are located at a same point (i.e. completely correlated), which yields that

$$B(n) < (n-1)!!$$

To evaluate the right-hand side of inequality (9.98), the following property is in need that is applicable to any BVP subject to general boundary conditions.

Given a BVP

$$\nabla \cdot \boldsymbol{\sigma}_0' + \boldsymbol{f} = 0 \quad \text{in } D \tag{9.99}$$

characterized with the elastic moduli $\boldsymbol{L}_0$, and the boundary conditions mixed of those of the slow-scale BVP (8.1–8.5) and the fast-scale BVP (8.13–8.18), i.e.

$$\boldsymbol{u}_0' = 0 \quad \text{on} \quad \partial D_u \tag{9.100}$$

$$\boldsymbol{\sigma}_0' \cdot \boldsymbol{n} = \tilde{\boldsymbol{t}} \quad \text{on } \partial D_t \tag{9.101}$$

the self-equilibrated stress $\boldsymbol{\sigma}_0' = \boldsymbol{L}_0\boldsymbol{\varepsilon}_0'$ is a fixed point for iterative operation of $\boldsymbol{\Gamma}$, i.e.

$$\boxed{\boldsymbol{\Gamma}\boldsymbol{L}_0\boldsymbol{\varepsilon}_0' = \boldsymbol{\varepsilon}_0'} \tag{9.102}$$

Proof: Construct a BVP as follows:

$$\begin{cases} \nabla\cdot\left[\boldsymbol{L}_0\nabla^s(\boldsymbol{u}' + \boldsymbol{u}_0)\right] + \boldsymbol{f} = 0 \text{ in } D \\ \boldsymbol{u}' = 0;\ \boldsymbol{u}_0 = \tilde{\boldsymbol{u}} \qquad \text{on } \partial D_u \\ \left(\boldsymbol{L}_0\nabla^s\boldsymbol{u}'\right)\cdot\boldsymbol{n} = 0; \left(\boldsymbol{L}_0\nabla^s\boldsymbol{u}_0\right)\cdot\boldsymbol{n} = \tilde{\boldsymbol{t}} \quad \text{on } \partial D_t \end{cases} \tag{9.103a, b, c}$$

Compared with the slow-scale BVP (8.1–8.5), the BVP (9.103) clearly has its solution $\boldsymbol{u}' + \boldsymbol{u}_0$ identical to the solution $\boldsymbol{u}_0$ of the slow-scale BVP (8.1–8.5) and therefore we simply have $\boldsymbol{u}' = 0$ in the BVP (9.103). Take $\boldsymbol{u}'$ as the unknown of (9.103) and its solution is expressed in terms of the Green function $\boldsymbol{G}$ of the fast-scale BVP as

$$\boldsymbol{u}' = \boldsymbol{G}\left(\nabla\cdot(\boldsymbol{L}_0\boldsymbol{\varepsilon}_0) + \boldsymbol{f}\right) \tag{9.104}$$

which, by using Gauss' divergence theorem, becomes

$$\boldsymbol{u}' = -(\nabla\boldsymbol{G})\boldsymbol{L}_0\boldsymbol{\varepsilon}_0 + \boldsymbol{G}(\tilde{\boldsymbol{t}} + \boldsymbol{f}) \tag{9.105}$$

and thereby

$$\boldsymbol{\varepsilon}' = -\boldsymbol{\Gamma}\boldsymbol{L}_0\boldsymbol{\varepsilon}_0 + \boldsymbol{\Lambda}(\tilde{\boldsymbol{t}} + \boldsymbol{f}) \tag{9.106}$$

with $\boldsymbol{\Lambda}$ defined in (7.34). Since $\boldsymbol{u}' = \boldsymbol{\varepsilon}' = 0$, it immediately follows that

$$(\nabla\boldsymbol{G})\boldsymbol{L}_0\boldsymbol{\varepsilon}_0 = \boldsymbol{G}(\tilde{\boldsymbol{t}} + \boldsymbol{f}) \tag{9.107}$$

$$\boldsymbol{\Gamma}\boldsymbol{L}_0\boldsymbol{\varepsilon}_0 = \boldsymbol{\Lambda}(\tilde{\boldsymbol{t}} + \boldsymbol{f}) \tag{9.108}$$

The right-hand sides of (9.107–9.108) correspond exactly to the solution of the mixed BVP (9.99–9.101), and thereby

$$(\nabla \boldsymbol{G})\boldsymbol{L}_0\boldsymbol{\varepsilon}_0 = \boldsymbol{u}_0' \tag{9.109}$$

$$\boldsymbol{\Gamma}\boldsymbol{L}_0\boldsymbol{\varepsilon}_0 = \boldsymbol{\varepsilon}_0' \tag{9.110}$$

By changing the boundary conditions of $\boldsymbol{u}_0$ in (9.103) to the mixed one (9.100–9.101), i.e. changing (9.103b) to

$$\boldsymbol{u}' = 0;\ \boldsymbol{u}_0 = 0 \qquad \text{on } \partial D_u \tag{9.111}$$

and repeating the derivation process from the beginning, the fixed point of the iterative operation is obtained as

$$\boxed{(\nabla \boldsymbol{G})\boldsymbol{L}_0\boldsymbol{\varepsilon}_0' = \boldsymbol{u}_0'} \tag{9.112}$$

$$\boxed{\boldsymbol{\Gamma}\boldsymbol{L}_0\boldsymbol{\varepsilon}_0' = \boldsymbol{\varepsilon}_0'} \quad \square$$

Apply (9.110) and (9.102) to the right-hand side of (9.98) and it yields

$$B(n)\nabla \boldsymbol{G}\boldsymbol{L}_\sigma(\boldsymbol{\Gamma}\boldsymbol{L}_\sigma)^{n-1}\boldsymbol{\varepsilon}_0 \leq B(n)C_V^{n-1}\nabla \boldsymbol{G}\boldsymbol{L}_\sigma\boldsymbol{\varepsilon}_0' \tag{9.113}$$

where the coefficient of variation

$$C_V = \max\left\{\frac{\sigma_\mu}{\bar{\mu}}, \frac{\sigma_\lambda}{\bar{\lambda}}\right\} \tag{9.114}$$

By further taking (9.112) into account, we have

$$B(n)C_V^{n-1}\nabla \boldsymbol{G}\boldsymbol{L}_\sigma\boldsymbol{\varepsilon}_0' \leq B(n)C_V^n\boldsymbol{u}_0' \tag{9.115}$$

In general, rise of $B(n)$ with n is slower than that of C_V^n when C_V is smaller than 1. In a geotechnical soil, the coefficient of variation is usually less than 1, and it is expected that the right-hand side of (9.115) and thereby the series (9.94) to converge quickly in most cases. Next it is numerically demonstrated that on a soil with the C_V of the Young's modulus being 40%, even the first non-zero term of (9.94) is sufficient to accurately capture the mean settlement.

9.3.3 Numerical example

In predicting settlement of a geotechnical foundation, random variation of the Young's modulus is a major factor to be considered in geotechnical design. In this example, a two-dimensional soil 30 meters deep and 120 meters wide is assumed subject to a plane strain condition (see Figure 9.5). A 10-meter-wide foundation at the top center of the soil imposes a pressure of 0.2 MPa onto the soil underneath. For the sake of simplicity, spatial variation of the Young's modulus is characterized as a one-dimensional lognormal random field along the depth only. In the MSFEM model, the bottom of the soil is fixed along the depth direction, and is free in the horizontal direction. Due to the symmetry, only the right half of the soil is modeled as shown in Figure 9.5, whereas a particular sample of the soil illustrates spatial variation of the Young's modulus.

The lognormal random field of the Young's modulus E is expressed in terms of a normalized Gaussian field $\tilde{Z}$ as

$$E(\boldsymbol{x},\vartheta) = \exp\left(\bar{Z} + \sigma_Z \tilde{Z}(\boldsymbol{x},\vartheta)\right) \tag{9.116}$$

Accordingly, the mean and variance of the Young's modulus are given, respectively, as

$$\bar{E} = \exp\left(\bar{Z} + \frac{\sigma_Z^2}{2}\right)$$

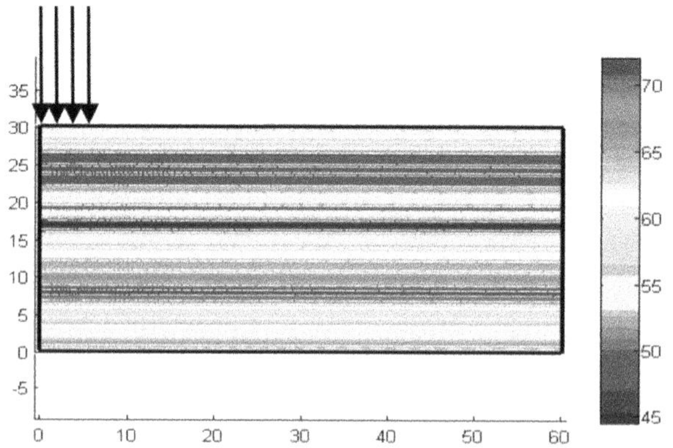

Figure 9.5 Right half of a symmetric lognormal random field sample for a heterogeneous soil under a pressure of a 10-meter-wide foundation at the center. (With kind permission from Springer Science+Business Media: *Front. Struct. Civ. Eng.*, Multiscale stochastic finite element method on random field modeling of geotechnical problems: A fast computing procedure, 9, 2015, 107–113, Xu, X.F. conveyed through Copyright Clearance Center, Inc.)

$$\sigma_E^2 = \exp\left(2\bar{Z} + \sigma_Z^2\right)\left(\exp\left(\sigma_Z^2\right) - 1\right)$$

and therefore $C_V = \sqrt{\exp\left(\sigma_Z^2\right) - 1}$. The mean and the coefficient of variation of the Young's modulus are given as 50 MPa and 40%, respectively, with the Poisson's ratio fixed at 0.3.

The correlation coefficient of the underlying Gaussian field $\tilde{Z}$ is assumed to be an exponential function

$$\rho_Z(\boldsymbol{x}) = e^{-\frac{|x|}{\ell_c}}$$

From (9.116), the correlation coefficient of the lognormal random field E is given in terms of ρ_z as

$$\rho_E(\boldsymbol{x}) = \frac{\exp\left(\rho_Z(\boldsymbol{x})\right) - 1}{\exp(\sigma_Z^2) - 1}$$

With the mean $\bar{E}$ = 50 MPa as the reference Young's modulus, the deterministic slow-scale BVP is solved by using the classical finite element method (Figure 9.6). The reference settlement of the foundation is found to be u_0 = 5.67 cm. This apparent settlement of the foundation corresponds to an unsafe geotechnical design considering the mean values of a soil's elastic moduli only.

In the fast-scale BVP, the second term on the right-hand side of (9.94) is calculated to provide an additional settlement as a correction to the above apparent settlement. To verify the accuracy of the fast-scale computation, a semianalytical result is derived by using a limiting case of this second term. When the correlation length is infinitely large, (9.96) becomes

$$\nabla \boldsymbol{G} \boldsymbol{L}_\sigma \boldsymbol{\Gamma} \boldsymbol{L}_\sigma \rho_E \boldsymbol{\varepsilon}_0 = \nabla \boldsymbol{G} \boldsymbol{L}_\sigma \boldsymbol{\Gamma} \boldsymbol{L}_\sigma \boldsymbol{\varepsilon}_0 \tag{9.117}$$

which, by using (9.110) and (9.112), becomes

$$\nabla \boldsymbol{G} \boldsymbol{L}_\sigma \boldsymbol{\Gamma} \boldsymbol{L}_\sigma \varepsilon_0 = C_V^2 \boldsymbol{u}_0' \tag{9.118}$$

Note that in this particular example the mixed BVP (9.99–9.101) is actually identical to the slow-scale BVP, and therefore $\boldsymbol{u}_0' = \boldsymbol{u}_0$.

In Table 9.1, numerical results of (9.96) computed with four meshes of finite element discretization are verified with the semianalytical result (9.118). The errors are approximately quadratic to the mesh size.

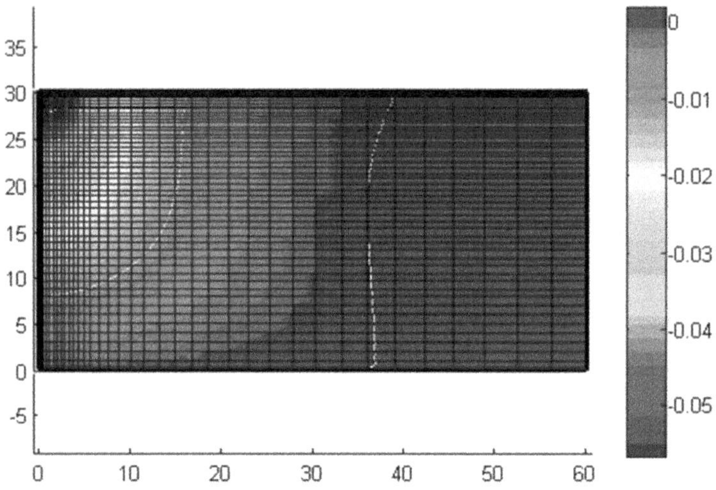

Figure 9.6 Finite element computation of the slow-scale BVP using a 36x48 nonuniform mesh. (With kind permission from Springer Science+Business Media: *Front. Struct. Civ. Eng.*, Multiscale stochastic finite element method on random field modeling of geotechnical problems: A fast computing procedure, 9, 2015, 107–113, Xu, X.F. Permission conveyed through Copyright Clearance Center, Inc.)

Table 9.1 Verification of the fast-scale BVP computation

Nonuniform mesh	*Mesh size (approx.)*	*Numerical result (cm)*	*Semianalytical result (cm)*	*Error (cm)*	*Relative error*
6 × 8	h	0.8857	$0.4^2 \times 5.67 =$ 0.9072	0.0215	2.4%
9 × 12	(2/3)h	0.8986		0.0086	0.95%
12 × 16	(1/2)h	0.9023		0.0049	0.54%
18 × 24	(1/3)h	0.9052		0.0020	0.22%

Source: With kind permission from Springer Science+Business Media: *Front. Struct. Civ. Eng.*, Multiscale stochastic finite element method on random field modeling of geotechnical problems: A fast computing procedure, 9, 2015, 107–113, Xu, X.F. Permission conveyed through Copyright Clearance Center, Inc.

With the above verification, the additional settlement of the fast-scale BVP in the case $\ell_C = 10$ m is solved by using the four meshes as shown in Table 9.2, which contributes to approximately further 14% increase. The total settlement is further verified with a Monte Carlo finite element simulation using 10,000 samples. As shown in Figure 9.7, the mean value becomes stabilized after the number of Monte Carlo samples reaches about 6,000. Since the autocorrelation function is much smoother than heterogeneities in a soil sample (see Figure 9.5), in computation of the fast-scale BVP a mesh much coarser than that of the Monte Carlo simulation (36 × 48) can

Table 9.2 MSFEM compared with Monte Carlo

Nonuniform mesh	*Slow-scale BVP (cm)*	*Fast-scale BVP (cm)*	*MSFEM total (cm)*	*Monte Carlo (10,000 samples) (cm)*	*Relative error*
6 × 8	–	0.74	5.67 + 0.74 = 6.41	–	1.2%
9 × 12	–	0.78	5.67 + 0.78 = 6.45	–	0.62%
12 × 16	–	0.79	5.67 + 0.89 = 6.46	–	0.46%
18 × 24	–	0.81	5.67 + 0.80 = 6.48	–	0.15%
36 × 48	5.67	–	–	6.49	–

Source: With kind permission from Springer Science+Business Media: *Front. Struct. Civ. Eng.*, Multiscale stochastic finite element method on random field modeling of geotechnical problems: A fast computing procedure, 9, 2015, 107–113, Xu, X.F. Permission conveyed through Copyright Clearance Center, Inc.

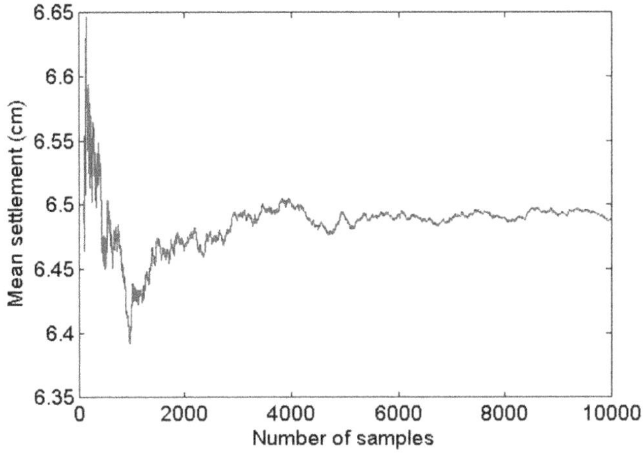

Figure 9.7 Mean settlement vs. number of Monte Carlo samples. (With kind permission from Springer Science+Business Media: Front. *Struct. Civ. Eng.*, Multiscale stochastic finite element method on random field modeling of geotechnical problems: A fast computing procedure, 9, 2015, 107–113, Xu, X.F. Permission conveyed through Copyright Clearance Center, Inc.)

be used to achieve the same level of accuracy. A comparison of computing time shows superiority of the perturbation-type MSFEM beyond traditional Monte Carlo finite element simulation; in the 9 × 12 mesh case the MSFEM takes only 3% of computing time consumed by the Monte Carlo simulation, whereas about 2,000 samples are demanded to reach the same level of accuracy. The source of the superior numerical efficiency of the MSFEM is identified as flexibility in reduction of a computing domain. More specifically, in computing (9.96), the location of the unknown displacement is focused at desired finite element nodes, like in this example, a single node at the center of the foundation, while in the Monte Carlo simulation computing resource is spent on computing of displacement all over the whole domain.

Appendix 1: MATLAB® code for the third-order SRC simulation

```
%3RD ORDER Short-Range-Correlation SIMULATION
%-----------------------------------------------------------------
% PART I - read target image and white noise initialization
%-----------------------------------------------------------------
t0=cputime;
N=256;
X=zeros(N);
XXX=imread('P4.jpg');
XX=sum(XXX,3);
X=-0.5*(abs(XX-500)-(XX-500))./(XX-500); %target image
FX=fft2(X);
Z=rand(N);
ZS=sort(reshape(Z,[1,N*N]));
Z=(Z-(ZS(FX(1,1)+1)+ZS(FX(1,1)))/2+abs(Z-(ZS(FX(1,1)+1)+ZS
(FX(1,1)))/2))/2./(Z-(ZS(FX(1,1)+1)+ZS(FX(1,1)))/2);
Z=1-Z;%white noise initialization
FZ=fft2(Z);
ZZ=Z;

%-----------------------------------------------------------------
% PART II - computation of triple correlation
%-----------------------------------------------------------------
L=15; % window size
LRX=zeros(L+1,2*L+1,L+1,2*L+1); %triple correlation of the
target image
LRZ=zeros(L+1,2*L+1,L+1,2*L+1); %triple correlation of the
simulated image
for i=1:N
   for j=1:N
      i
    if X(i,j)==1
        for m1=1:L+1
```

```
            im1=i-L-1+m1-N*floor((i-L-1+m1-1e-6)/N);
             for n1=1:2*L+1
             jn1=j-L-1+n1-N*floor((j-L-1+n1-1e-6)/N);
             LRX(:,:,m1,n1)=X(im1,jn1)*X(i-L-1+(1:L+1)
             -N*floor((i-L-1+(1:L+1)-1e-6)/N),j-L-1
             +(1:2*L+1)-N*floor((j-L-1+(1:2*L+1)-1e-6)/N))
             +LRX(1:L+1,1:2*L+1,m1,n1);
             end
          end
       end
       if Z(i,j)==1
         for m1=1:L+1
            im1=i-L-1+m1-N*floor((i-L-1+m1-1e-6)/N);
             for n1=1:2*L+1
             jn1=j-L-1+n1-N*floor((j-L-1+n1-1e-6)/N);
             LRZ(:,:,m1,n1)=Z(im1,jn1)*Z(i-L-1+(1:L+1)
             -N*floor((i-L-1+(1:L+1)-1e-6)/N),
             j-L-1+(1:2*L+1)-N*floor((j-L-1+(1:2*L+1)
             -1e-6)/N))+LRZ(1:L+1,1:2*L+1,m1,n1);
             end
          end
       end
    end
end
t1=cputime-t0;
t0=cputime;
LRZZ=LRZ;
NRX=sqrt(sum(sum(sum(sum(LRX.^2)))));
error=sqrt(sum(sum(sum(sum((LRX-LRZ).^2))))); % norm of the
difference between two triple correlations

%-------------------------------------------------------------
% PART III - Metropolis simulator
%-------------------------------------------------------------
K=3;
t=zeros(K,1);
Y=zeros(N,N,K);
nf=zeros(K,1);
err=zeros(K,1);
err0=error;
for k=1:K
    [A,IXY]=sort(rand(N^2,1));
     IX=ceil(IXY/N);
     IY=IXY-(IX-1)*N;
    for n=1:N^2
       i=IX(n);j=IY(n);
         ZZ(i,j)=1-ZZ(i,j);
         a1=[i-L:i]-N*floor(([i-L:i]-1e-6)/N);
         b1=[j-L:j+L]-N*floor(([j-L:j+L]-1e-6)/N);
         a2=[i+L:-1:i]-N*floor(([i+L:-1:i]-1e-6)/N);
```

```
        b2=[j+L:-1:j-L]-N*floor(([j+L:-1:j-L]-1e-6)/N);
        for p=1:L+1
            for q=1:2*L+1
                p1=i-L-1+p-N*floor((i-L-1+p-1e-6)/N);
                q1=j-L-1+q-N*floor((j-L-1+q-1e-6)/N);
                a3=a2-L-1+p-N*floor((a2-L-1+p-1e-6)/N);
                b3=b2-L-1+q-N*floor((b2-L-1+q-1e-6)/N);
                p2=i+L+1-p-N*floor((i+L+1-p-1e-6)/N);
                q2=j+L+1-q-N*floor((j+L+1-q-1e-6)/N);
                a4=[p2-L:p2]-N*floor(([p2-L:p2]-1e-6)/N);
                b4=[q2-L:q2+L]-N*floor
                (([q2-L:q2+L]-1e-6)/N);
                LRZZ(1:L+1,1:2*L+1,p,q)=LRZ(1:L+1,1:2*L+1,p,
q)+(2*ZZ(i,j)-1)*(ZZ(p1,q1)*ZZ(a1,b1)+ZZ(a2,b2).*ZZ(a3,b3)+ZZ
(p2,q2)*ZZ(a4,b4));
                LRZZ(L+1,L+1,p,q)=LRZ(L+1,L+1,p,q)+(2*ZZ
(i,j)-1)*(ZZ(p1,q1)+ZZ(p2,q2));
                LRZZ(p,q,p,q)=LRZ(p,q,p,q)+(2*ZZ
(i,j)-1)*(ZZ(p1,q1)+ZZ(p2,q2));
            end
         end
        LRZZ(1:L+1,1:2*L+1,L+1,L+1)=LRZ(1:L+1,1:2*L+1,L+1,L+1)
+(2*ZZ(i,j)-1)*(ZZ(a1,b1)+ZZ(a2,b2));
        LRZZ(L+1,L+1,L+1,L+1)=LRZ(L+1,L+1,L+1,L+1)+(2*ZZ
(i,j)-1);
        error1=sqrt(sum(sum(sum(sum((LRX-LRZZ).^2)))));
        if error1>=error ZZ(i,j)=1-ZZ(i,j); LRZZ=LRZ;
        else nf(k)=nf(k)+1;LRZ=LRZZ;
           error=error1
        end
    end
   if nf(k)==0 break
   end
   err(k)=error;
    Y(:,:,k)=ZZ;
   t(k)=cputime-t0;
   t0=cputime;
end
FZZ=fft2(ZZ);
cputime-t0
figure, pcolor(X),shading interp
figure, pcolor(ZZ),shading interp
```

Appendix II: Free space Green function and modified Green function

3D ELASTICITY

Denote μ_0 the shear modulus and ν_0 the Poisson's ratio of an isotropic elastic moduli tensor. The free space Green function is analytically available in real space and Fourier space, respectively, as

$$G_{ij}^{\infty}(\boldsymbol{x}) = \frac{1}{16\pi\mu_0(1-\nu_0)|\boldsymbol{x}|}\left((3-4\nu_0)\delta_{ij} - \frac{x_i x_j}{|\boldsymbol{x}|^2}\right) \tag{A2.1}$$

$$\hat{G}_{ij}^{\infty}(\boldsymbol{\xi}) = \frac{1}{\mu_0|\boldsymbol{\xi}|^2}\left(\delta_{ij}|\boldsymbol{\xi}|^2 - \frac{1}{2(1-\nu_0)}\xi_i\xi_j\right) \tag{A2.2}$$

where $|\boldsymbol{x}| = \sqrt{x_1^2 + x_2^2 + x_3^2}$ and $|\boldsymbol{\xi}| = \sqrt{\xi_1^2 + \xi_2^2 + \xi_3^2}$. By substituting (A2.1) into (3.22), the modified Green function is obtained in real space as

$$\begin{aligned}
\Gamma_{ijkl}^{\infty}(\boldsymbol{x}) = \frac{1}{4\pi\mu_0|\boldsymbol{x}|^3}\Bigg[&\left(\frac{\delta_{ik}\delta_{jl} + \delta_{il}\delta_{jk}}{2} - \frac{3}{4}\frac{\delta_{ik}x_j x_l + \delta_{jk}x_i x_l + \delta_{il}x_j x_k + \delta_{jl}x_i x_k}{|\boldsymbol{x}|^2}\right) \\
&- \frac{1}{4(1-\nu_0)}\Bigg(\delta_{ij}\delta_{kl} + \delta_{ik}\delta_{jl} + \delta_{il}\delta_{jk} \\
&- 3\frac{\delta_{ij}x_k x_l + \delta_{kl}x_i x_j + \delta_{ik}x_j x_l + \delta_{jk}x_i x_l + \delta_{il}x_j x_k + \delta_{jl}x_i x_k}{|\boldsymbol{x}|^2} \\
&+ \frac{15 x_i x_j x_k x_l}{|\boldsymbol{x}|^4}\Bigg)\Bigg]
\end{aligned} \tag{A2.3}$$

Given the inverse Fourier transform

$$\boldsymbol{G}^{\infty}(\boldsymbol{x}) = \frac{1}{(2\pi)^3}\int_{-\infty}^{+\infty} \hat{\boldsymbol{G}}^{\infty}(\boldsymbol{\xi})e^{i\boldsymbol{\xi}\cdot\boldsymbol{x}}\,d\boldsymbol{\xi} \tag{A2.4}$$

it yields the Fourier version of (3.22)

$$\hat{\Gamma}_{ijkl}(\boldsymbol{\xi}) = \frac{1}{4}\left(\hat{G}_{ik}(\boldsymbol{\xi})\xi_j\xi_l + \hat{G}_{jk}(\boldsymbol{\xi})\xi_i\xi_l + \hat{G}_{il}(\boldsymbol{\xi})\xi_j\xi_k + \hat{G}_{jl}(\boldsymbol{\xi})\xi_i\xi_k\right) \tag{A2.5}$$

and thereby the modified Green function in Fourier space

$$\hat{\Gamma}^{\infty}_{ijkl}(\boldsymbol{\xi}) = \frac{1}{\mu_0|\boldsymbol{\xi}|^4}\left[\frac{|\boldsymbol{\xi}|^2}{4}\left(\delta_{ik}\xi_j\xi_l + \delta_{il}\xi_j\xi_k + \delta_{jk}\xi_i\xi_l + \delta_{jl}\xi_i\xi_k\right) - \frac{1}{2(1-\nu_0)}\xi_i\xi_j\xi_k\xi_l\right] \tag{A2.6}$$

2D ELASTICITY

In the plane strain case, the free space Green function is analytically available as

$$G^{\infty}_{ij}(\boldsymbol{x}) = \frac{1}{8\pi\mu_0(1-\nu_0)}\left(\frac{x_i x_j}{|\boldsymbol{x}|^2} - (3-4\nu_0)\delta_{ij}\log|\boldsymbol{x}|\right) \tag{A2.7}$$

with subscripts being 1 or 2. In Fourier space, the plane strain free space Green function is identical to (A2.2) with subscripts $i, j = 1$ or 2, and accordingly the modified Green function follows (A2.6). The plane stress Green function and the modified Green function are obtained from those in the plane strain case by replacing ν_0 with $\nu_0/(1+\nu_0)$ and keeping μ_0 unchanged.

2D AND 3D CONDUCTIVITY

In case of conductivity, given K_0 the conductivity of a homogeneous medium, the free space Green function is analytically available in real and Fourier space, respectively, as

$$G^{\infty}(\boldsymbol{x}) = \frac{1}{K_0}\frac{1}{4\pi|\boldsymbol{x}|} \tag{A2.8}$$

$$\hat{G}^{\infty}(\boldsymbol{\xi}) = \frac{1}{K_0}\frac{1}{|\boldsymbol{\xi}|^2} \tag{A2.9}$$

By substituting (A2.8) into the definition of the modified Green function

$$\Gamma_{ij}(\boldsymbol{x}) = \frac{\partial^2 G(\boldsymbol{x})}{\partial x_i \partial x_j} \tag{A2.10}$$

and its Fourier version

$$\hat{\Gamma}_{ij}(\boldsymbol{\xi}) = \hat{G}_{ij}(\boldsymbol{\xi})\xi_i\xi_j \tag{A2.11}$$

the free space modified Green function is obtained in real and Fourier space, respectively, as

$$\Gamma_{ij}^{\infty}(\boldsymbol{x}) = \frac{1}{K_0}\frac{|\boldsymbol{x}|^2\delta_{ij} - 3x_i x_j}{4\pi|\boldsymbol{x}|^5} \tag{A2.12}$$

$$\hat{\Gamma}_{ij}^{\infty}(\boldsymbol{\xi}) = \frac{1}{K_0}\frac{\xi_i\xi_j}{|\boldsymbol{\xi}|^2} \tag{A2.13}$$

Appendix III: Fourth-rank Eshelby tensor in elasticity (Mura, 1987)

Any other component of the rank-four Eshelby tensor not listed here has a zero value:

$$
\begin{aligned}
&S_{ijkl} = S_{jikl} = S_{ijlk} \\
&S_{1111} = \frac{3}{8\pi(1-\nu)} a_1^2 I_{11} + \frac{1-2\nu}{8\pi(1-\nu)} I_1 \\
&S_{1122} = \frac{3}{8\pi(1-\nu)} a_2^2 I_{12} - \frac{1-2\nu}{8\pi(1-\nu)} I_1 \\
&S_{1133} = \frac{3}{8\pi(1-\nu)} a_3^2 I_{13} + \frac{1-2\nu}{8\pi(1-\nu)} I_1, S_{1112} = S_{1223} = 0 \\
&S_{1212} = \frac{a_1^2 + a_2^2}{16\pi(1-\nu)} I_{12} + \frac{1-2\nu}{16\pi(1-\nu)} (I_1 + I_2)
\end{aligned}
\tag{A3.1}
$$

All other components are obtained by the cyclic permutation of (1,2,3).

Sphere $(a_1 = a_2 = a_3 = a)$

$$
I_1 = I_2 = I_3 = \frac{4\pi}{3}, \quad I_{11} = I_{22} = I_{33} = I_{12} = I_{23} = I_{31} = \frac{4\pi}{5a^2} \tag{A3.2}
$$

Elliptic Cylinder (a_1 & $a_2 << a_3$)

$$
\begin{aligned}
&I_1 = \frac{4\pi a_2}{(a_1 + a_2)}, I_2 = \frac{4\pi a_1}{(a_1 + a_2)}, I_3 = 0 \\
&I_{12} = \frac{4\pi}{(a_1 + a_2)^2}, 3I_{11} = \frac{4\pi}{a_1^2} - I_{12}, 3I_{22} = \frac{4\pi}{a_2^2} - I_{12}, I_{13} = I_{23} = I_{33} = 0
\end{aligned}
\tag{A3.3}
$$

Disk-like or Penny-shaped ($a_1 = a_2 >> a_3$)

$$
\begin{aligned}
&I_1 = I_2 = \frac{\pi^2 a_3}{a_1}, I_3 = 4\pi - 2I_1, I_{11} = I_{22} = I_{12} = I_{21} = \frac{3\pi^2 a_3}{4a_1^3} \\
&I_{13} = I_{23} = I_{31} = I_{32} = 4\left(\frac{\pi}{a_1^2} - I_{12}\right), I_{33} = \frac{4\pi}{3a_3^2}
\end{aligned}
\tag{A3.4}
$$

Spheroid ($a_1 = a_2 = a_3/\eta = a$)

$$
\begin{aligned}
&I_1 = I_2 = \frac{2\pi a_1^2 a_3}{(a_3^2 - a_1^2)^{3/2}}\left[\frac{a_3}{a_1}\left(\frac{a_3^2}{a_1^2} - 1\right)^{1/2} + g\left(\frac{a_3}{a_1}\right)\right] \\
&I_3 = 4\pi - 2I_1, I_{13} = I_{23} = \frac{I_1 - I_3}{a_3^2 - a_1^2}, I_{12} = \frac{\pi}{a_1^2} - \frac{1}{4}I_{13} \\
&I_{11} = I_{22} = I_{12}, 3I_{33} = \frac{4\pi}{a_3^2} - 2I_{13}
\end{aligned}
\tag{A3.5}
$$

Appendix IV: Second-rank Eshelby tensor in conductivity (Mura, 1987)

Any other component of the rank-two Eshelby tensor not listed has a zero value.

Sphere ($a_1 = a_2 = a_3 = a$)

$$S_{11} = S_{22} = S_{33} = \frac{1}{3} \tag{A4.1}$$

Cylinder with an Elliptic Cross Section $(a_3 \to \infty)$

$$S_{11} = \frac{a_2}{a_1 + a_2}, S_{22} = \frac{a_1}{a_1 + a_2}, S_{33} = 0 \tag{A4.2}$$

Needle-like $(\eta \to \infty)$

$$\begin{aligned} S_{11} = S_{22} &= \frac{1}{2} + \frac{1}{2}\left(1 - \log 2 + \log\frac{1}{\eta}\right)\frac{1}{\eta^2} + O\left(\frac{1}{\eta^4}\right) \\ S_{33} &= 1 - 2S_{11} \end{aligned} \tag{A4.3}$$

Disk-like or Penny-shaped $(\eta \to 0)$

$$\begin{aligned} S_{11} = S_{22} &= \frac{\pi}{4}\eta - \eta^2 + \frac{3\pi}{8}\eta^3 + O(\eta^4) \\ S_{33} &= 1 - 2S_{11} \end{aligned} \tag{A4.4}$$

Spheroid $(a_1 = a_2 = a_3/\eta = a)$

$$S_{11} = S_{22} = \frac{a_1^2 a_3}{2\left(a_3^2 - a_1^2\right)^{3/2}}\left[\frac{a_3}{a_1}\left(\frac{a_3^2}{a_1^2} - 1\right)^{1/2} + gic\left(\frac{a_3}{a_1}\right)\right] \tag{A4.5}$$

$$S_{33} = 1 - 2S_{11}$$

Noted that the *gic* function is given as (6.8) in Chapter 6.

References

Adler, P.M. 1992. *Porous media: Geometry and transports.* Oxford: Butterworth-Heinemann.

Aifantis, E.C. 1984. On the microstructural origin of certain inelastic models. *Trans. ASME J. Engrg. Mat. Tech.*, 106: 326–330.

Auriault, J.-L., C. Boutin, and C. Geindreau. 2009. *Homogenization of coupled phenomena in heterogeneous media.* London: ISTE & Wiley.

Avellaneda, M., A.V. Cherkaev, K.A. Lurie, and G.W. Milton. 1988. On the effective conductivity of polycrystals and a three-dimensional phase-interchange inequality. *J. Appl. Phys.*, 63: 4989–5003.

Balberg, I., N. Binenbaum, and N. Wagner. 1984. Percolation thresholds in the three-dimensional sticks system. *Phys. Rev. Lett.*, 52: 1465–1468.

Bazant, Z.P. and G. Pijaudier-Cabot. 1988. Nonlocal continuum damage, localization instability and convergence. *J. Appl. Mech.*, 55: 287–293.

Ben-Amoz, M. 1966. Variational principles in anisotropic and nonhomogeneous elastokinetics, *Quart. Appl. Math.*, XXIV: 82–86.

Bensoussan, A., J.L. Lions, and G. Papanicolaou. 1978. *Asymptotic analysis for periodic structures.* Amsterdam: North-Holland.

Benveniste, Y. 1987. A new approach to the application of Mori–Tanaka's theory in composite materials. *Mech. Mater.*, 6: 147–157.

Beran, M.J. 1965. Use of the variational approach to determine bounds for the effective permittivity in random media. *Nuovo Cimento*, 38: 771–782.

Beran, M.J. and J. Molyneux. 1966. Use of classical variational principles to determine bounds for the effective bulk modulus in heterogeneous media. *Q. Appl. Math.*, 24: 107–118.

Berk, N.F. 1991. Scattering properties of the leveled-wave model of random morphologies. *Phys. Rev. A*, 44(8): 5069–5079.

Bristow, J.R. 1960. Microcracks, and the static and dynamic elastic constants of annealed and heavily cold-worked metals. *Br. J. Appl. Phys.*, 11: 81–85.

Broadbent, S. and J. Hammersley, J. 1957. Percolation processes: I. Crystals and mazes. *Proc. Cambridge Philos. Soc.*, 53: 629–641.

Budiansky, B. and R.J. O'Connell. 1976. Elastic moduli of a cracked solid. *Int. J. Solids Structures*, 12: 81–97.

Cahn, J.W. 1965. Phase separation by spinodal decomposition in isotropic systems. *J. Chem. Phys.*, 42(1): 93–9.

Cambanis, S. and E. Masry. 1978. On the reconstruction of the covariance of stationary Gaussian processes observed through zero-memory nonlinearities. *IEEE Trans. Inform. Theory IT*, 24(4): 485–494.

Cho, S.W. and I. Chasiotis. 2007. Elastic properties and representative volume element of polycrystalline silicon for MEMS. *Exp. Mech.*, 47: 1741–2765.

Copeland, A.C., G. Ravichandran, and M.M. Trivedi. 2001. Texture synthesis using gray-level co-occurrence models: Algorithms, experimental analysis, and psychophysical support. *Opt. Eng.*, 40(11): 2655–2673.

Corson, P.B. 1974. Correlation functions for predicting properties of heterogeneous materials: II. Empirical construction of spatial correlation functions for two-phase solids. *J. Appl. Mech.*, 45: 3165–3170.

Dederichs, P.H. and R.Z. Zeller. 1973. Variational treatment of the elastic constants of disordered materials. *Physik*, 259: 103–116.

Deodatis, G. and M. Shinozuka. 1991. Weighted integral method: II. Response variability and reliability. *J. Eng. Mech. ASCE*, 117: 1865–1877.

Deodatis, G. and R.C. Micaletti. 2001. Simulation of highly skewed non-Gaussian stochastic processes, *J. Eng. Mech. ASCE*, 127: 1284–1295.

Der Kiureghian, A. and J.B. Ke. 1988. The stochastic finite element method in structural reliability. *Probab. Eng. Mech.*, 3: 83–91.

Drugan, W.J. and J.R. Willis. 1996. A micromechanics-based nonlocal constitutive equation and estimates of representative volume element size for elastic composites. *J. Mech. Phys. Solids*, 44: 497–524.

Elishakoff, I., Y.J. Ren, and M. Shinozuka. 1996. Variational principles developed for and applied to analysis of stochastic beams. *J. Eng. Mech.*, ASCE 122: 559–565.

Eringen, A.C. 1983. Theories of nonlocal plasticity. *Int. J. Eng. Sci.*, 21: 741–751.

Eringen, A.C. 1999. *Microcontinuum field theories: I. Foundations and solids.* New York: Springer-Verlag.

Eshelby, J.D. 1957. The determination of the elastic field of an ellipsoidal inclusion. *Proc. R. Soc. London A*, 241: 376–396.

Evans, A.G. and J.W. Hutchinson. 2009. A critical assessment of theories of strain gradient plasticity. *Acta Materialia*, 57: 1675–1688.

Eyre, D.J. and G.W. Milton. 1999. A fast numerical scheme for computing the response of composites using grid refinement. *Eur. Phys. J. AP*, 6: 41–47.

Fleck, N.A. and J.W. Hutchinson. 1997. Strain gradient plasticity. *Adv. in App. Mech.*, 33: 295–361.

Fokin, G. 1982. Iteration method in the theory of nonhomogeneous dielectrics. *Physica Status Solidi. B Basic Research*, 111: 281–288.

Francfort, G. and F. Murat. 1986. Homogenization and optimal bounds in linear elasticity. *Arch. Rat. Mech. Anal.*, 94: 307–334.

Frisch, H.L. 1965. Statistics of random media. *Trans. Soc. Rheol.*, 9: 293–312.

Gagalowicz, A. 1981. A new method for texture fields synthesis: Some applications to the study of human vision. *IEEE Trans. Patt. Anal. Machine Intell.*, 3: 520–533.

Gagalowicz, A. and S.D. Ma. 1985. Sequential synthesis of natural textures. *Computer Vision, Graphics, and Image Processing*, 30: 289–315.

Garboczi, E.J., K.A. Snyder, J.F. Douglas, and M.F. Thorpe. 1995. Geometrical percolation threshold of overlapping ellipsoids. *Phys. Rev. E.*, 52: 819–828.

Geman, S. and D. Geman. 1984. Stochastic relaxation, Gibbs distribution and the Bayesian restoration of images. *IEEE Trans. Patt. Anal. Machine Intell.*, 6(6): 721–741.

Germain, P. 1973. Method of virtual power in continuum mechanics: 2. Microstructure. *SIAM J. Appl. Math.*, 25(3): 556–575.

Ghanem, R.G. and P.D. Spanos. 1991. *Stochastic finite element: A spectral approach.* New York: Springer-Verlag.

Gibiansky, L.V. and S. Torquato. 1996. Bounds on the effective moduli of cracked materials. *J. Mech. Phys. Solids*, 44: 233–242.

Gibiansky, L. and O. Sigmund. 2000. Multiphase elastic composites with extremal bulk modulus. *J. Mech. Phys. Solids*, 48: 461–498.

Gilbert, E.N. 1961. Random plane networks. *SIAM J. Appl. Math.*, 9(4): 533–543.

Giona, M. and A. Adrover. 1996. Closed-form solution for the reconstruction problem in porous media. *AIChE. J.*, 42: 1407–1415.

Graham-Brady, L. and X.F. Xu. 2008. Stochastic morphological modeling of random multiphase materials. *J. Appl. Mech.*, 75: 061001. doi:10.1115/1.2957598

Grigoriu, M. 1984. Crossings of non-Gaussian translation processes, *J. Eng. Mech. ASCE*, 110: 610–620.

Grigoriu, M. 1998. Simulation of stationary non-Gaussian translation processes. *J. Eng. Mech. ASCE*, 124: 121–126.

Gueguen, Y., T. Chelidze, and M. Le Ravalec. 1997. Microstructures, percolation thresholds, and rock physical properties. *Tectonophysics*, 279: 23–35.

Hammersley, J.M. and P. Clifford. 1971. *Markov field on finite graphs and lattices.* Retrieved from http://www.statslab.cam.ac.uk/~grg/books/hammfest/hamm-cliff.pdf

Haralick, R.M. and L.G. Shapiro. 1992. *Computer and robot vision* (Vol. I). Boston: Addison-Wesley.

Hashin, Z. 1965. On the elastic behavior of fiber reinforced materials of arbitrary transverse phase geometry. *J. Mech. Phys. Solids*, 13: 119–134.

Hashin, Z. 1988. The differential scheme and its application to cracked materials. *J. Mech. Phys. Solids*, 36: 719–734.

Hashin, Z. and S. Shtrikman. 1962. A variational approach to the theory of the effective magnetic permeability of multiphase materials. *J. Appl. Phys.*, 33: 3125–3131.

Hashin, Z. and S. Shtrikman. 1963. A variational approach to the theory of the elastic behavior of multiphase materials. *J. Mech. Phys. Solids*, 11: 127–140.

Hazlett, R.D. 1997. Statistical characterization and stochastic modeling of pore networks in relation to fluid flow. *Math. Geol.*, 29: 801–822.

Helsing, J. and A. Helte. 1991. Effective conductivity of aggregates of anisotropic grains. *J. Appl. Phys.*, 69: 3583–3588.

Hernandez, Y.R., A. Gryson, F.M. Blighe et al. 2008. Comparison of carbon nanotubes and nanodisks as percolative fillers in electrically conductive composites. *Scr. Mater.*, 58: 69–72.

Hien, T.D. and M. Kleiber. 1990. Finite element analysis based on stochastic Hamilton variational principle. *Comput. Struct.*, 37(6): 893–902.

Hill, R. 1952. The elastic behavior of crystalline aggregate, *Proc. R. Soc. London A*, 65: 349–354.

Hill, R. 1964. Theory of mechanical properties of fiber-strengthened materials: I. Elastic behavior. *J. Mech. Phys. Solids*, 12: 199–212.

Hill, R. 1965. A self-consistent mechanics of composite materials. *J. Mech. Phys. Solids*, 13: 213–222.

Hori, M. and S. Munasinghe. 1999. Generalized Hashin-Shtrikman variational principle for boundary-value problem of linear and non-linear heterogeneous body. *Mech. Mater.*, 31: 471–486.

Horii, H. and S. Nemat-Nasser. 1985. Elastic fields of interacting inhomogeneities. *Int. J. Solids Struct.*, 21: 731–745.

Johnson, M.E. 1987. *Multivariate statistical simulation.* New York: John Wiley & Sons.

Jona-Lasinio, G. 2001. Renormalization group and probability theory. *Phys. Rep.*, 352: 4–6.

Ju, J.W. and T.-M. Chen. 1994. Effective elastic moduli of two-dimensional brittle solids with interacting cracks, Part I: Basic formulations. *J. Appl. Mech.*, 61: 349–357.

Julesz, B. 1962. Visual pattern discrimination. *IRE Trans. Inf Theory IT*, 8: 84–92.

Julesz, B., E.N. Gilbert, L.A. Shepp et al. 1973. Inability of humans to discriminate between visual textures that agree in second-order statistics–revisited. *Perception*, 2: 391–405.

Kachanov, M. 1993. Elastic solids with many cracks and related problems. In *Advanced applied mechanics* (Vol. 30), eds. J.W. Hutchinson and T. Wu, 258–445. Cambridge: Academic Press.

Kotz, S., N. Balakrishnan, and N.L. Johnson. 2000. *Continuous multivariate distributions* (Vol. 1): *Models and applications.* New York: John Wiley & Sons.

Kröner, E. 1972. *Statistical continuum mechanics.* Wien: Springer-Verlag.

Kröner, E. 1977. Bounds for effective elastic moduli of disordered materials. *J. Mech. Phys. Solids*, 25: 137.

Kunin, I.A. 1983. *Elastic media with microstructure* (Vol. 2.): *Three-dimensional models.* Berlin: Springer-Verlag.

Li, S.T. and J.L. Hammond. 1975. Generation of pseudo-random numbers with specified univariate distributions and correlation coefficients. *IEEE Trans. Systems Man. Cybern.*, 5: 557–561.

Lin, Y.K. 1976. *Probabilistic theory of structural dynamics.* Melbourne: Robert E. Krieger.

Liu, W.K., G.H. Bestfield, and T. Belytschko. 1988. Variational approach to probabilistic finite elements. *J. Eng. Mech. ASCE*, 114: 2115–2133.

Liu, B. and D.C. Munson. 1982. Generation of a random sequence having a jointly specified marginal distribution and autocovariance. *IEEE Trans. Acoust. Speech Signal Process. ASSP*, 30: 973–983.

Liu, C. 2005. On the minimum size of representative volume element: An experimental investigation. *Exp. Mech.*, 45: 238–243.

Luciano, R. and J.R. Willis. 2005. FE analysis of stress and strain fields in finite random composite bodies. *J. Mech. Phys. Solids*, 53: 1505–1522.

Lukkassen, D. 1999. A new reiterated structure with optimal macroscopic behavior. *SIAM J. Appl. Math.*, 59: 1825–1842.

Mal, A.K. and L. Knopoff. 1967. Elastic wave velocities in two-component systems. *J. Inst. Maths Appl.*, 3: 376–387.

Mardia, K.V. 1970. Translation family of bivariate distribution and Frechet's bounds. *Sankhya Ser*, A 32: 119–122.

Markov, K.Z. 1998. On the cluster bounds for the effective properties of microcracked solids. *J. Mech. Phys. Solids*, 46: 357–388.

Matheron, G. 1967. *Elements pour un théorie des milieux poreux*. Paris: Masson.

Metropolis, N., A.W. Rosenbluth, M.N. Rosenbluth et al. 1953. Equations of state calculations by fast computational machine. *J. Chem. Phys.*, 21: 1087–1091.

Michel, J.C., H. Moulinec, and P. Suquet. 1999. Effective properties of composite materials with periodic microstructure: A computational approach. *Comput. Methods Appl. Mech. Eng.*, 157: 109–143.

Milton, G.W. 1981. Bounds on the electromagnetic, elastic, and other properties of two-component composites. *Phys. Rev. Lett.*, 46: 542–545.

Milton, G.W. 1982. Bounds on the elastic and transport properties of two-component composites. *J. Mech. Phys. Solids*, 30: 177–191.

Milton, G.W. 1992. Composite materials with poisson's ratios close to - 1. *J. Mech. Phys. Solids*, 40(5): 1105–1137.

Milton, G.W. 2002. *The theory of composites*. Cambridge: Cambridge University Press.

Milton, G.W. and N. Phan-Thien. 1982. New bounds on effective elastic moduli of two-component materials. *Proc. R. Soc. London A*, 380: 305–331.

Mindlin, R.D. 1964. Micro-structure in linear elasticity. *Arc. Ration. Mech. Anal.*, 16: 51–78.

Mori, T. and K. Tanaka. 1973. Average stress in matrix and average elastic energy of materials with misfitting inclusions. *Acta Metallurgica*, 21: 571–574.

Moulinec, H. and P. Suquet. 1998. A numerical method for computing the overall response of nonlinear composites with complex microstructure. *Comput. Methods Appl. Mech. Eng.*, 157: 69–94.

Mura, T. 1987. *Micromechanics of defects in solids*. Dordrecht: Martinus Nijhoff.

Nemat-Nasser, S. and M. Hori. 1999. *Micromechanics: Overall properties of heterogeneous materials*. Amsterdam: Elsevier.

Nikias, C.L. and A.P. Petropulu. 1993. *Higher-order spectra analysis, a nonlinear signal processing framework*. Englewood Cliffs: PTR Prentice Hall.

Nix, D. and H. Gao. 1998. Indentation size effects in crystalline materials: A law for strain plasticity. *J. Mech. Phys. Solids*, 46: 411–425.

Oden, J.T. and J.N. Reddy. 1976. *Variational methods in theoretical mechanics*. New York: Springer-Verlag.

Ostoja-Starzewski, M. 2008. *Microstructural randomness and scaling in mechanics of materials*. London: Chapman & Hall/CRC.

Pan, H.H. and G.J. Weng. 1995. Elastic moduli of heterogeneous solids with ellipsoidal inclusions and elliptic cracks. *Acta Mechanica*, 110: 73–94.

Pan, Y., G.J. Weng, S.A. Meguid et al. 2011. Percolation threshold and electrical conductivity of a two-phase composite containing randomly oriented ellipsoidal inclusions. *J. Appl. Phys.*, 118: 065101. https://doi.org/10.1063/1.4928293

Papanicolaou, G.C. and S.R.S. Varadhan. 1981. Boundary value problems with rapidly oscillating random coefficients, in *Random fields, Vol. I, II (Esztergom, 1979), volume 27 of Colloq. Math. Soc. János Bolyai*, ed. J. Fritz, J.L., Lebowitz, and D. Sza´sz, 835–873. Amsterdam: North-Holland.

Papoulis, A. 1989. *Probability random variables & stochastic processes: Sanitary & water resources engineering*. New York: McGraw-Hill.

Ponte Castañeda, P. and J.R. Willis. 1995. The effect of spatial distribution on the effective behavior of composite materials and cracked media. *J. Mech. Phys. Solids*, 43: 1919–1951.

Ponte Castañeda, P. 2002. Second-order homogenisation estimates for nonlinear composites incorporating field fluctuations. *I. Theory. J. Mech. Phys. Solids*, 50: 737–757.

Price, R. 1958. A useful theorem for nonlinear devices having Gaussian inputs. *IRE Trans. Inform. Theory IT*, 4: 69–72.

Puig, B., F. Poirion, and C. Soize. 2002. Non-Gaussian simulation using Hermite polynomial expansion: Convergences and algorithms. *Probab. Eng. Mech.*, 17: 253–264.

Quiblier, J.A. 1984. A new three-dimensional modeling technique for studying porous media. *J. Colloid Interface Sci.*, 98: 84–102.

Ren, Z.Y. and Q.S. Zheng. 2002. A quantitative study of minimum sizes of representative volume elements of cubic polycrystals–numerical experiments, *J. Mech. Phys. Solids*, 50: 881–893.

Reuss, A. 1929. Berechnung der fließgrenze von mischkristallen auf grund der plastizitätsbedingung für einkristalle. *ZAMM*, 9: 49–58.

Roberts, A.P. and M.A. Knackstedt. 1996. Structure-property correlations in model composite materials. *Phys. Rev. E.*, 54: 2313–2328.

Roberts, A.P. and M. Teubner. 1995. Transport properties of heterogeneous materials derived from Gaussian random fields: Bounds and simulation. *Phys. Rev. E.*, 51: 4141–4154.

Rogula, D. 1982. Introduction to nonlocal theory of material media, In *Nonlocal theory of material media, CISM courses and lectures*, ed. D. Rogula, 125–222. Wien: Springer.

Sahimi, M. 2003. *Heterogeneous materials I: Linear transport and optical properties*. New York: Springer.

Sanchez-Palencia, E. 1980. *Non-homogeneous media and vibration theory*. Berlin: Springer-Verlag.

Shen, L. and X.F. Xu. 2010. Multiscale stochastic finite element modeling of random elastic heterogeneous materials. *Comput. Mech.*, 45: 607–621.

Silnutzer, N. 1972. Effective constants of statistically homogeneous materials (doctoral dissertation). Philadelphia: University of Pennsylvania.

Sinai, Y.G. 1982. *Theory of phase transitions: Rigorous results*. Oxford: Pergamon Press.

Slepian, D. 1972. On the symmetrized Kronecker power of a matrix and extensions of Mehler's formula for Hermite polynomials. *SIAM J. Math. Anal.*, 3: 606–616.

Suquet, P. 1993. Overall potentials and extremal surfaces of power-law or ideally plastic materials. *J. Mech. Phys. Solids*, 41: 981–1002.

Talbot, D.R.S. and J.R. Willis. 1997. Bounds of third order for the overall response of nonlinear composites. *J. Mech. Phys. Solids*, 45: 87–111.

Torquato, S. 2002. *Random heterogeneous materials, microstructure and macroscopic properties*. New York: Springer.

Toupin, R.A. 1962. Elastic materials with couple-stresses. *Arch. Rational Mech. Anal.*, 11: 385–414.

Tuck, C. 1999. *Effective medium theory*. Oxford: Oxford University Press.

Vanmarcke, E.H. and M. Grigoriu. 1983. Stochastic finite element analysis of simple beams. *J. Eng. Mech. ASCE*, 109: 1203–1214.

Voigt, W. 1887. Theoretische Studien über die Elasticitätsverhältnisse der Krystalle. Abh.Kgl.Ges.Wiss.Göttingen. *Math.Kl.*, 34: 3–51.

Walpole, L. 1966. On bounds for the overall elastic moduli of inhomogeneous systems. *I. J. Mech. Phys. Solids*, 14: 151–162.

Wiener, O. 1912. Die Theorie des Mischkörpers für das Feld der Stationären Strömung. *Abh. Math. Phys. K1 Königl. Sächs. Ges.*, 32: 509–604.

Willis, J.R. 1977. Bounds and self-consistent estimates for overall properties of anisotropic composites. *J. Mech. Phys. Solids*, 25: 185–202.

Willis, J.R. 1981. Variational and related methods for the overall properties of composites. In *Advances in applied mechanics*, ed. C.-S. Yih, 1–78. London: Elsevier.

Willis, J.R. 2002. Lectures on mechanics of random media. In *Mechanics of random and multiscale structures, CISM lecture notes*, eds. D. Jeulin and M. Ostoja-Starzewski, 221–267. Wein/New York: Springer.

Xu, X.F. 2005. Morphological and multiscale modeling of stochastic complex materials (doctoral dissertation). Baltimore: Johns Hopkins University.

Xu, X.F. 2009. Generalized variational principles for uncertainty quantification of boundary value problems of random heterogeneous materials. *J. Eng. Mech. ASCE*, 135: 1180–1188.

Xu, X.F. 2011a. On the third-order bounds of effective shear modulus of two-phase composites. *Mech. Mater.*, 43: 269–275.

Xu X.F. 2011b. Stochastic computation based on orthogonal expansion of random fields. *Comput. Methods Appl. Mech. Eng.*, 200: 2871–2881.

Xu, X.F. 2012a. Ellipsoidal bounds and percolation thresholds of transport properties of composites. *Acta Mech.*, 223: 765–774.

Xu, X.F. 2012b. Ellipsoidal bounds of elastic composites. *J. of Appl. Mech.*, 79: 021016.

Xu, X.F. 2012c. Optimal percolation thresholds of two- and three-dimensional engineering composites. *J. Eng. Mater. Technol.*, 134: 031008.

Xu, X.F. 2012d. Quasi-weak and weak formulation of stochastic finite elements on static and dynamic problems: A unifying framework. *Probab. Eng. Mech.*, 28: 103–109.

Xu, X.F. 2012e. *Convolved orthogonal expansions on dynamics and random media problems*. Proc. of the 5th Asian-Pacific Symposium on Structural Reliability and its Applications. P262.

Xu, X.F. 2014. *Probabilistic resonance and variance spectra*. Proceedings of Vulnerability, Uncertainty, and Risk: Quantification, Mitigation, and Management.

Xu, X.F. 2015. Multiscale stochastic finite element method on random field modeling of geotechnical problems: A fast computing procedure. *Front. Struct. Civ. Eng.*, 9: 107–113.

Xu, X.F. and I.J. Beyerlein. 2016. Probabilistic strength theory of carbon nanotubes and fibers. In *Advanced computational nanomechanics*, ed. N. Silvestre, 123–146. Chichester: John Wiley & Sons.

Xu, X.F. and X. Chen 2009. Stochastic homogenization of random multi-phase composites and size quantification of representative volume element. *Mech. Mater.*, 41: 174–186.

Xu, X.F., X. Chen, and L. Shen. 2009. A Green-function-based multiscale method for uncertainty quantification of finite body random heterogeneous materials. *Comput. Struct.*, 87: 1416–1426.

Xu, X.F., G. Dui, and Q. Ren. 2014. A note on scale-coupling mechanics. In *Multiscale modeling and uncertainty quantification of materials and structures*, eds. M. Papadrakakis, G. Stefanou, 159–169. New York: Springer.

Xu, X.F. and L. Graham-Brady. 2005. A stochastic computation method for evaluation of global and local behavior of random elastic media. *Comput. Methods Appl. Mech. Engrg.*, 194: 4362–4385.

Xu, X.F. and L. Graham-Brady. 2006. Computational stochastic homogenization of random media elliptic problems using Fourier Galerkin method. *Finite Elem. Anal. Des.*, 42: 613–622.

Xu, X.F. and Y.X. Jie. 2014a. Third-order bound of nonlinear composites and porous media under hydrostatic deformation. *Mech. Mater.*, 68: 137–146.

Xu, X.F. and Y.X. Jie. 2014b. Variational approach to percolation threshold of nanocomposites considering clustering effect. *J. Nanomech. Micromechanics*, 4: A4013009.

Xu, X.F. and G. Stefanou. 2011. Variational bounds on effective elastic moduli of randomly cracked solids. *Int. J. Multiscale Comput. Eng.*, 9: 347–363.

Xu, X.F. and G. Stefanou. 2012a. Explicit bounds on elastic moduli of solids containing isotropic mixture of cracks and voids. *Fatigue Fract. Eng Mater. Struct.*, 35: 708–717.

Xu, X.F. and G. Stefanou. 2012b. Convolved orthogonal expansions for uncertainty propagation: Application to random vibration problems. *Int. J. Uncertainty Quantification*, 2: 383–395.

Yamazaki, F. and M. Shinozuka. 1988. Digital generation of non-Gaussian stochastic fields. *J. Eng. Mech. ASCE*, 114(7): 1183–1197.

Yeong, C.L.Y. and S. Torquato. 1998. Reconstruction random media. *Phys. Rev. E.*, 57: 495–506.

Zimmerman, R.W. 1985. The effect of microcracks on the elastic moduli of brittle materials. *J. of Mat. Sci. Letters*, 4: 1457–1460.

Zimmerman, R.W. 1991. *Compressibility of sandstones*. Amsterdam: Elsevier.

Index

D

E

O

P

Q

R

Milton Keynes UK
Ingram Content Group UK Ltd.
UKHW020314111024
449327UK00040B/1094

9 780367 65702